Y. N. KORMILITSIN, O. A. KHALIZEV

THEORY OF

SUBMARINE DESIGN

Theory of Submarine Design
Published by Riviera Maritime Media
www.rivieramm.com
ISBN number 0-9541446-0-0

2001

Professor Yuri N.Kormilitsin
Professor Oleg A.Khalizev
Saint-Petersburg State Maritime Technical University
Lotsmakskaya, 3
190008 Saint-Petersburg
Russia

A catalogue record for this book is available from the British Library
Library of Congress cataloguing in publication data

Kormilitsin, Yuri N.
Khalizev, Oleg A.

Theory of submarine design.

This work is subject to copyright. All rights are reserved, whether the whole or part of the material is concerned, specifically the rights of translation, reprinting, reuse of illustrations, recitation, broadcasting, reproduction on microfilm or in any other way, and storage in data banks. Duplication of this publication or parts thereof is permitted only under the provisions of the Great Britain Copyright Law in its current version, and permission for use must always be obtained from the authors. Violations are liable for prosecution under the Great Britain Copyright Law.

© Yuri N.Kormilitsin, Oleg A.Khalizev

Printed in Great Britain

The use of general descriptive names, registered names, trademarks, etc. in this publication does not imply, even in the absence of a specific statement, that such names are exempt from the relevant protective laws and regulations and therefore free for general use.

From the Authors

This book has been written based on the textbook "Submarine Design", which has been recommended by the Ministry of Education of the Russian Federation for students of institutions of higher education, specialising in "Shipbuilding". It is published in the year of a centenary anniversary of submarine professional design in Russia.

The book is published with the permission of the authors and the consent of the St.Petersburg State Maritime Technical University.

The book explores methodological issues, theory of submarine design, general methods of submarine displacement and its trimming determination, architectural aspects and determination of principal particulars as well as some other issues related to submarine specific features.

It is emphasised in the book that in our century of extensive use of computer aids for designing complicated engineering objects, and submarines undoubtedly refer to such objects, it is especially important for a designer to have a deep understanding of interrelation of submarine individual parameters and characteristics as well as an understanding of the intricacies of the submarine design process.

The authors hope that the book will be useful for everyone who dedicates their life to these most complicated but most interesting ships of the present time.

Authors express special thanks to Lidiya A.Krutitskaya, Anna B.Scorokhodova, Aleksei N.Umarov, Gennady D.Serditov.

Yury Kormilitsin was born on 1 July 1932 in Khabarovsk, USSR. He spent his childhood in Vladivostok. In 1956 he graduated from the Leningrad Shipbuilding Institute.

Since 1956 he has worked in the CDB ME «Rubin» and has gradually risen from a design engineer of the Design Department to the post of the General Designer on conventional submarines and deep-water submersibles with non-nuclear power plants. He is State Prize Winner (1984 and 1998), a Doctor of Sciences (Engineering) and a Member of the St.Petersburg Engineering Academy. He has been the Chief Designer of four generations of diesel-electric submarines: Projects «Foxtrot», «Tango», «Kilo», «Amur» and their various modifications.

Yury Kormilitsin has developed and practically implemented a concept of submarine series development that is based on using a set of common architecture and arrangement features, ensuring continuously high combat efficiency throughout subsequent equipment and sensor upgrades at the least possible displacement and low acoustic signatures. He is the author of many engineering and design developments, scientific publications and inventions. He is a professor of the Ship Design Department of the St.Petersburg State Maritime Technical University.

Oleg Anatolievich Khalizev was born on November 18, 1944 in Kazan. He grew up in Leningrad and in 1970 graduated from the Naval Architecture Faculty of the Leningrad Shipbuilding Institute.

Since 1970 he has worked at the Ship Design Department of the Leningrad Shipbuilding Institute, the Faculty of Naval Architecture, starting as an engineer in a research team and gradually rising to the position of Professor of the Ship Design Department. He is a Master of Technical Sciences and since 1974 has been lecturing on problems of submarine and deepwater submersible designs to students attending major courses at the St.Petersburg State Maritime Technical University. Since 1987 he has also been the head of the «Underwater Shipbuilding» course at the Faculty of Naval Architecture and Offshore Engineering.

His scientific interests are associated with various design aspects of submarines and submersible vehicles. He has supervised a number of research projects in support of practical submarine design work.

Oleg Khalizev is the author of over 100 scientific publications and inventions, 19 textbooks and methodological instruction documents.

WELCOME TO THE SMTU

The St.-Petersburg State Marine Technical University is a well-known highly ranked institution providing educational services in a wide spectrum of marine - related subjects and also in up-to-date branches of general engineering, business and social studies.

We are offering a choice of B Sc and M Sc degree programs as well as over 30 vocational courses leading to the Diplomas of Economist and Manager, Sociologist and Secondary School Teacher. With over five thousand full-time students, we are currently expanding a number of the new courses and programs within and beyond engineering. The high standards of teaching at the SMTU are confirmed by accreditation of five engineering courses by Institute of Marine Engineers (U.K.).

The students from 18 countries are an integral part of our community and we invite new applicants to join our full-time and short-term courses covering sciences, engineering, language and social studies.

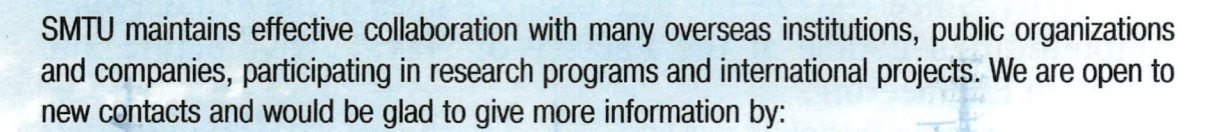

SMTU maintains effective collaboration with many overseas institutions, public organizations and companies, participating in research programs and international projects. We are open to new contacts and would be glad to give more information by:

The St.-Petersburg State Marine Technical University is headed by Professor K.P.Borisenko.

Contact numbers
tel: +7 812 114 2923, 7 812 114 4168, tel/fax: +7 812 318 5227
email: xmas@peterlink.ru
Prof. Kirill Rozhdestvensky is responsible for International Science and Education.

CONTENTS

Authors 4

Contents 7

Introduction 12

1. Submarine Design Management and Methods
   1.1. Submarine Design Management 16
   1.2. Submarine Design Methods 23
   1.3. Computer Application in Submarine Design 29

2. Submarine Displacement
   2.1. Submarine Displacement Types 37
   2.2. Displacement (Load) Breakdown
       into Standard Groups 39
   2.3. Load Breakdown at Early Design Stages.
       Weight Indices 42

3. Weights of Individual Groups and Load Items As Functions of Principal Characteristics of the Submarine
   3.1. The «Hull» Load Group 46
   3.2. Loads for «Hull Gears, Fittings»
       and «Hull Systems» Groups 65
   3.3. Loads for the «Mechanical Equipment,
       Pipelines and Systems of Power Plants» Group 68
   3.4. Load for the «Electric Equipment, Cables of Electric
       Power Systems, Electric Networks
       and Radioelectronic Equipment» Group 90
   3.5. Load for the «Weapons and
       Their Supporting Systems» Group 91

3.6. Load for «Stocks and Complement», «Displacement Margin» and «Solid Ballast» Groups........93

4. The Load Equation As a Function of the Displacement. Determination of the Displacement

4.1. Load Equation Derivation and Solution for the First Approximation 97

4.2. Load Equation for the Second Approximation 99

4.3. Effects of Variations in Submarine Particulars and Independent Weights Upon the Load Balance. The Differential Form of the Load Equation As a Function of the Displacement. The Normand's Number 102

4.4. Load Control..104

5. Submarine Constant Buoyant Volume (CBV)

5.1. The Constant Buoyant Volume and Its Relationship with the Submarine Normal Load....................................111

5.2. Relationships among CBV Components Considered at Early Design Stages 114

5.3. Determination of the Pressure Hull Volume. Volume Indices 116

5.4. The Equation of Volumes. Submarine Volume Ratios 152

6. Submarine Trimming

6.1. Approximate Estimation of Weight Levers 157

6.2. Approximate Estimations of CBV Component Levers 159

6.3. Submarine Trimming 160

6.4. Solid Ballast Weight and Longitudinal Profile Updating 161

7. Compensation of Submarine Variable Weights

7.1. General Issues of Variable Weight Compensation 165

7.2. Weapon Weight Compensation 170

7.3. Fuel and Oil Weight Compensation 174

7.4. Compensation for Submarine Buoyancy
Variations Due to Hydrological Effects 175

7.5. Variable Weight Compensation Analysis 176

7.6. Estimation of the Required Volume of Auxiliary
Ballast Tanks 179

8. **Submarine Hullform Design**

8.1. Selection of the Architectural and Structural Type 185

8.2. External Hullform and Appendages 190

8.3. Effects of and Ratios on Submarine
Performance and Manoeuvring Qualities 195

8.4. Effects of Relative Lengths of
the Parallel Middlebody, Fore and Aft Ends
upon Submarine Performance 202

8.5. Outer Hullform Factors 208

8.6. Sail Shape Selection 209

8.7. Shaft Number Effects upon Submarine Hullform
and Hydrodynamic Qualities 213

8.8. Effects of Layout Considerations upon
the Choice of Outer Hullform Parameters 220

9. **Determination of Submarine Principal Dimensions**

9.1. General 223

9.2. Submarine Hull Length Estimations 224

9.3. Submarine Hull Beam Estimations 226

9.4. Submarine Hull Height Estimations 227

9.5. Reserve Buoyancy. Approximate
Draught Estimations Based on the Specified
Reserve Buoyancy 228

9.6. Application of the Stability Equation in Estimations
of Main Dimensions 239

10. Submarine Lines Drawing
- 10.1. General 243
- 10.2. Phase-Wise Specifics of Submarine Lines Drawing Design 247
- 10.3. Methods for Lines Design and Specification 251
- 10.4. Submarine Sail and Fin Lines Drawings 266
- 10.5. Modern Computer Technologies for Lines Drawing Generation 270

11. Design of Submarine Control Surfaces
- 11.1. Submarine Control Surfaces: Types and Purposes 273
- 11.2. Determination of «Hull + Control Surfaces» Hydrodynamic Characteristics 278
- 11.3. Criteria Used in the Control Surface Design 285
- 11.4. Methods Used in Submarine Control Surface Design 289

12. Military-and-Economic Analysis and Its Role in Submarine Design
- 12.1. The Subject of Military-and-Economic Analysis 299
- 12.2. Basic Notions of Military-and-Economic Analysis 302
- 12.3. The Military-and-Economic Efficiency Evaluation Model 308
- 12.4. The Submarine Combat Employment Model 310
- 12.5. Determination of the Submarine Task Force Strength. Operational Strain and Operational Utilisation Coefficients 316
- 12.6. The Concept of Military-and-Economic Analysis Dynamic Problems 318
- 12.7. Estimation of Submarine Technical-and-Economic Indices 322

References 332

1901-2001

100 YEARS ON THE WAY OF PROGRESS & QUALITY

**CENTRAL DESIGN BUREAU
FOR MARINE ENGINEERING**
90 Marata, Saint-Petersburg,
191119, Russia
Phone: (812) 113 5132
Fax: (812) 164 3749
E-mail: neptun@ckb-rubin.spb.su

RUBIN

INTRODUCTION

The Evolution of the Submarine Design Theory

As with the situation with surface ships, we understand the submarine design theory as a scientific discipline that studies the creative process of producing a submarine and, first of all, evolving principal particulars of this submarine and justifying optimum design choices.

The physical meaning of some formulae of the basic ship design theory, like equations of mass, capacity and others, is as valid for submarines as for surface ships. Nevertheless, submarines are such specific engineering products that any direct transfer of surface ship analytical tools and techniques is, as a rule, unacceptable due to the very structure and physical essence of the involved relationships. Therefore, we believe that the submarine design theory has a right for independence and it should not be regarded as just a particular case of ship design theory.

The progress of underwater shipbuilding over the past 30 years has already resulted in its own diversification of design approaches. For example, there are distinctive specifics in design features of submersible vehicles, passenger submarines, etc.

The evolution of the submarine design general theory into an independent discipline started fairly recently and still is an on-going process.

Early developments in this field were initiated in 1939-1941 by B.M.Malinin [55], a prominent submarine designer and a professor of the Shipbuilding Institute. Based on evaluations of submarine designs, he identified individual groups of weights, which constitute the load, as functions of the displacement and other parameters of a submarine. Those formulae have enabled to use the weight (mass) equation for finding the displacement. He also derived approximate formulae for stability characteristics, the wetted area and other values suitable for initial design stages. In 1940

A.N.Shcheglov published a book titled «Design of Submarines» [92] that described methods for determining main dimensions and the displacement of submarines. In 1949 another publication appeared - «A Reference Book of Submarine Designer» - a two volume work edited by professor B.M.Malinin, summarising the experience of underwater shipbuilding in Russia from 1927 to 1945 [80]. In the same year S.A.Yegorov, Z.A. Deribin et al. completed their book on submarine design methods. Unfortunately, that work has never been published. They suggested a method to determine submarine displacement analytically by formulating and separately solving two separate equations: the equation of weight (mass) and the equation of buoyant volumes, and then to match the obtained results.

All those achievements had been based on pre-war experience of underwater shipbuilding. In 1950s the development of submarines with new structural and technical characteristics rendered a lot of earlier derived formulae inapplicable.

In 1952 A.N.Donchenko analysed weight loads of the first submarines built after the war and found new numerical coefficients, as well as new relationships between the displacement and individual components of the submarine load balance. It should be noted that since the 1950s various institutions (mainly of the Navy and of the Shipbuilding Ministry) performed a large number of studies associated with particular tasks at the nexus of the submarine design theory and other ship science disciplines. In this respect one should mention work by D.P.Skobov, L.V.Kalacheva, S.S.Zolotov, V.V.Rozhdestvenskiy in the field of submarine dynamics, Yu.A.Shemanskiy, N.S.Solomenko, V.T.Tomashevskiy in the field of submarine structural strength, A.A.Pravdin, M.K.Glozman, N.L.Sivers in the field of hull design, and many others.

It was during the same period that design theory issues reached textbooks. In 1954 there was a handbook by V.N.Kvasnikov and M.V.Saveliev [35] and in 1959 a textbook by K.P.Yefremov was published [31].

A considerable contribution to the development of the submarine design theory was made by professor S.A.Bazilevskiy [9]. In his scientific publications, as well as in lectures at the Naval Academy, he deliberated on techniques of submarine design aiming towards wider application of the similarity theory and error analysis.

Since the mid 1950s combat efficiency and capabilities of submarines have been continuously growing, as a result of the evolvement of new, more efficient means of maritime warfare. As submarines are fitted with more and more state-of-the-art tools of war, the number of possible design solutions becomes greater and greater. Under such conditions increasing numbers of design decisions have to be made at initial design stages when, based on evaluations of the efficiency of weapons, sensors and equipment, taking into account anticipated costs of developing and operating the submarine, they formulate design specifications. This stage of ship design is known as the Conceptual Design (CD). The CD development was stimulated by the progress of the submarine design theory in the direction of wider and deeper substantiation of the submarine design analytical arsenal and introduction of computers into CD procedures.

Among numerous studies aimed to support the Conceptual Design, one should note publications by L.B.Breslav, K.B.Malinin, N.N.Grigoriev, L.Yu.Khudiakov.

In the first of the above-mentioned investigations, the application of modern mathematical statistics methods has made it possible to refine the structure of approximate formulae for masses and volumes of the designed submarine. The second one considered various aspects associated with the use of analytical methods for generating tentative sketches of hull lines. The third one offered a method to find the normal displacement from masses and volumes, while the fourth of these publications dealt with methodological aspects of the Conceptual Design.

Crucially important for modern shipbuilding are military and economic evaluations of submarine designs. These issues happen to rest at the nexus of a whole bunch of sciences: design theory, tactics, economics, specific problems of mathematical analysis.

Many publications on such issues were made by L.B.Breslav [11], A.A.Narusbaev [58], I.G.Zakharov [66], L.Yu.Khudiakov [87] and some other authors.

As has been already mentioned above, the submarine design theory is closely related to other naval architecture disciplines: hydrodynamics, ship theory and structural mechanics, to construction technologies, economics and a number of other disciplines. The design theory extensively uses scientific tools of these disciplines for tasks associated with, for example, propulsive performance, manoeuvrability or strength of the designed submarine. However, here these tools are used in a different way and approached from different positions. This is due to the following specific features of design-oriented applications.

Firstly, unlike other disciplines, the design theory solves not direct but inverse tasks. For example, while ship theory calculations serve to determine the speed of an actual submarine, the aim of design efforts is to establish particulars of the designed submarine that would guarantee the desired speed.

Secondly, in contrast to all other disciplines that consider various abilities and parameters of the submarine individually, without relating them to each other, the submarine design theory can be characterised as a comprehensive system approach to the submarine as a single product in which everything is interrelated and interdependent. Therefore, it is necessary to consider how this or other decisions may influence not just a certain quality intended for improvement but the submarine as a whole, i.e. all other qualities and performances. Consequently it is not always that recommendations developed based on the submarine design theory agree with similar recommendations of other disciplines [14].

At the same time it should be mentioned that, in spite of the apparent multitude of investigations, the submarine design theory is not yet sufficiently fully developed even in such classical domains as design procedures. Relevant work performed by different agencies and organisations has not yet been truly systematised and generalised.

Evidently, international references contain virtually nothing on the submarine design theory. The only exceptions are a detailed publication [96] made in Great Britain that describes a number of particular problems of submarine design and a monograph [23] on general issues of ship design management.

1. SUBMARINE DESIGN MANAGEMENT AND METHODS

1.1. Submarine Design Management

Naval Force Development Programmes for certain periods of time are generated on the grounds of *long-term military-and-economic planning* that is itself based on forecasted national foreign-policy targets and tasks of military forces, particularly of the Navy, in achieving these goals. Other significant input comes from predictions of developments in weapons and other military hardware in both the subject and other countries.

Long-term military-and-economic planning is a hierarchic procedure performed at different levels, starting with the overall fleet configuration and going down to individual classes of ships. In particular, long-term military-and-economic plans should outline what types of submarines are to be constructed or designed within the considered period. The final result of these efforts is the Naval Force Development Programme that usually covers a period of about 10 years. Construction of any new type of a ship is included into such programmes only when initial design stages are already completed (usually after they have formulated and approved Submarine Design Specifications (SDS)). While being implemented, the programme may be corrected due to changes in the political situation, in production and economic resources and capabilities, in military equipment developments.

Naval long-term military-and-economic planning and programme development are performed using contributions from Navy Research Institutes, various organisations of the shipbuilding and other involved industries, Navy specialists and ship designers.

It is generally known that there is no way to design a sophisticated engineering product in one step. The process has to consist of several successive phases or stages of design efforts (Fig.1.1).

A flow chart of traditional phased submarine design process adopted in Russia [23] may be seen in Fig.1.1.

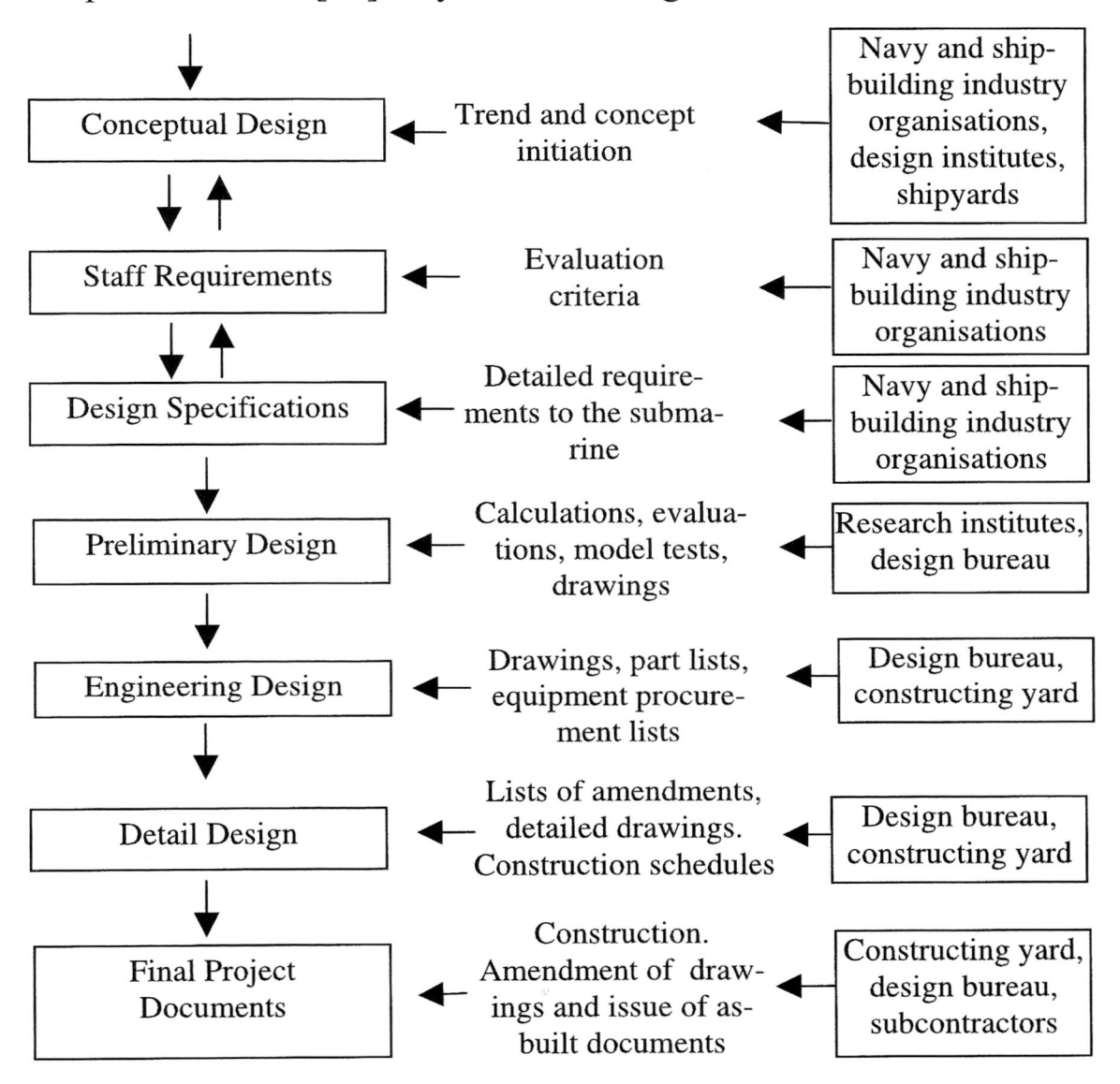

Phase	Inputs/Outputs	Organisations
Conceptual Design	Trend and concept initiation	Navy and shipbuilding industry organisations, design institutes, shipyards
Staff Requirements	Evaluation criteria	Navy and shipbuilding industry organisations
Design Specifications	Detailed requirements to the submarine	Navy and shipbuilding industry organisations
Preliminary Design	Calculations, evaluations, model tests, drawings	Research institutes, design bureau
Engineering Design	Drawings, part lists, equipment procurement lists	Design bureau, constructing yard
Detail Design	Lists of amendments, detailed drawings. Construction schedules	Design bureau, constructing yard
Final Project Documents	Construction. Amendment of drawings and issue of as-built documents	Constructing yard, design bureau, subcontractors

Fig.1.1. Design Phases

For the greater part of ships, submarines included, the design process starts with Conceptual Design which generalises the outcome of various research work relevant for design applications. Such studies are continuously carried out by advanced concept design teams from design and research establishments of both the shipbuilding industry and the Navy.

17

At this stage they investigate issues pertaining to the application of advanced equipment and weapon packages, new materials and design solutions for submarines; they evaluate effects of wide variations in tactical and technical particulars (e.g.: speed, diving depth, etc.), in the displacement and the architecture of submarines, perform feasibility studies in terms of cost and construction capabilities; investigate promising types of submarines. Conceptual Design is a multi-variant process involving extensive application of military-and-economic analysis in order to justify chosen decisions and utilising analytical methods to determine principal particulars of submarines.

With the start of a new Naval Construction Programme conceptual design efforts are re-directed to feasibility studies on the chosen types of submarines and formulation of requirements to their service and tactical characteristics. Actually, they often tend to see conceptual design as a comparatively narrow task and regard it as feasibility study and customer's requirements phase [87]. Conceptual Design is a synthesis of strategies and tactics of the Navy, of science and technology of shipbuilding and other involved industries.

The Staff Requirements (SR; Russian name: Operational and Tactical Assignment) are generated by the Naval Command and research organisations belonging to the Navy. This process also involves contributions from design bureaus of the shipbuilding industry.

In the Staff Requirements they:
− define the designation and missions of the submarine;
− specify and analyse data on the potential adversary, evaluate enemy countermeasures taking into account possible improvements in such capabilities, assess potential areas of combat operations, base locations and conditions, and availability of repair facilities, etc.

Based on the SR, they formulate Operational and Tactical Requirements that are tentative requirements of the Navy to configuration and main characteristics of the weapon package, protection features, endurance, speed, sea range, diving depth and seakeeping abilities of the future submarine.

The main goal of work on the SR is to evolve the most rational, in terms of both military and economic aspects, combination of tactical characteristics that would enable the submarine to fulfil the assigned missions in the best possible way.

In order to match tactical and technical features of the submarine and to make sure, in the first approximation, that they fit together, the

Conceptual Design phase includes design studies on the subject submarine under rather wide variations of tactical inputs. For every considered variant they determine the displacement, main dimensions, the power plant capacity and the approximate construction cost. Several, so-called «basic», options are studied in greater detail, including making a sizeable bulk of calculations and drawings. Upon the completion of Conceptual Design work, they evaluate the efficiency of considered variants with the help of military–and–economic analysis methods. Then the variants are compared and the recommended version of the SR is submitted to the Naval Command for approval [75].

The generation of the Staff Requirements should, together with its supporting Conceptual Design efforts, be considered as an initial stage of the submarine design process.

Development of the Technical Proposal and Submarine Design Specifications (SDS).

The Technical Proposal for a submarine design (this design stage was earlier called «tentative» or «feasibility» design) is prepared by a design organisation from the shipbuilding industry based on the approved Staff Requirements. The aim of this design stage is to justify the advisability and to check the feasibility of creating a submarine to the approved SR.

Tasks to be solved during the development of the Technical Proposal may be grouped as follows:

– to check whether the SR fits available technical and economic capabilities; to find principal technical solutions necessary for achieving the specified tactical performances. For these purposes they carry out basic shipbuilding calculations required to determine submarine particulars, draw general arrangement plans, prepare an explanatory note; estimate submarine design, construction and operation costs;

– to find first-iteration solutions for administrative issues pertaining to the creation of the submarine, to compile lists of weapons and major equipment (power plant, machinery, instrumentation); to select manufacturers and suppliers of existing equipment, as well as organisations that will develop new equipment; to define the scope of research and development work necessary for developing new equipment and validating new technical solutions; to estimate time–frames for each stage of the submarine development project;

– to assess the technical level of the intended submarine from the point of view of national and international achievements in science and technology and to provide military and economic justifications for the project; to establish the number of such submarines required to achieve missions assigned to the Navy and to look into other issues of military–and–economic analysis. This group of tasks is as a rule assigned to research institutions of the Navy.

Essentially, the generation of the Technical Proposal may be considered to be the final stage of the Conceptual Design. It differs from the previous phase by a considerably smaller number of evaluated variants, but it involves a much greater scope of calculation and drawing work on each of the variants. At the Technical Proposal phase they mostly consider alternatives for key technical solutions, e.g., hull architectural type, propulsion plant type and other features affecting principal particulars of the intended submarine. This is the stage when they make first approximations of the displacement and main dimensions, as well as of other principal particulars of the submarine depending on alternative choices of the weapon package, the speed, etc.

Based on the solution of the above tasks, they select the optimum variant of the Technical Proposal and it serves as the basis for formulating Submarine Design Specifications (SDS; the Russian term is «Tactical and Technical Assignment»).

SDS set out detailed requirements of the customer (the Navy) for the intended submarine and usually contain the following data:

Designation of the submarine;

Weapons (missile, torpedo, mine) and sensors (sonar, radar, communication, computers, etc.);

Requirements for protection and stealth features (submarine signature intensity levels and platform noise affecting onboard sonar performance);

Platform features (tentative displacement, speed and sea range, diving depth, manoeuvring qualities, endurance, etc.);

Habitability conditions;

Extent of automation;

Power plant (type and key parameters);

Additional requirements set to the submarine in general or to some special features relevant for obtaining this or other additional quality.

Submarine Design Specifications are reviewed and approved by various authorities of the shipbuilding industry and of the Navy, and after their approval the project is included in the Naval Construction Programme.

Preliminary Design is based on the approved SDS and is actually the main design stage. While developing the Preliminary Design, it is necessary to cover the following issues which determine whether the SDS implementation is practically feasible:

a) Determination of submarine displacement and main dimensions, as well as surface and submerged performance and manoeuvring qualities, speed and sea range;

b) Generation of block diagrams of systems and gears;

c) Resolution of issues associated with equipment arrangement in compartments and with the overall configuration of the submarine;

d) Selection of main machinery and major items of equipment;

e) Resolution of key issues of construction technology and management.

The Preliminary Design phase includes submarine model tank tests. Based on the outcome of these tests, the lines drawing of the submarine, hull appendages and propeller geometry are finalised.

They also carry out wind tunnel model tests in order to determine manoeuvrability characteristics of the submarine. During the Preliminary Design it may be found that it is impossible to meet certain SDS requirements. In this case specifications have to be reconciled. Preferably, the Preliminary Design should be as detailed as npossible to avoid any major changes at later design stages. The Preliminary Design package contains so-called «to-be-submitted» and «not-to-be-submitted» documents, as well as design data and documents from subcontractors [25]. The final scope of to-be-submitted documents, i.e. drawings, calculations, schematics and other technical documents, is established depending on the type and particulars of the subject submarine.

Engineering Design work is based on the approved Preliminary Design taking into account the changes and amendments introduced during its review and approval. Such adjustments should not imply changes in basic particulars of the subject submarine.

The aim of the Engineering Design stage is to make preparations for submarine construction, as well as for production and procurement of hull materials and equipment, weapons and sensors to be delivered by subcontractors. For this purpose they prepare sets of sufficiently detailed drawings, calculations and other technical documents. The Engineering Design work involves detailed consideration of all technical issues and should confirm tactical and technical features of the submarine. At the same time submarine construction specifications and part lists are compiled. The scope of to-be-submitted documents and the total scope of design work at this stage are approximately three times greater than at the Preliminary Design.

In order to optimise equipment and compartment-wise arrangements, General Arrangement drawings are made to a rather large scale (usually 1:10).

A special emphasis is placed at the Engineering Design stage on the construction technology that is developed so as to suit the actual shipyard, with regard to submarine construction schedules and financial aspects.

The Preliminary and the Engineering Design packages are reviewed and approved by the customer (the Navy) and shipbuilding industry authorities.

Detail (Workshop) Design. The main aim of this phase is to generate and issue the full set of detailed drawings of hull, mechanical and electromechanical components necessary for submarine construction. Accordingly, the main part of the Detail Design work is the generation of detailed drawings in numbers ranging from 6 to 10 thousands. Another crucial aspect of Detail Design efforts is the development of detailed construction technology. The Detail Design package also includes compartment-wise equipment installation (mounting) drawings, bills of materials, technical requirements and specifications for the subject submarine construction.

The Detail Design stage is completed with the issue of «to-be-executed» documents which are the Engineering Design documents corrected to comply with detailed drawings. E.g., there is a finalised record load calculation to check the position of the centre of gravity, etc. A large part of these documents are operating instructions for various equipment on the submarine.

Submarine Construction. During the construction the designer's functions are to perform daily supervision and monitoring of the construction progress to ensure that the yard meets design requirements (e.g., maintains the weight discipline) and to provide technical assistance to the shipyard. The designer's representatives participate in acceptance tests and trials of the submarine.

Preparation of As-built Documents. After the completion of the construction, the last stage of the designer's work is to produce as-built drawings and documents. The goal of this task is to provide the commissioned submarine, Naval bases, technical educational institutions and other relevant organisations with drawings and other technical documents that accurately describe the built submarine in terms of structures, equipment layout and all tactical and technical particulars and characteristics. The issue of as-built drawings and documents essentially means incorporation of corrections into the Detail Design package to cater for any deviations from detailed drawings made during construction and for actual values of tactical and technical performances measured during tests and trials of the lead submarine.

1.2. Submarine Design Methods

It is well known that the common approach to the design of sophisticated engineering products is the convergence method, often known internationally as the «trial-and-error method». This is to the full extent applicable to submarines. The essence of the method is that since it is impossible to reach the target in a single step, there should be several successive stages of design efforts (Fig.1.1).

At initial design stages the variation method is applied extensively. This method enables a solution to be found that would in the best possible way fit the designation assigned to the intended submarine.

Each step of the convergence method differs from the previous one by more complete and detailed substantiation of selected parameters, and more thorough deliberations on key tasks. Elements of the design are updated and supported by more accurate calculations. The scope of design documents grows and the number of drawings developed at every step becomes larger. The contents of documents under the same task titles also change step by step. Documents incorporate updated data on equipment and weaponry developed simultaneously with the submarine design. Accordingly, the scope of the load balance and constant buoyant volume (CBV) calculations at Preliminary and Detailed Design stages is quite different, calculation methods for various parameters and characteristics of the developed design respectively become more and more complicated.

Under the convergence method they do not attempt to reconcile all parameters and particulars of the submarine at every step. Otherwise, it would overcomplicate both the preparation for construction and the design process itself, which are quite lengthy (from 3 to 10 years) [16]. Therefore, at certain design stages some parameters and characteristics should be preferably finalised and fixed. Normally, certain technical characteristics of the submarine adopted in the Preliminary Design should not drastically change at the Engineering Design stage. Among these characteristics are main dimensions, hullform, architecture and structural configuration of the submarine, although to a certain extent these parameters inevitably do undergo some changes.

The design phase sequence may vary depending on the Submarine Design Specifications and on how much the subject boat differs from previous ones. In real life there have been cases when a project started at the Engineering Design stage (e.g., Project 641 «FOXTROT» since it was largely a follow-on development and improvement of Project 611 «ZULU») (Figs.1.2 and 1.3) [43], [91], [95].

A crucially important design technique is the use of prototypes, i.e. data from earlier designs and statistic information about parameters and characteristics of already constructed submarines. However, utilising this approach one should keep in mind that for all practical purposes there is never a prototype that would fit all design requirements. This is obvious because the availability of such a prototype would render the new design meaningless.

Nevertheless, the use of prototypes in the design of such complex things as submarines is both reasonable and necessary. The trick is that the very notion of «prototype» should be understood in broad terms. There may be several existing designs used as prototypes in the development of a new one. Thus, one of them may serve for hullform selection, another may be used for developing the general arrangement, the third one may help in calculations on strength, loads and volumes required for equipment layout, etc. Still, even using prototypes in this wide interpretation, it should be remembered that any prototype represents the past. Some time has already passed since its development, science and engineering have achieved new advances, and therefore the new submarine design should take into account the changes in requirements to design work introduced during the elapsed period.

In other words, data of the chosen prototype have to be corrected in a certain way. The same is applicable to statistic information. When using any statistical data in design work it is first of all necessary to be well aware of their origins, to know when, how and for what types of submarines those data have been obtained. At the same time it is necessary to remember that statistics and prototypes are nothing but experience accumulated over a given period of time. Developments in equipment and new materials, changes of requirements to design work, introduction of new structural configurations, new technological solutions — all these things influence various weights and volumes, parameters and characteristics.

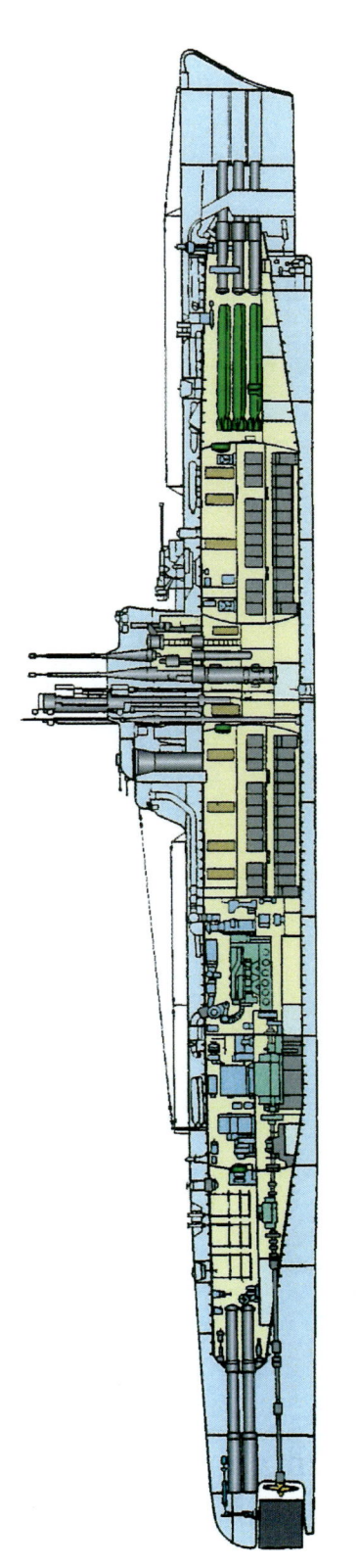

Torpedo tubes	10
Torpedoes (calibre, mm)	22 (533)
Normal displacement, m³	1831
Main dimensions, m: length overall beam overall	 90.5 7.5
Full surface speed, kn	17
Full submerged speed, kn	16
Submerged range at 2.1 kn, miles	440
Maximum diving depth, m	200
Endurance, days	75
Type of MPP	DEP
Full speed power, h.p. (r.p.m.) diesel engines electric motors	 3x2,000 (500) 1x2,700(540) 2x1350 (440)
Shafts	3
Complement	72

Fig.1.2 Project 611 "ZULU" Submarine

25

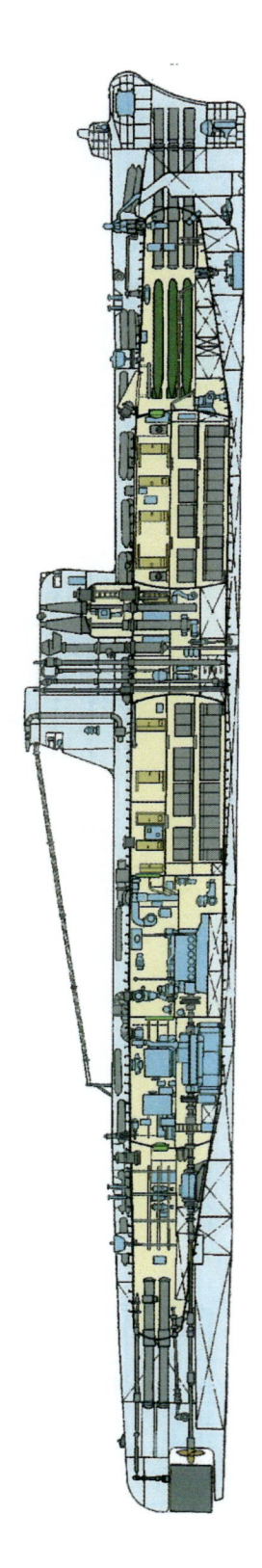

Torpedo tubes	10
Torpedoes (calibre, mm)	22 (533)
Normal displacement, m³	1952
Main dimensions, m:	
length overall	91.3
beam overall	7.5
Full surface speed, kn	16.8
Full submerged speed, kn	16
Submerged range at economic speed, miles	400
Maximum diving depth, m	280
Endurance, days	70
Type of MPP	DEP
Full speed power, h.p. (r.p.m.)	3x2,000 (500)
diesel engines	1x2,700(540)
electric motors	2x1350 (440)
Shafts	3
Complement	70

Fig.1.3 Project 641 "FOXTROT" Submarine

Further to the above, it should follow that utilisation of experience accumulated by previous generations of designers is justified and required, but it will help to successfully solve the specified task only if approached creatively [17].

Let us now consider specifically the early design stages, i.e. the Technical Proposal and the Submarine Design Specifications phases. In terms of creative work these stages are professionally the most attractive for designers and at the same time they are the most difficult ones since the designer has only a list of operational requirements and a clean sheet of paper to start with. At this stage almost everything is vague and uncertain. There are only ideas, some outlines of considerations that should be somehow implemented in the future design. Compared to other stages, at this point designing is an art rather than a science. The success to a large extent depends on the broad-mindedness and experience of the designer and the design team. An experienced designer familiar with many submarine projects may by applying various coefficients, weights and volumes at an early stage obtain rather accurate values for basic particulars of the new boat.

What methods can be used in submarine design at the very early stages?

Theoretically, all methods recommended by the ship design theory [6], [60], [61] are applicable. However, special features inherent to submarines to a large extent make the application of many of these methods quite difficult. These include the need for pressure and outer hulls to be arranged with respect to each other, the need for painstaking co-ordination of the load balance with the constant buoyant volume, the troubles of trimming under various service conditions – this is by far an incomplete list of submarine design specifics. Under such conditions even observing the law of Archimedes, which is Law No. 1 for submariners, is rather a complicated task. Besides, it should be kept in mind that unlike all other ships, space ones included, the submarine has to suffer numerous variations in external pressures from 1 to 100 atmospheres, and some special types even up to 1,000 atmospheres.

The early-stage choice of the submarine design method also depends on the nature of design specifications. Thus, in a case where the submarine is designed for an existing specified power plant, a specified package of weapons and sensors, it is somewhat easier for the designer since weights and sizes of at least some equipment are known from the very beginning. However, such specifications do not appear very often. Usually, the greater bulk of submarine components has yet to be selected and the designer has yet to optimise their numerous parameters: output power, weight, volume, consumed power, cooling requirements, etc.

All currently used methods are based on experience gained in the process of practical design work. Great contributions to the development of these design methods were made by well-known designers of the national submarine fleet: I.G.Bubnov, P.P.Pustyntsev, N.N.Isanin, B.M.Malinin, V.N.Peregudov, G.N.Chernyshov and many others [53]. In real life every school of designers prefers to use its own traditional methods for early-stage design work [82].

The Drawing or Graphic Method

Under this method the design work begins with graphic exercises on equipment arrangement in pressure hull compartments and a general study on the architectural outlook of the submarine. After that, based on the data of the graphic study, they perform first-approximation calculations on the load balance and the constant buoyant volume, and then check the boat trimming in terms of moments and forces. If the performed design study proves to be unsatisfactory from the point of view of these issues, it is modified and the results are checked again.

This method can be used, obviously, only when at the very beginning of the design the submarine equipment is defined (or specified), e.g., the main power plant and weights and dimensions of the larger part of other equipment items are known. This design method only allows for a limited number of options to be studied because the procedure is rather time-consuming.

Nevertheless, modern computers with suitable databases make this method quite promising.

The Graphoanalytical Method

The basic idea of this method is that to determine the displacement, main dimensions and other particulars of the submarine both graphic exercises and design formulae are used.

E.g., with the help of the mass equation they find the submarine displacement and select the power plant. Then they update the mass equation, re-calculate the displacement, make a graphic study on the arrangement of compartments and approximately estimate the reserve of buoyancy. They also make the first approximation of the constant buoyant volume and evaluate submarine trimming. As a rule, they generate several equipment layout options. One should, however, remember that as long as the pressure hull is selected, the potential variety of alternative task solutions becomes limited. In some cases they may suggest several alternatives for the pressure hull too.

The Analytical Method

Contrary to both previous methods, this one is based not on graphic deliberations but exclusively on various analytical formulae. Thus, taking equations of mass, volume, stability, propulsion, etc., expressed as functions of target figures of the Submarine Design Specifications, displacement and other design parameters, one may solve these equations jointly and obtain the sought displacement, main dimensions and other submarine design values. The analytical method is widely used in Conceptual Design. It does not require labour-consuming graphic studies and enables a large number of alternative solutions for the given task to be considered. This method allows optimum task solutions based on a chosen set of criteria to be obtained. Obviously, the validity of these obtained results in many respects depends both on the chosen criteria and on variation ranges adopted for subject variables.

At the same time, the design quality depends as much on the designer's experience as on the designer's command of tools available from other disciplines relevant for submarine design purposes and on all achievements in other branches of science and engineering that may be utilised to enhance the quality of the design.

1.3. Computer Application in Submarine Design

A characteristic feature of modern submarine design is the ever-increasing utilisation of computer and graphic hardware.

Initially (1960–70's) computers were used just to perform the most standard ship theory calculations. Their procedures normally could be derived from «manual» calculation routines without any need significant modifications. These, primarily, included calculations of loads, ship statics and dynamics, strength of standard structures, etc. [39], [97].

However, the progress in computer hardware has stimulated further improvement of calculation methods, e.g., thanks to the opportunity to drop various simplifications and assumptions. New utility methods have rapidly found applications in design work, including the finite element method which made it possible to make more accurate predictions of stress fields, thus improving both the reliability of structures and material savings [81].

The domain of computer applications has expanded and covered new fields of engineering and design activities [61], [98].

At present, in addition to making the above-mentioned calculations, computers compile material bills, part lists, generate drawings, etc. First of all, computers are applied for the most time-consuming and routine jobs, they successfully cope with laying out equipment in compartments and with the preparation of general arrangement drawings. A real breakthrough is that computers are now assigned to find optimum solutions.

The goals and purposes of computer applications in design work, to a large extent, depend on the stage of the design development (Staff Requirements, Submarine Design Specifications, Technical Proposal, Preliminary, Engineering or Detail Design).

Thus, in the process of work on the Staff Requirements it is necessary to select and justify requirements for tactical and technical features of the submarine, to compile lists of major weapons, equipment, etc. The main aim of using computers at this stage is to find a combination of submarine characteristics that would enable the boat to fulfil the assigned tasks in the best possible way.

At the Technical Proposal stage, same as during initial steps of Preliminary Design work, the main task is to evolve the architectural and structural outlook of the submarine, to update the equipment and system package, to determine principal particulars like normal displacement, main dimensions, manoeuvring characteristics, etc.

At these stages the task of computers is not just to determine some particulars but to optimise submarine particulars and the whole configuration aiming, e.g., at the least normal displacement or the lowest cost to satisfy SDS requirements.

During the later stages the design work becomes much more labour-consuming. This happens due to the geometric–progression growth of the amount of information that needs to be recorded, corrected and used or relayed to other departments of the design bureau. Considering that all design documents must be perfectly co-ordinated, it is irrational to do all this work manually because of both the huge labour and time costs and because of unavoidable errors associated with the human factor. In this task the computer is indispensable but its function becomes different: to reduce human labour content and to enhance design work quality.

In this respect it is most interesting to consider early design stages because they involve extensive application of design theory methods to get practical results.

Computerisation of the design process for any kind of engineering products, including submarines, assumes the availability of a mathematical model and corresponding software implementing the algorithm of that model.

Let us formulate a submarine design task as an extremum problem. Let C ($C_i...C_j$) be the vector of SDS targets, data from prototypes, etc. (e.g., diving depth, speed).

Vector X ($X_i...X_n$) is the vector of variables to be optimised, i.e. submarine particulars (displacement, main dimensions).

The vector X components are restricted from both sides as:

$$(X_i)_{min} \geq X_i \leq (X_i)_{max}, i=1...n \quad (1.1)$$

These constraints follow, e.g., from service and construction limitations.

Requirements to the submarine are formulated as:

$$B_j(X,C) \oplus A_j(C), j=1...m \quad (1.2)$$

where B_j is an estimation of the j-th quality of the subject design variant;

$\oplus$ is the relator (>, <, etc.); $A_j(C)$ are requirements to the j-th quality.

In this case any variant (any set of ї) that satisfies (1.1) and (1.2), is acceptable.

To select the best variant, we need to introduce the efficiency criterion, i.e. an index of design perfection:

$$Z(X, C) \rightarrow extr \quad (1.3)$$

It is assumed that Z (X, C) is a steady function of the design quality.

Thus, the task of submarine design is to find out such an X vector with which criterion (1.3) reaches extreme values at known C and satisfied (1.1) and (1.2).

A submarine mathematical model representing the totality of functions B_j A_j and ё is shown in Fig.1.4 [82].

```
       Input                                        Output
┌─────────────────────┐  ┌─────────────┐   ┌──────────────────────┐
│    SR or SDS        │  │  Criteria   │   │  Major architectural │
│                     │  └──────┬──────┘   │  and acoustic char-  │
│                     │         ↓          │  acteristics of the  │
│  Design solutions   │→ │ Mathematical │→ │      submarine       │
│                     │  │    model     │   │                      │
│                     │  │ of the sub-  │   │                      │
│  Scientific guide-  │  │   marine     │   │  Architectural and   │
│      lines          │  └──────↑───────┘   │  structural concept  │
│                     │  │ Restraints B │   │                      │
│  Material resources │  │ to seakeeping│   │  Construction cost   │
│                     │  │   abilities  │   │                      │
│  Technical resources│  │ to particulars│  │  Requirements to     │
│                     │  │    other     │   │  subsystems, load    │
│  Service conditions │  │              │   │                      │
└─────────────────────┘  └──────────────┘   └──────────────────────┘
```

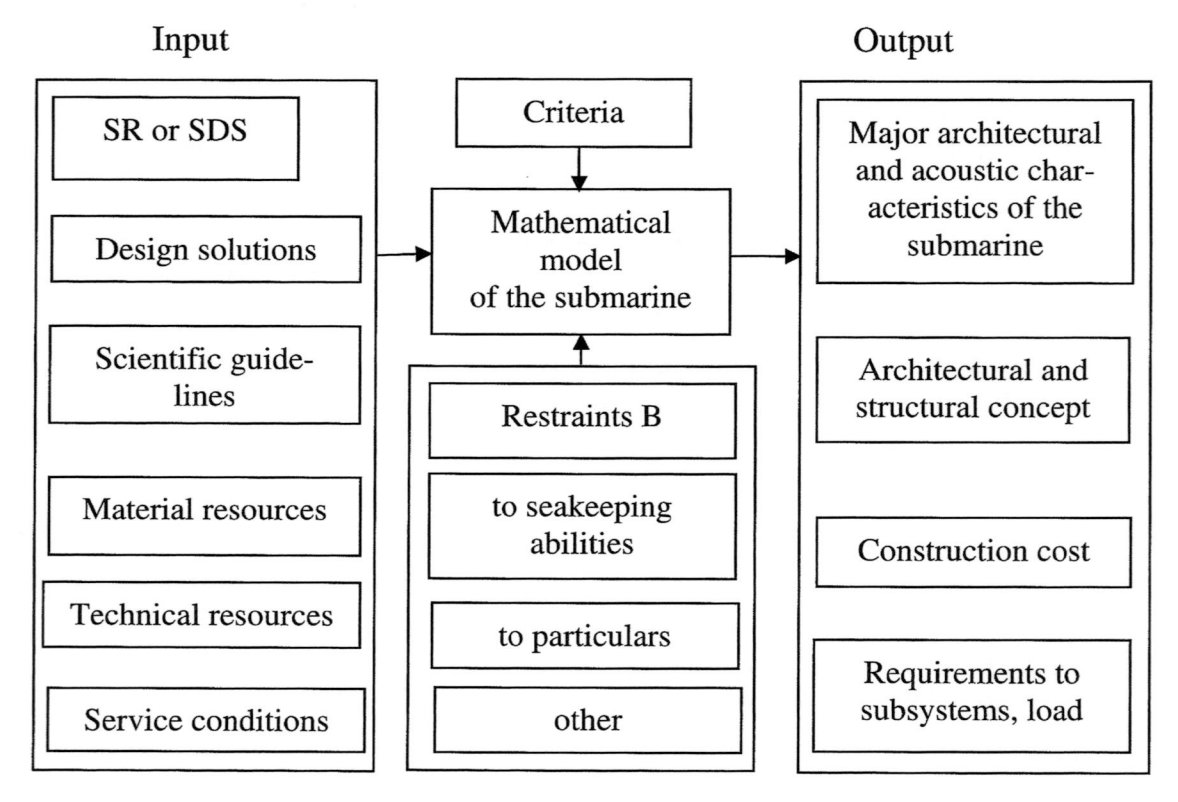

Fig. 1.4 Mathematical Model of a Submarine

The main requirement to any submarine mathematical model is its adequacy, i.e. theoretical solutions obtained by means of this model should be validated by available practical data.

This is achieved by:

– generating the submarine geometrical model allowing it to describe different shapes of the outer and the pressure hulls, as well as of the appendages;

– developing algorithms representing the physical essence of the described relationships, as well as algorithms based on calculation procedures;

– increasing the extent of detailing of the mathematical description of the design process;

– using direct calculation methods of naval architecture.

Mathematical models used in design bureaus are developed mainly for the sake of solving the following design tasks:

– assessments of practical achievability of targets set by the customer, checks of compatibility and consistency of these requirements, justification of submarine characteristics outlined in design specifications;

– determination (updating) of principal particulars of the submarine, of the architectural and structural concept.

The first task is relevant mostly to the Technical Proposal stage (Fig.1.5) and includes generation of different variants of the submarine, evaluation of their particulars and characteristics, and selection of the preferable variant for the future design work.

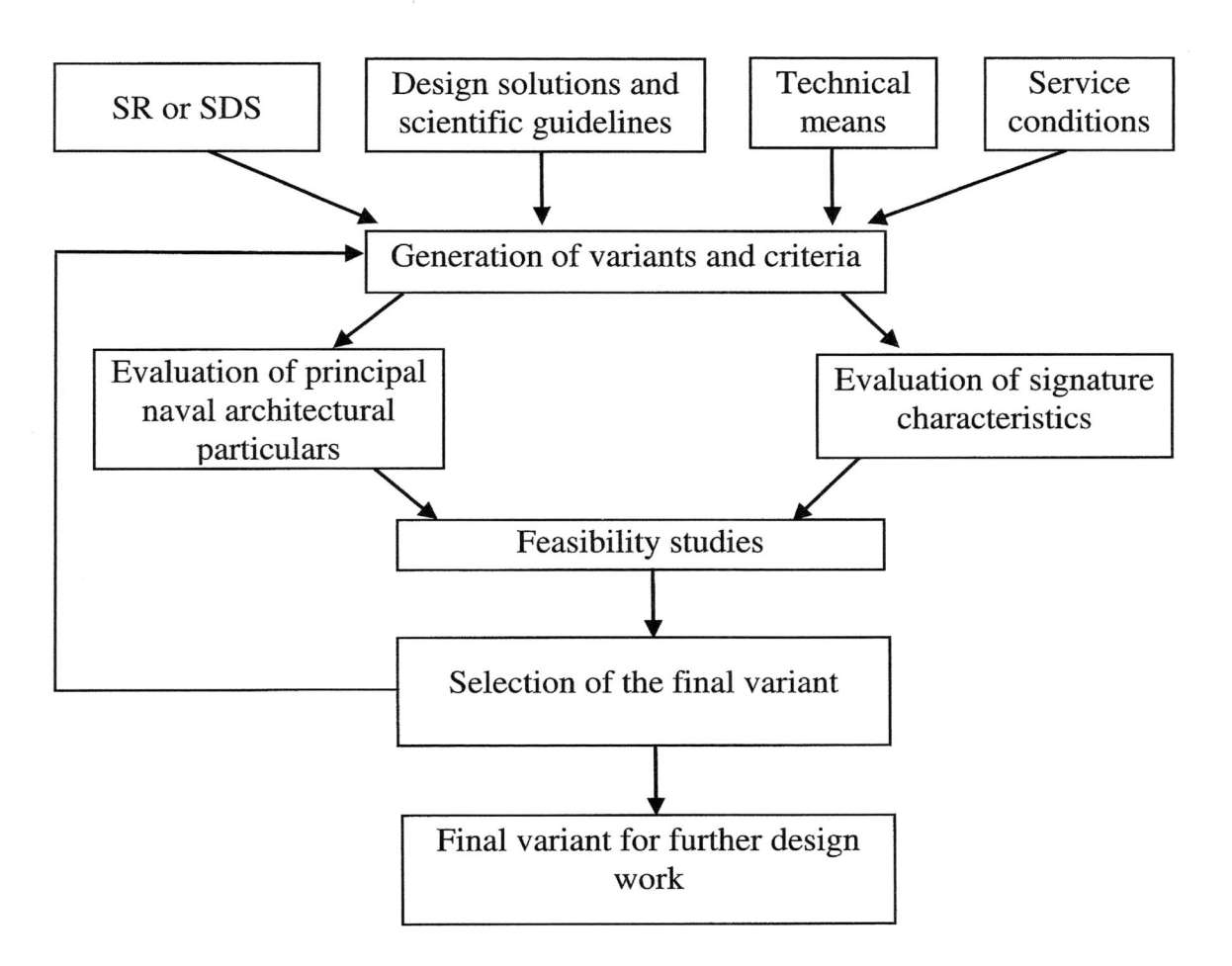

Fig. 1.5. The Traditional Task Solution Flowchart of the Technical Proposal Stage

Fig.1.6 shows a possible configuration of a computer-aided design (CAD) system intended for the problems of the first type.

The second task logically follows from the first one. At this stage they refine the description of the outer and the pressure hulls, update the arrangement, perform calculations on submarine statics and dynamics, and prepare information for further analysis by other departments.

At the same time it should be noted that although a computer is a powerful tool with many abilities, it is only a tool in the hands of a designer and enables the latter to solve specified tasks faster, in more depth and in a broader sense. Just so. The results obtained with computer assistance largely depend on how well the designer understands the physical basis of this or other phenomenon and can analyse the

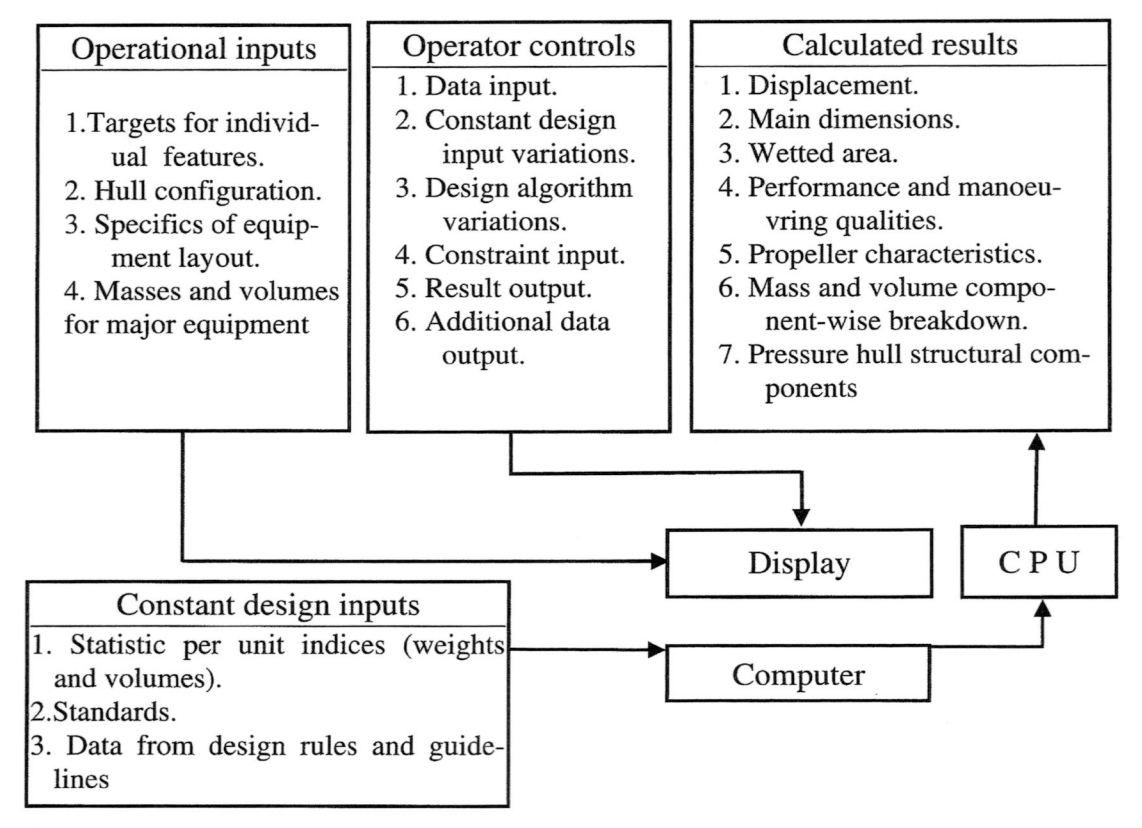

Operational inputs	Operator controls	Calculated results
1. Targets for individual features. 2. Hull configuration. 3. Specifics of equipment layout. 4. Masses and volumes for major equipment	1. Data input. 2. Constant design input variations. 3. Design algorithm variations. 4. Constraint input. 5. Result output. 6. Additional data output.	1. Displacement. 2. Main dimensions. 3. Wetted area. 4. Performance and manoeuvring qualities. 5. Propeller characteristics. 6. Mass and volume component-wise breakdown. 7. Pressure hull structural components

Constant design inputs
1. Statistic per unit indices (weights and volumes). 2. Standards. 3. Data from design rules and guidelines

Display — CPU — Computer

obtained results.

Fig. 1.6. An example of CAD system architecture

This is an appropriate point to mention that any design work, including that for the most sophisticated products like submarines, is an art. The designer should possess and continuously improve his skills and knowledge, perfecting them through hand drawing and analysis, never trusting the baby solely to a machine [50].

CENTRAL RESEARCH INSTITUTE OF SHIPBUILDING TECHNOLOGY

- Basic high scientific input resource-saving technologies employed in construction, repair and scrapping of ships.
- Automation of reproduction engineering in ship construction.  
- Design, modernization and technical refurbishment of shipyards and shops.
- Ecology and safety of labour.
- Manufacture of production equipment, jigs and fixtures.  
- Design of fishing and transport vessels, offshore area development facilities.
- Design and manufacture of valves and fittings for ship pipelines.

CRIST, 7, Promyshlennaya Street, St Petersburg, 198095, Russia
Telephone (812) 186 0401; 186 0429. Fax: (812) 186 0459. E-mail: cnits@telegraph.spb.ru

Federal State Unitary Enterprise
Central Research Institute of Structural Materials

PROMETEY

Central Research Institute of Structural Materials Prometey is a well-known in scientific and business circles multi-functional enterprise, working effectively in the field of creating unique metallic, polymeric and composite materials, technologies for its manufacturing and processing.

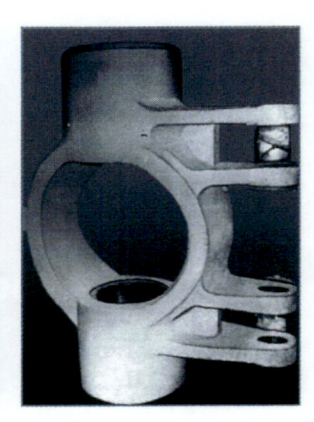

STEEL
TITANIUM
ALUMINUM
NON-METALLIC COMPOSITE MATERIALS

- Development of new materials.
- Development of industrial production technology of materials on metallic and non-metallic base, welding, methods and means of non-destructive suspension.
- Carrying out engineering works and examination of projects, selection of materials for different operating conditions.
- Development of methods and means of surface treatment (coatings, treatment with concentrated energy flows, applying or varnishes, paints, heat and chemical treatment).
- Materials serviceability and lifetime prediction, delivery of materials and welded structures.
- Transfer of experience and "know-how", sale of licenses.
- Technological audit, attestation and certification of materials in accordance with the international requirements.

The materials developed by the Institute meet all specific requirements to brand-new materials which are dictated by the conditions of exploitation of naval and nuclear equipment.

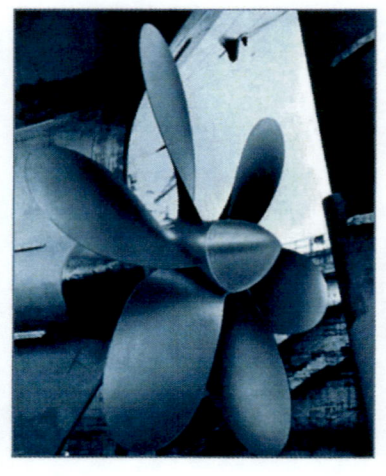

49, Shpalernaya Str., St.-Petersburg, 193015, Russia.
Phone: +7-812-2741619, fax: +7-812-2741707
E-mail: vv@prometey2.spb.su   http://www.prometey.nw.ru

2. SUBMARINE DISPLACEMENT

2.1. Submarine Displacement Types

Unlike for surface ships, for submarines there is only one natural load condition: the normal load. The reason is that in accordance with the law of Archimedes the submarine weight has to be equal to the submerged buoyancy while the latter is definitely dictated (at water specific weight $\gamma = \rho g = $ const) by the submerged constant buoyant volume.

By «normal load» we understand the total weight and the position of the centre of gravity of a completely outfitted submarine with the complement, machinery, systems and gears fully ready for operation, with issued special-purpose equipment and normal stocks of consumables (fuel, water, provisions, etc.) and trimmed with solid ballast [74].

The submarine normal load corresponds to the normal displacement D_0. The normal displacement is equal to the product of the design water density into the volume displacement. The notion of constant buoyant volume V_0 is used when considering the submerged submarine; for the surface condition they apply volume displacement V_{SFB} that corresponds to the full-buoyancy waterline. These two volumes are equal in value but formulated differently.

$$D_0 = \rho g V_0 \quad D_0 = \rho g V_{SFB} \tag{2.1}$$

In underwater shipbuilding they also use several other displacement notions that are necessary to account for the weight of the water flooding hull volumes that are not included into the constant buoyant volume (CBV): those of the main ballast tanks (MBT) V_{MBT} and of the permeable structures V_{MBT}.

Submerged displacements D_S and V_S take into account the weight and the volume of water in the net MBT volume.

Submerged displacements D_S and V_S represent the total watertight volume that remains watertight until the submarine dives.

$$D_S = D_0 + \rho g \sum V_{MBT}$$
$$V_S = V_0 + \sum V_{MBT} = V_0 \cdot (1+\varepsilon) \qquad (2.2)$$

where $\varepsilon = \dfrac{\sum V_{MBT}}{V_0}$ – is the relative buoyancy reserve.

The total submerged displacements D_{FS} and V_{FS} include the weight and the volume of water in MBT and in all permeable parts of the hull.

$$D_{FS} = D_0 + \rho g \sum V_{MBT} + \rho g \sum V_{PEP}$$
$$V_{FS} = V_0 + \sum V_{MBT} + \sum V_{PEP} \qquad (2.3)$$

The total submerged volume is calculated by the outer surface of the hull plating including volumes of all appendages. It should also include any external coating applied to the hull.

The greater part of V_{FS} is the bare hull volume V_{BH} calculated from the moulded surface (ignoring plate thickness, external coating on the outer hull and appendages).

The above-listed displacement notions are applicable to any floating underwater object.

Diesel-electric submarines can take extra fuel capacity into specially fitted fuel-ballast tanks. In order for the submarine to stay submerged, the residual positive buoyancy of the fuel has to be balanced by an adequate amount of additional stocks (oil, water, provisions) and water in auxiliary ballast tanks (ΔP). For the case of sailing with extra fuel, they use the notion of «displacement with excessive fuel capacity» D_{EC}, V_{EC} found as:

$$D_{EC} = D_0 + \rho_F g \sum V_{FBT} + \Delta P = D_0 + \rho g \sum V_{FBT}$$
$$V_{EC} = V_0 + \sum V_{FBT} \qquad (2.4)$$

where V_{FBT} is the net volume of fuel-ballast tanks.

From formulae (2.4) one may see that all these displacements are modifications of the normal displacement as in every case we just extend the «added weight» method, which is commonly used in naval architecture, to some flooded volumes of the hull.

Each of the above-listed displacement notions has a certain physical meaning and is applied within its own range of tasks. Thus, the normal displacement determines the submarine weight in air without water in MBTs and permeable structures. Derivatives of this displacement are used in tasks associated with construction technologies, submarine transportation, and in economic calculations (launching weight, transported weight). Actually, it is the normal displacement that they usually state in SDS.

The normal displacement is a basic, initial parameter. Nevertheless, it is not what defines the hullform, the main dimensions or such major submarine particulars like surface and submerged performance and manoeuvrability. These qualities respectively govern the surface and the total submerged displacements because only they are characteristics of the hullform, main dimensions, total moving weight and wetted surface of the submarine. This is one of the major specific features of submarine design. Considering the above, as well as the fact that signature levels depend on the total submerged displacement, when ordering a submarine it should be reasonable to specify requirements to this particular displacement.

2.2. Displacement (Load) Breakdown into Standard Groups

In order to properly manage load calculations, to reduce the probability of errors, as well as to make it possible to compare loads of different submarine designs (prototypes) and utilise such data in design work, load balance calculation results should be presented in a uniform format. Therefore, load calculations for submarines, same as for surface ships, are regulated by relevant industrial standards that contain some general rules for calculations, define the breakdown of component weights constituting the submarine load and the format of load balance calculation result tables.

All weights collectively constituting the normal weight load of the submarine are broken down into groups, subgroups and types. A numeral code is assigned to each weight. Weight groups and group codes correspond to the «Classifier of the Unified Design Document System, Class 36 (ships, ship equipment)» [33] following which they generate submarine detailed drawings at the Detail Design stage.

Table 2.1 demonstrates the format for logging load balance components.

Depending on the design stage, the submarine load balance is calculated with different extents of detailing and with different methods. For the Technical Proposal, loads are usually subdivided only into groups while at the Preliminary Design phase they are further segregated into subgroups. In such cases weights are mostly re-calculated from prototypes using approximate formulae, or are determined based on some steady statistic regularities derived from the analysis of loads of earlier constructed or designed submarines. Generally, the accuracy of load and displacement component calculations with these methods is not high (about 10%) because initial data may vary within wide ranges. At the later design stages, when they need higher accuracy, the use of such methods becomes very limited.

Table 2.1

Tabular Log for Load Calculation Results

Load component code	Load component description	P, t	Levers			Moments			Notes
			X, m	Y, m	Z, m	Mx, tm	My, tm	Mz, tm	

At the Engineering Design phase the load is broken down into subgroups and components. Weights and levers are determined more accurately due to the more detailed weight breakdown and rather detailed structural and general arrangement drawings, system and electrical equipment schematics, etc.

At the Detail Design stage weights and levers are calculated based on detailed drawings, more detailed general arrangement drawings and specifications for subcontracted equipment.

As a result of these efforts the so-called record load balance (as calculated from detailed drawings) is issued.

In addition to levers and moments in terms of the height above the base plane and the distance from the midship section, at Engineering and Detail Design stages they calculate moments of weight with respect to the centreplane in order to make sure that there is no static heel and, if necessary, to eliminate it by structural modifications or ballast.

The load balance calculation should contain a clear indication of the abscissa reference plane position with respect to the nearest actual frame and a table of frame offsets from this reference plane.

Table 2.2 shows the standard breakdown of submarine weights into groups [16], [33], [65], [69].

The standard weight breakdown is a convenient tool both for load calculations while generating submarine drawings and for weight management during construction.

Table 2.2

Submarine Load Component Groups

Group Number	Description	Diesel-electric submarines	Nuclear torpedo submarines	Nuclear missile submarines
000	Displacement margin	0.5-1.5	1.0-1.5	1.0-1.5
100	Hull	37-38	38-39	39-40
200	Hull gears, fittings	3-4	3-4	2-3
300	Furniture and equipment of spaces, paint, insulation, special coatings, protectors, spare parts and supplies	8	6-7	6.5-7.5
400	Mechanical equipment, pipelines and systems of power plants	8-9	16-18	13-14
500	Hull systems	8.0-8.5	8.0-8.5	6-7
600	Electric equipment and cables of electric power systems, electric networks and radioelectronic equipment	16-20	7-8	6-8
700	Weapons and their supporting systems	4-5	5	11-16
800	Stocks and complement	3-5	4-5	2-3
*	Total per load groups without ballast Solid ballast Normal load of the submarine at $\rho = t/m^3$	88.5-99.0	88-96	87-100

In conceptual design, when basic particulars of the boat are yet to be determined, one may notice certain drawbacks of this standard breakdown because its groups combine weights dictated by different characteristics and particulars of the submarine.

In this regard it is necessary to make an important comment. Since the appearance of computers their abilities are widely used for calculating all load components: weights, moments and their sums. This is absolutely natural. However, if, especially at early stages, designers neglect «manual» and comparative checks of weight and moment calculations, mistakes are inevitable and their results may be deplorable.

2.3. Load Breakdown at Early Design Stages. Weight Indices

The equation of weights, which determines the displacement of a submarine, can be solved only when relationships between load components and submarine characteristics and particulars are already established. Therefore, at early design stages the weight breakdown differs from the procedure prescribed by the standard.

First, they segregate those submarine component weights that can be derived accurately enough from the data of design specifications. These include weights of weapons, stocks, complement, power plants, etc.

Secondly, all other weights are grouped according to the degree of their dependence on principal characteristics and yet unknown particulars of the submarine.

This load breakdown approach (let us for the sake of convenience call it «design» principle) allows to rather accurately calculate, even at early design stages, all specified weights and to establish for the rest of the weights some more obvious physical relationships than those implied in the standard [74].

Let us see how submarine load groups change when we apply the design breakdown instead of the standard one.

First of all we put together all structures subjected to the full external pressure: the pressure hull proper (plates, frames, stiffeners) and equistrong structures (end bulkheads, the sail, pressure tanks, etc.). Let us designate these weights as P_{PH} and P_{EST}.

The rest of the standard Group 100 «Hull» components become a new category called «Light Hull» P_{LH}. When we need more accurate calculations, this item can be further subdivided into a number of sub-items.

Submarine hull systems and gears (Groups 200 and 500) are combined under a single title P_{GS}, though, similarly to P_{LH}, they can be calculated more accurately.

In Group 300, it is advisable to segregate «Coatings» P_{COAT} as their weight is determined, mostly, by signature control requirements and can be calculated separately.

In diesel-electric submarine (SSDE) design, Group 400 as a rule consists of the following items:
 – diesel plant P_{DP},
 – electric propulsion plant P_{EPM},
 – shafting P_{SH} with associated auxiliary machinery and gears.

For a nuclear submarine (SSN), this group consists of the main power plant P_{MPP} and the auxiliary power plant P_{APP}.

It is advisable to represent Group 600 as three items:
 – storage battery system P_{SB},
 – general-purpose electric equipment and cables P_{EEG};
 – radioelectronic aids P_{REA}.

For nuclear submarines, P_{SB} may be considered together with the weight of the main power plant.

Group 800 should be split into:
 – fuel and oils P_{FO};
 – complement provision and fresh water stocks P_{SCR};
 – transported weights P_{TRC};
 – trimming and trapped water P_{TTW}.

Group 000 covers reserves of displacement left for future upgrading P_{UDM} and for design and construction of the submarine P_{CDM}. The solid ballast P_{BAL} is presented as a separate item.

Depending on special features of the subject submarine, on the scope of initial data inputs and on the availability of a close prototype, this design load breakdown may be modified. In particular, it may be made more detailed or, vice verse, more general. Sometimes it may be more convenient to group weights by similar designations or by submarine weight modules.

Formulae in which weight indices are used for approximate estimations of submarine weights may be divided into two groups.

The first group: weights are presented as functions of those characteristics and principal particulars of the submarine that determine her geometric configuration

$$P_i = p_i f_i(D_{FS}, L, B, H, \delta \ldots \vartheta, H_{LIM}, R) \qquad (2.5)$$

where p_i – weight index of the 1st group;
f_i – functions of principal particulars;
L – hull length;
B – hull breadth;
H – midship hull height;
δ – block coefficient;
ϑ – submarine speed;
H_{LIM} – maximum diving depth;
R – sea range under subject conditions (surface, snorkel, submerged).

Similar functions are commonly used in the design of surface ships and vessels. It is more difficult to use such formulae in submarine design because weight calculations are made for the normal displacement that does not determine the main dimensions and the outer hullform. The relationship between D_0 and D_{FS}, as will be shown below, is not very stable even within any single architectural type of submarine, and the use of D_{FS} in formulae for weight estimations leads to higher errors. Therefore, weight estimation relationships of the first group are, in submarine design, usually formulated like:

$$P_i = p_i f_i(D_0, \vartheta_1, H_{LIM}, R \ldots) \qquad (2.6)$$

The second group of formulae serves to find weights of individual structures from their known dimensions or volumes

$$P_i = g_i \varphi_i(V_i, l_i, b_i, h_i \ldots) \qquad (2.7)$$

where g_i – weight index of the 2nd group;
φ_i – functions of characteristic dimensions of the subject structure;
V_i – volume of the subject structure;
l_i, b_i, h_i – length, breadth and height of the subject structure.

Formulae of the second group are used for more detailed calculations when the general arrangement schematics are already available but strength calculations and detailed drawings of structures have yet to be made.

Weight indices may be derived from the following documents:
- documents of prototype submarine designs (one or several);
- documents of submarine designs for which load calculations have been made in sufficient detail;
- subcontractors' documents from which it is possible to find, e.g., weight indices of the power plant, electrical equipment, etc.;
- special design studies made as preliminary feasibility checks for some new design solutions suggested for the submarine;
- predicted weight indices assumed in the Conceptual Design based on the analysis of future engineering developments.

In practical design tasks it is crucial to select weight indices corresponding to the level of the considered task.

When we calculate the weight of a whole group, we should apply weight indices derived for this entire load group; if we consider a subgroup weight, we should use subgroup weight indices; and finally, when we come to the weight of an individual structure, we should apply weight indices computed from a characteristic dimension or parameter of that very structure.

We should, however, note that while using weight indices or other values derived from above-mentioned documents, the designer has to remember that any direct transfer of these data to a newly designed submarine may turn into just a way to conserve the technical level of the prototype. Therefore, the application of weight indices and documents of prototype designs should be preceded by creative analysis and corrections for the latest engineering achievements and special features of the subject design.

3. WEIGHTS OF INDIVIDUAL GROUPS AND LOAD ITEMS AS FUCTIONS OF PRINCIPAL CHARACTERISTICS OF THE SUBMARINE

3.1. The «Hull» Load Group

There are a number of formulae for the weight of surface ship hulls. These formulae relate the weight to the ship displacement and main dimensions, including the hull depth and the block coefficient.

It is impossible to use these formulae for submarines for two reasons:

– first, submarine main dimensions are determined exclusively by the total submerged displacement;

– secondly, there are two hulls (pressure and light) with their separate functions. Therefore, in weight calculations for the «Hull» group we have to consider these hulls separately.

Pressure Hull and Equistrong Structures

The Pressure Hull Weight P_{PH} depends on the design pressure (diving depth), geometrical characteristics, physical and mechanical properties of material. Besides, it depends on the structural features, e.g., the framing system (internal or external), on the chosen frame spacing, on rules and procedures applied to structural strength analysis [1], [64], [77], [90].

At the stage when principal particulars of the submarine are being determined, when dimensions of pressure hull (PH) compartments are not yet established, compartment weights are estimated with elementary formulae:

$$P_{PH} = p_{PH} D_0 \quad (3.1)$$

$$P_{PH} = g_{PH} V_{PH}, \qquad (3.2)$$

where V_{PH} – the PH moulded volume as measured on the internal surface of the plating.

Formula (3.1) may be used to estimate submarine displacement. It represents the simplest (and at the same time the least accurate) relationship between the PH weight and the submarine displacement. The P_{PH} index (the PH relative weight), same as all other indices of this kind, is dimensionless.

Formula (3.2) is used at the early design stages quite widely. It includes the dimensional index g_{PH} (t/m³) sometimes called «specific weight of the PH», that shows the weight per 1 m³ of the buoyant volume provided by the PH. This index has a definite physical meaning. The difference

$$\eta_{PH} = \rho - g_{PH} \qquad (3.3)$$

may be regarded as an efficiency index of the pressure hull as a source of buoyancy. The η_{PH} value determines what payload can be accommodated by the PH per 1 m³ of its volume.

Comparing formulae (3.1) and (3.2) we can derive a formula to relate indices P_{PH} and g_{PH}. It shows that index

$$P_{PH} = g_{PH} \frac{1}{\rho} \left(\frac{V_{PH}}{V_0} \right) \qquad (3.4)$$

should be less steady as it depends on g_{PH} and on $\left(\frac{V_{PH}}{V_0}\right)$, which, as will be described below, may vary within rather a wide range. Thus, e.g., at the same value $g_{PH} = 0.18$ t/m³ the PH relative weight varies within $P_{PH} = 0.13$ to 0.17.

Let us consider the same aspects associated with practical application of (3.2).

To escape effects of individual features of any actual submarine PH configuration, which is usually a combination of cylindrical and tapered components, let us introduce an arbitrary analogue of the pressure hull: a cylinder with a constant radius along the entire length, stiffened by equirigid frames without any cutouts or local reinforcements. This «analogue cylinder» has the same as the real PH volume and length, is built of a material of the same grade and designed to the same external load using the same calculation procedure (Fig.3.1)

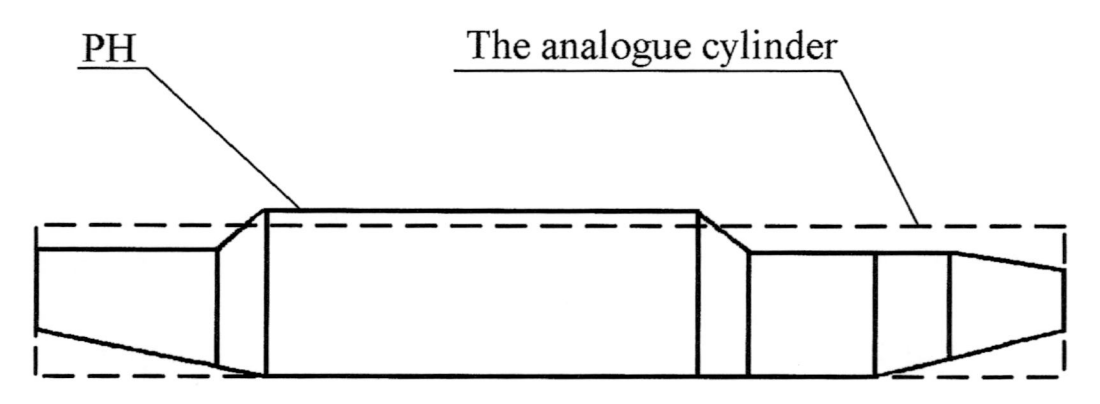

Fig. 3.1. The Pressure Hull Analogue Schematic

Let us call the combined weight of plates and frames of this analogue cylinder the «theoretical weight» of the PH

$$P_{PH}^T = g_{PH}^T V_{PH}$$

where g_{PH}^T – the PH theoretical weight index.

In order to be able to consider the weight of the actual hull, let us introduce two static coefficients k_{SSR} and k_{LSR}:

$$P_{PH} = k_{SSR} \cdot k_{LSR} \cdot g_{PH}^T V_{PH} \qquad (3.5)$$

Coefficient k_{SSR} accounts for the fact that the real hull has small distributed stiffeners and reinforcements for plating, framing and tapered portions while some plates have cutouts and some parts of PH frames are removed where they would have crossed pressure tanks (PT) equistrong with the pressure hull. If the value of the index found from the specified thickness of the plating and volumes of equistrong tanks is 2 to 3% of D_0, this coefficient is $k_{SSR} \approx 1.02$ to 1.04.

Coefficient $k_{LSR} = 1.05$ to 1.22 accounts for large concentrated reinforcements: bossings of propeller shafts, coamings of missile silos, etc. and is re-calculated from a prototype sufficiently close in terms of the type and scope of hull reinforcements.

Comparing (3.2) and (3.5) we can find a relation between indices of the actual and the theoretical PH weights.

$$g_{PH} = k_{LSR} k_{SSR} g_{PH}^T \qquad (3.6)$$

Since the real submarine PH is quite close to the analogue cylinder and the factors determining their weights are identical, we may investigate these factors using the g_{PH}^T index as an example.

The general functional dependence for g_{PH}^T can be formulated as:

$$g_{PH}^T = f(P_D, \rho_{PH}, \sigma_T, E, d_{PH}, \ell_{COM}...) \qquad (3.7)$$

where: P_D — design load;
ρ_{PH} — hull material density;
σ_T — hull material yield strength;
E — modulus of elasticity;
d_{PH} — PH diameter;
ℓ_{COM} — PH compartment length.

For the analogue cylinder the g_{PH}^T index can be determined from a single frame spacing (Fig.3.2).

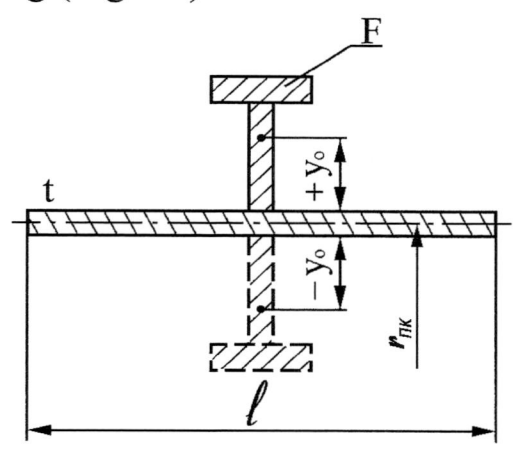

Fig. 3.2 The Frame Spacing Design Model.

$$g_{PH}^T = \frac{P_{PH}^T}{V_{PH}} = \frac{P_{PL} + P_{FR}}{V_{PH}} = \frac{2\pi r_{PH} t \ell \rho_{PH} + 2\pi (r \pm y_0) F \rho_{PH}}{\pi r_{PH}^2 \ell}$$

After relevant manipulations we arrive at:

$$g_{PH}^T = 2\rho_{PH} \frac{t + \frac{F}{\ell}(1 \pm \frac{y_0}{r_{PH}})}{r_{PH}} = 2\rho_{PH} \frac{t_{RFT}}{r_{PH}} \qquad (3.8)$$

where: t — PH plating thickness;
ℓ — frame spacing;
F — frame area;
y_0 — offset of the frame section neutral axis («+» for external frames);
t_{RPT} — reduced plating thickness.

Let us expand the g_{PH}^T index formula (3.7) into functions of governing parameters.

Structural particulars of the PH (t, F and ℓ) in formula (3.8) are determined by PH strength analysis in terms of stresses and buckling resistance of plates and frames. It is rather difficult to obtain a sufficiently accurate analytical expression for the g_{PH}^T index that would account for all conditions of PH calculations, even for such a simplistic structure like a stiffened cylinder. Assuming that the plate thickness is determined based only on the strength considerations, we can re-calculate g_{PH}^T from a prototype with the same grade of material:

$$g_{PH}^T = g_{PH}^T \frac{H_{LIM}}{H_{LIM_0}} \cdot \frac{\sigma_{T_0}}{\sigma_T} \qquad (3.9)$$

where H_{LIM_0} – extreme diving depth for the chosen prototype submarine;

σ_{T_0} – prototype hull material yield strength.

This formula can be used for a fast approximate evaluation of the PH weight index at different H_{LIM} and σ_T when their variations are small because the neglect of buckling conditions and other assumptions may lead to considerable errors. Therefore, they use results of systematic strength calculations for circular cylinders simulating submarine pressure hulls with variations of initial data: the material (ρ_{PH}, σ_T and E), the design load (P_D), the PH geometry (r_{PH} and ℓ_{COM}). Values of g_{PH}^T obtained by such calculations are plotted as curves.

Let us consider the effects of geometric parameters: the PH diameter d_{PH} and the compartment length ℓ_{COM} upon g_{PH}.

The diving depth and the material grade will be assumed to be invariable.

Under a strict task formulation, for the sake of result definiteness with every variant of d_{PH} and ℓ_{COM} we should look for such a combination of t, F, ℓ that would minimise g_{PH}^T, i.e. would show the least PH weight.

In Fig. 3.3 one can see plots of the following functions:

$$(g_{PH}^T)\min = f_1(d_{PH}, \ell_{COM} = \text{const})$$
$$(g_{PH}^T)\min = f_2(d_{PH}, \ell_{COM}/d_{PH} = \text{const}) \qquad (3.9)$$

In the first case g_{PH}^T considerably decreases with the growth of d_{PH}. This is explained by the increasing effect of bulkheads becoming relatively closer to each other.

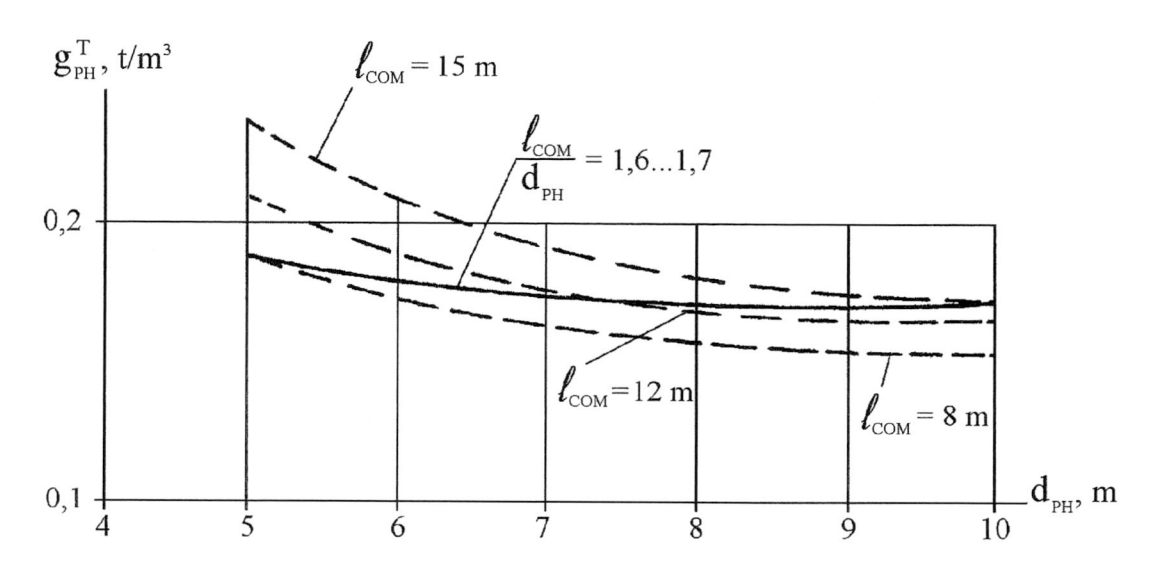

Fig. 3.3. The Pressure Hull Weight Index Versus the PH Diameter and the Compartment Length

In the second case the d_{PH} influence is much less than in the first case. Within the range of diameters and relative lengths of compartments $\frac{\ell_{COM}}{d_{PH}} = 1.6$ to 1.7 typical for modern submarines, it may be regarded as virtually non-existent.

Hence, at constant P_D and σ_T the PH diameter optimisation aiming at the absolute minimum of the PH weight at $V_{PH} = $ const is meaningless. Transverse dimensions and the PH length-to-diameter ratio are in most cases dictated by the general arrangement of the submarine taking into account performance-wise requirements to the hull-form, construction technology considerations and other factors but never by PH weight reduction considerations.

Let us now consider g_{PH}^T curves plotted as functions of H_{LIM} and σ_T for steel (Figs.3.4 and 3.5). Similar curves are routinely used in submarine design work.

Examining Figs.3.4 and 3.5 we may notice that at a constant design load (diving depth) the increase in σ_T results in a gradual reduction of the useful effect: benefits in terms of index values and PH weights drop and at higher σ_T values even tend to zero. In Fig.3.5 the line connecting points with conventionally extreme (for the effect upon the PH weight) values of σ_T is shaded.

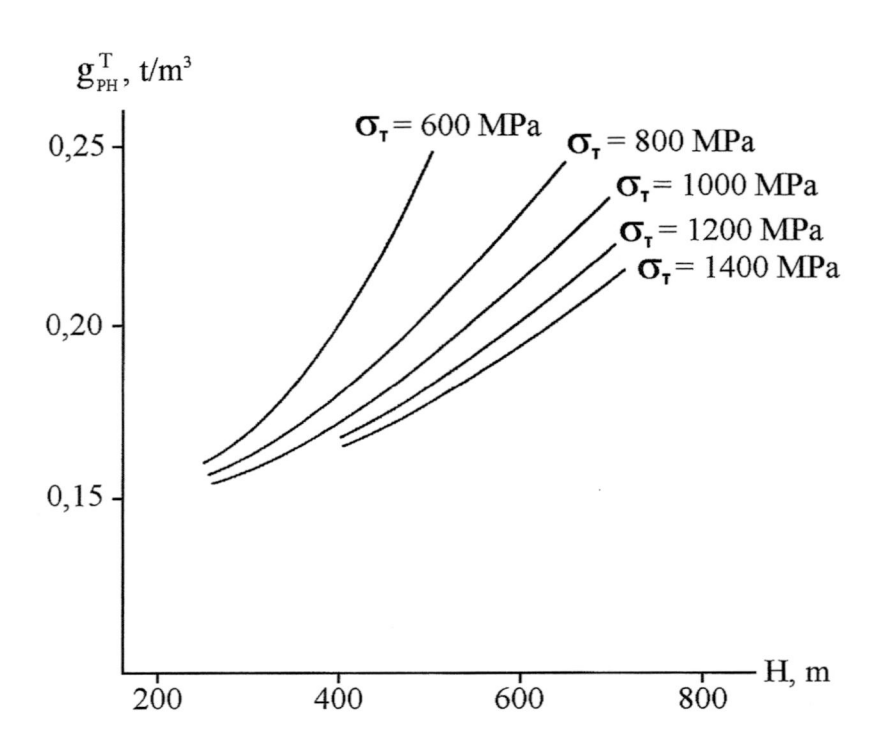

Fig. 3.4. The g_{PH}^{T} Index As a Function of the Diving Depth

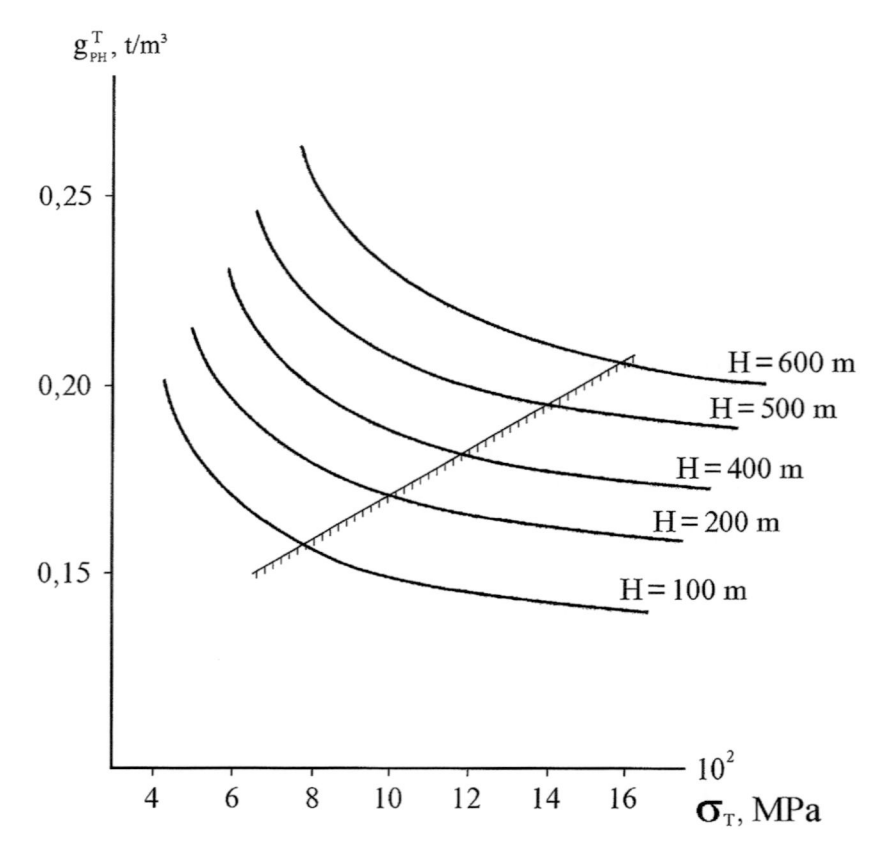

Fig. 3.5. The g_{PH}^{T} Index As a Function of the Material σ_T

The design load (diving depth) increase expands the domain of effective influence of the yield strength on the PH weight.

This pattern of σ_T effect upon the PH weight is explained by the buckling resistance factor because σ_T variations do not change the modulus of normal elasticity E, and therefore it is impossible to reduce plate thickness without changing the frame spacing length. Strains in the material will be below the allowable ones and its high mechanical properties will not be utilised to the full extent.

Strictly speaking, there is some reduction of the g_{PH}^T index with the increase in σ_T even when the plate thickness is determined by buckling resistance conditions, but this reduction is so small that it is irrelevant for our general conclusions.

Fig.3.6 gives an idea of the material type effect upon the PH weight index. We may see that new high-yield materials can considerably reduce the PH weight or, accordingly, increase the diving depth of the submarine.

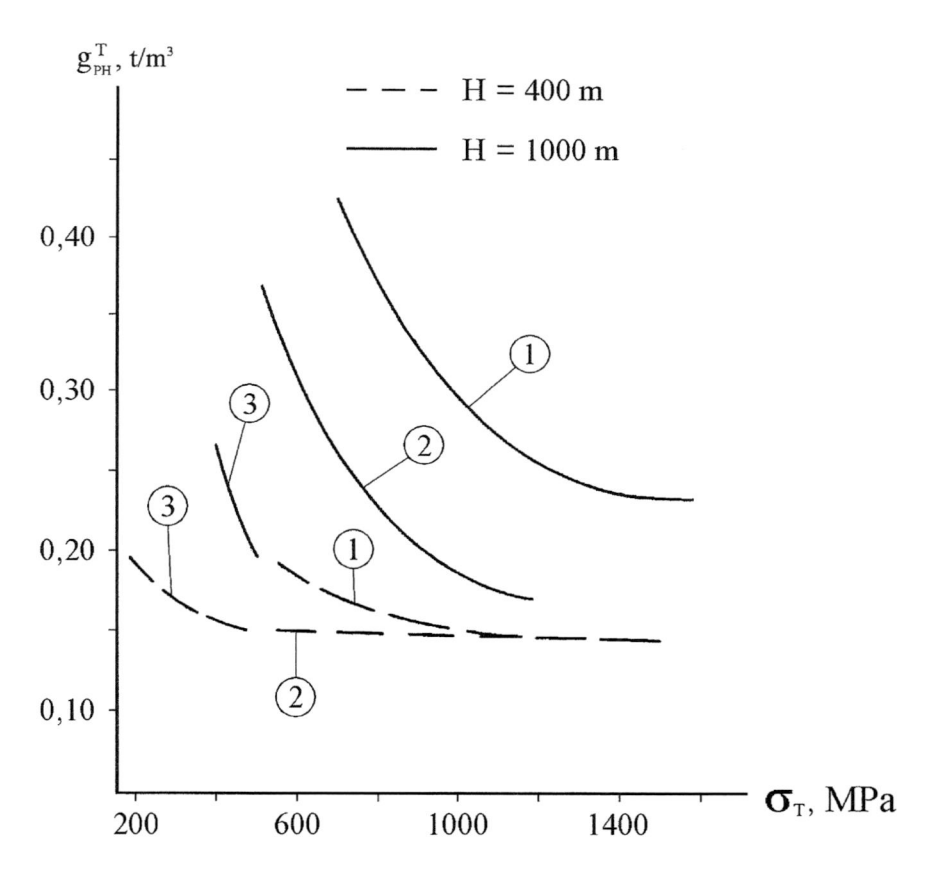

Fig. 3.6. Material Type Effect Upon
the Submarine Pressure Hull Weight Index
1 - steel, 2 - titanium, 3 - aluminium-magnesium alloy.

The PH weight also depends on the framing system. External frames increase the g_{PH}^T index compared to internal ones:

at d_{PH} < 8.0 m approximately by 1.5 to 3.0% at $d_{PH} \approx$ 8 to 11 m approximately by 3.0 to 5.0%

Nevertheless, it should be kept in mind that when frames are mounted outside the PH they not only increase its weight but add extra buoyancy as well. Comparing the increase in the weight with the additional buoyancy, M.K.Glozman [1] has formulated a condition that, if satisfied, would balance (exactly or in excess) the increase in g_{PH}^T by the additional buoyancy.

$$r_{PH} \geq \frac{2\rho_{PH} - \rho}{\rho} y_0 \qquad (3.11)$$

where: r_{PH} — pressure hull radius;
$\quad\quad\quad\;\; y_0$ — offset of the frame neutral axis from the pressure hull axis.

For steel hulls (ρ_{PH} = 7.85 t/m³) the (3.11) condition is: $r_{PH} \geq 14.7\, y_0$.

In practical design the (3.11) condition as a rule is observed, though it should be noted that the choice between internal and external framing is actually more complicated as it is also necessary to consider general arrangement requirements, hull configuration (the double-hull system), service requirements, etc.

When PH dimensions are established, the g_{PH}^T index is a function of the frame spacing length ℓ, which can be selected in such a way that together with other structural elements of the PH – t, F – it would minimise the g_{PH}^T value.

However, this is a task of «internal» optimisation, it pertains to hull structure design and falls beyond the scope of present considerations.

Besides, the influence of the frame spacing on the PH weight is relatively minor and when choosing its length one should also consider its impact on labour requirements and, hence, duration and cost of the submarine construction [104].

Listed below are approximate methods for PH weight estimations that may be applied depending on the available initial information and on the problem to be solved. The PH volume is assumed to be known.

1. There is a close (in terms of the pressure hull) prototype.

a) material grades and design depths of the subject project and of the prototype are the same: $g_{PH}^T = g_{PH_0}^T$.

$$P_{PH} = P_{PH_0} \frac{V_{PH}}{V_{PH_0}} \qquad (3.12)$$

b) material grades and design depths of the subject project and of the prototype are different: functions $g_{PH}^T (P_D; \sigma_T)$ (see Fig.3.4), which act as extrapolators are used to scale the prototype g_{PH_0} to design P_D and σ_T and then to find the PH weight.

$$g_{PH}^T = g_{PH_0} \frac{g_{PH}^T(P_D; \sigma_T)}{g_{PH}^T(P_D; \sigma_{T_0})} \qquad (3.13)$$

$$P_{PH} = g_{PH} V_{PH}$$

2. There is no sufficiently close prototype.
The theoretical index $g_{PH}^T (P_D; \sigma_T)$ is found from Fig.3.4.
The PH weight is calculated with (3.5).
3. Configurations and dimensions of PH compartments are available.
Using an applicable standard procedure, they make tentative strength calculations for each PH compartment to find t_i and F_i (the frame spacing is assumed to be uniform along the entire hull). Then they apply the formula suggested by E.A.Gorigledzhan to determine the PH weight:

$$P_{PH} = (P'_{PL} \cdot k_{PL} + P'_{FR} \cdot k_{FR}) k_{WD} + P_{SR} \qquad (3.14)$$

where: $P'_{PL} = \sum_{1}^{n} P'_{PL}$ — weight of the plating (ignoring cutouts and reinforcements) found as the total of compartment plating weights;

$P'_{FR} = \sum_{1}^{n} P''_{PL} - \Delta P_{FR}$ — weight of the framing by compartments less frame portions removed because of pressure tanks (Fig.3.7);

$\sum_{1}^{n} P''_{PL}$ — weight of the framing by compartments without the subtraction of portions removed not to cross pressure tanks;

ΔP_{FR} — weight of frame portions removed not to cross pressure tanks;

n — number of PH compartments;

$k_{PL} = 1.02$ — coefficient accounting for plating cutouts and reinforcement;

$k_{PL} = 1.03–1.05$ — coefficient accounting for framing reinforcement and stiffening, e.g., in way of removable plates

$k_{WD} = 1.02$ — coefficient accounting for the metal of welded joints;

P_{SR} — weight of large concentrated reinforcements (coamings of silos, etc.).

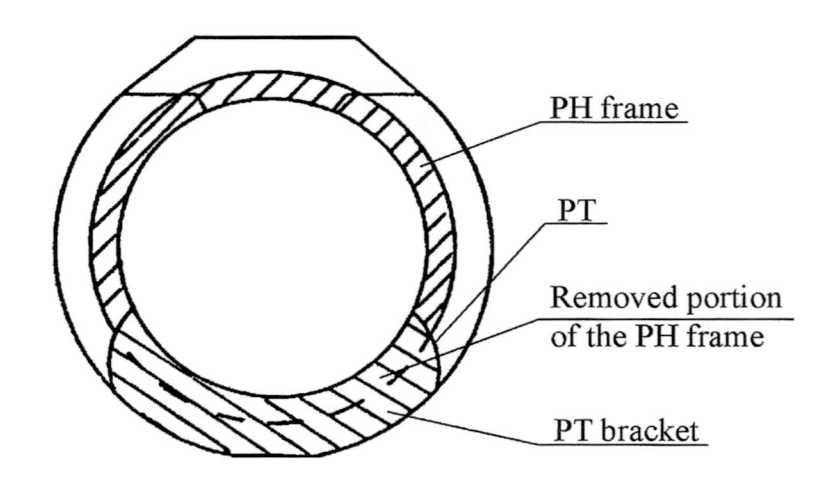

Fig. 3.7. The Pressure Hull Model for Weight Calculations with Formula (3.14)

There is a number of submarine hull structures that are calculated in the same way and for the same design load as the pressure hull. Their weights depend on same parameters as the PH weight.

These structures, collectively called «equistrong» (with the PH) include:
– end bulkheads of the pressure hull;
– pressure tanks;
– pressure superstructures.

For a tentative estimation of the total weight of these structures we can assume that:

$$P_{EST} = p_{EST} D_0 \qquad (3.15)$$

Design load and material grade changes compared to the prototype are covered by re-calculating the p_{EST} index:

$$p_{EST} = \frac{P_D}{P_{D_0}} \cdot \frac{\sigma_{T_0}}{\sigma_T} \qquad (3.16)$$

Formulae (3.15) and (3.16) do not reflect all factors that determine the weight of equistrong structures but their share in the load is not very considerable and this inaccuracy in their estimation does not result in any noticeable error in the submarine weight.

Index $p_{EST} = 0.03$ to 0.04 and with the increase of D_0 its value somewhat reduces.

The weight of end bulkheads has a steady share of about 0.01 of the displacement (0.03 to 0.04 of the PH weight). The major portion of the weight of all equistrong structures belongs to pressure tanks. The weight contribution of pressure superstructures on modern submarines is minor and never exceeds 0.2 to 0.3%.

When the designer already has compartment and tank schematics enabling him to establish characteristic dimensions of the PH, weight calculations for equistrong structures should be differentiated using either g_i indices or results of preliminary strength calculations for these structures.

End Bulkheads:

$$P_{ENB} = P_{ENP}^F + T_{ENP}^{AF} = g_{ENP}^F S_F + g_{ENP}^{AF} S_{AF} \tag{3.17}$$

where $S_{F(AF)} = \dfrac{\pi}{4}(d_{ENB}^{F(AF)})^2$ – cross-section area covered by the bulkhead.

Indices $g_{ENB}^{F(AF)}$ found from prototypes largely depend on bulkhead diameter (increasing with its growth) and structural configuration (plane or spherical bulkhead). In spite of the fact that the spherical bulkhead works in buckling resistance, its weight is 1.05 to 1.3 times less than that of the plane one.

Pressure Tanks:

$$P_{EGT} = \sum_1^n g_{EGT_i} V_{EGT_i} \tag{3.18}$$

It is also possible to use the 1st group index related to g_{EGT} as:

$$p_{EGT} = \dfrac{P_{EGT}}{D_0} = g_{EGT} \dfrac{1}{\rho} \left(\dfrac{\Sigma V_{EGT}}{V_0} \right) \tag{3.19}$$

where g_{EGT} – averaged weight index for equistrong tanks.

From (3.19) it follows that the share of pressure tanks in the submarine load balance is a function of the relative volume of these tanks. The weight of equistrong tanks always includes transverse bulkheads separating them from other tanks. The weight of bulkheads and pressure tank plating is comparatively high, and therefore, the g_{EGT} index is considerably larger than the PH index g_{PH} (sometimes 3 to 4 times higher). Due to these factors tanks with larger volumes have lower g_{EGT} indices.

The g_{EGT} index values are notably unsteady and can vary within 0.4 to 0.8 t/m³, the lower end of the range typically belonging to internal tanks.

The Light Hull

Following the design weight breakdown approach, the light hull (LH) in this case includes all structures left in Group 100 after we have separated the pressure hull and equistrong structures.

The LH share in the total load balance depends on the architecture of the submarine and on how extensive the outer hull (OH) is.

For double-hull submarines with large buoyancy reserves satisfying present requirements to surface trim floodability*, the LH weight is up to 45 to 50% of the weight of the entire hull (Group 100). The outer hull of such submarines contributes up to 60% of the LH weight.

The light hull includes structures of different types, different functions, different design loads and different materials. To calculate the weight of such a multipurpose system like the light hull in general (Fig.3.8), at the zero level of detail elaboration, we may use the correlation function (3.20) correcting it against the prototype.

$$P_{LH} = A_{LH}D_0^n + B_{LH} \qquad (3.20)$$

where B_{LH} is the weight of light hull structures that does not depend on the displacement.

The statistic data analysis has shown that quite satisfactory results could be obtained with the simplest formula for the light hull weight as a function of the normal displacement:

$$P_{LH} = p_{LH}D_0 \approx 0{,}16 \div 0{,}18 D_0 \qquad (3.21)$$

Considering the huge share of the light hull in the submarine load balance, to obtain more accurate results its weight calculations should be elaborated in greater detail.

Let us consider the LH weight calculation at the first level of detail elaboration. Since at this level all components of the LH can be expressed approximately as functions of the displacement, Table 3.1 provides approximate values of $p_i = P_i / D_0$ indices.

*The submarine should remain afloat with one PH compartment and two adjacent main ballast tanks flooded [30], [89].

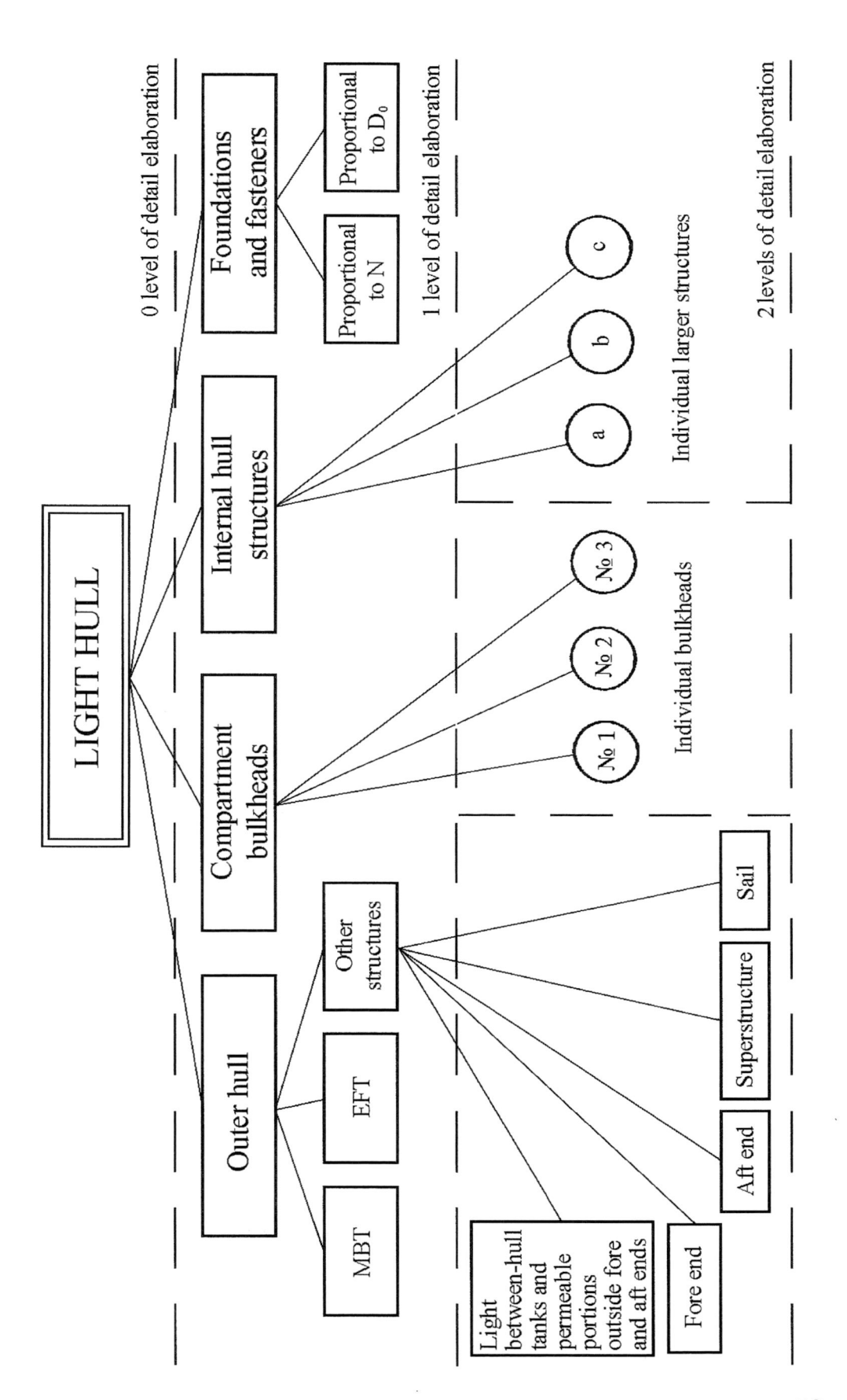

Fig.3.8. Light Hull Breakdowns for Different Levels of Detail Elaboration

59

Dimensions of the outer hull are the overall dimensions for the submarine, and therefore from the physical point of view it would be more appropriate to apply formulae containing the submerged displacement or the main dimensions. Nevertheless, formula $P_{OH} = p_{OH}D_0$ gives quite good results.

Table 3.1

Weight Indices of Light Hull Structures Referred to the Normal Displacement

Weight	Symbol	Index value
Outer hull	p_{OH}	0.10–0.11
Compartment bulkheads*	p_{CBHD}	0.020–0.025
Internal hull structures	p_{IHS}	0.02–0.03
Foundations and fasteners	p_{HF}	0.020–0.025
Total for the light hull	p_{LH}	0.16–0.115

* When a submarine is designed with an escape compartment, its restricting bulkheads are equistrong with the pressure hull. In this case their weight can be found from (3.17).

The weight of the outer hull can be refined if formulated as:

$$P_{OH} = P_{MBT} + P_{EFT} + P_{OTH}$$

where: P_{MBT} — weight of main ballast tanks;

P_{EFT} — weight of external fuel tanks;

P_{OTH} — weight of other parts of the OH, mainly permeable structures (permeable end structures, the superstructure, the sail, etc.).

On diesel-electric submarines, MBTs and external fuel tanks (EFT) occupy almost the entire between-hull volume and if there are end tanks, they partially take up end spaces as well. On nuclear submarines, due to the fact that the between-hull space is used for a large variety of purposes (biological shielding tanks, various special-purpose bays) the MBT share is less.

MBT and EFT weights can be expressed through some characteristics and particulars of the submarine. Let us formulate the weight of these tanks using indices of both the first and the second groups:

$$P_{MBT} = p_{MBT}D_0 \qquad (3.22)$$

$$P_{MBT} = g_{MBT} V_{MBT}^{GR} \qquad (3.23)$$

$$P_{EFT} = p_{EFT}D_0^{2/3} \qquad (3.24)$$

$$P_{EFT} = g_{EFT} V_{EFT}^{GR} \qquad (3.25)$$

where V_{MBT}^{GR} and V_{EFT}^{GR} – total gross volumes of tanks.

To consider net volumes we introduce coefficients k_{MBT} and k_{EFT} averaged for the subject groups of tanks and accounting for framing in tanks, PH plating, trapped water, etc. Besides, let us introduce another coefficient k_{NFC} to show what portion of the normal fuel capacity is allotted to the external tanks. Then

$$V_{MBT}^{GR} = k_{MBT} V_{MBT}^{NET} \qquad (3.26)$$

$$V_{EFT}^{GR} = k_{EFT} \, k_{NFC} \, V_{EFT}^{NET} \qquad (3.27)$$

From (3.22) and (3.23) we can, taking into account (3.26), derive formula for the MBT weight as a function of the displacement and of the relative reserve buoyancy (tentative):

$$P_{MBT} = p_{MBT}D_0 = g_{MBT} \frac{k_{MBT}}{\rho} \left(\frac{V_{MBT}^{NET}}{V_0}\right)D_0 = g_{MBT} \frac{k_{MBT}}{\rho} \varepsilon D_0 \qquad (3.28)$$

where ε – relative reserve buoyancy.

From (3.24) and (3.25) with account for (3.27) we can derive the EFT weight formula:

$$P_{EFT} = g_{EFT} \cdot k_{NFC} \cdot k_{EFT} \frac{V_{EFT}^{NET}}{D_0^{2/3}} = g_{EFT} \frac{k_{NFC} \cdot k_{EFT}}{\rho_F} \cdot \frac{P_F}{D_0^{2/3}} \qquad (3.29)$$

where P_F – full fuel capacity in tanks (useful capacity + trapped fuel in tanks).

From the general case of ship design it is known that:

$$P_F = g_F(1 + k_F)N\frac{R}{\vartheta_i} = g_F(1 + k_F)\frac{\vartheta_i^2}{C_i} R_i D_0^{2/3} \qquad (3.30)$$

where g_F – specific fuel consumption;

ϑ_i and C_i – speed and Admiralty coefficient of the submarine speed condition for which they specify the sea range under diesel engines R (usually this is economic snorkel cruising);

$k_F \approx 0.02$ – coefficient accounting for trapped fuel left in tanks.

Substituting (3.29) and (3.30) into (3.24) we obtain an expression for the fuel tank weight.

$$P_{EFT} = g_{EFT} \frac{k_{NFC}k_{EFT}(1+k_F)}{\rho_F C_{S\,NORT}} \vartheta^2_{SNORT} R_{SNORT} D_0^{2/3} \qquad (3.31)$$

Coefficients and indices for formulae (3.28) through (3.31) are established from prototypes. Some approximate values are:

$$g_{MBT} = 0.13 \sim 0.16 \text{ t/m}^3;$$

$$g_{MBT} = 0.20 \sim 0.22 \text{ t/m}^3;$$

$$k_{MBT} = 1.05 \sim 1.10;$$

$$k_{EFT} = 1.03 \sim 1.05.$$

Weights of other structures are found from:

$$P_{OTH} = P'_{OTH} D_0 = (0.04 - 0.05) D_0 \qquad (3.32)$$

If all foundations are placed in the «Hull» group, their weight can be refined already at the first level of detail elaboration by formulating it as a sum of two components:

$$P_{HF} = p'_{HF} \sum_1^n N + p''_{HF} D_0 \qquad (3.33)$$

where $p'_{HF} \sum_1^n N$ – accounts for foundations of the power plant (PP) and the shaftline

$p''_{HF} D_0$ – accounts for foundations of general-purpose machinery and equipment together with fasteners.

At the second level of detail elaboration of the outer hull weight they calculate weights titled «other structures» (Fig.3.8).

Approximate formulae and indices for weights of outer hull structures may be found in Table 3.2.

Indices g_i and weights of bossings P_{BOS} are derived from prototypes with similar structures.

Table 3.2

Approximate Formulae for Weights of the Outer Hull Structures

Structures	Weight calculation formulae	Weight indices
Light between-hull structures (BHS)	$P_{BHS} = g_{BHS} V_{BHS}$, where g_{BHS} - averaged value for all light between-hull structures.	$g_{BHS}= 0.13\sim0.15$ t/m^3
Fore end (FE)	$P_{FE}=g_{FE}V^{2/3}_{FE}$ or $P_{FE}=g'_{FE}S_{FE}$, where $V_{FE}=\delta_{FE}(lbh)_{FE}$ — gross volume; S — wetted surface of the end; $\delta_{FE} = 0.55\sim0.60$ for stem type; $\delta_{FE} = 0.70$ for body of revolution;	$g_{FE}= 1.15\sim1.40$ t/m^3 $g'_{FE}= 0.22\sim0.28$ t/m^2
Aft end (AFE)	$P_{AFE}=g_{AFE}V_{AFE} + g_{ST}S_{ST} + P_{BOS}$ or $P_{AFE}=g'_{AFE}S_{AFE} + g_{ST}S_{ST} + P_{BOS}$ $\delta_{AFE} = 0.5$ for AE of stabilising type; $\delta_{AFE}=0.35$ for body of revolution;	$g'_{AFE} = 0.20\sim0.25$ t/m^2 for stabilising type of AE; $g'_{AFE} = 0.25\sim0.30$ t/m^2 for body of revolution
Superstructure	$P_{SST}=g_{SST}l_{SST}(b_{SST}+2c_{SST})$ or $P_{SST}=g'_{SST} S_{SST}$, where $l_{SST}, b_{SST}, c_{SST}$ - superstructure dimensions	$g'_{SST} = 0.06\sim0.08$ t/m^2
Sail	$P_{SAIL}=g_{SAIL}V^{2/3}_{SAIL}$ or $P_{SAIL}=g'_{SAIL} S_{SAIL}$ $\delta_{SAIL}=0.60\sim0.70$ for foil-type sails; $\delta_{SAIL}=0.57\sim0.59$ for larger-volume sails with vertical walls; $\delta_{SAIL}=0.47\sim0.48$ for larger-volume sails with slant walls.	$g_{SAIL} = 0.90\sim1.10$ t/m^3 $g'_{SAIL} = 0.13\sim0.15$ t/m^2

If the aft end weight has to be found while control surface areas are not known yet, they can be estimated by re-calculations from the prototype proportionally to $V^{2/3}_{FS}$.

When stabiliser and plane/rudder areas are established, weights of these structures can be found with the help of indices from Table 3.3.

For formulae containing volumes of structures, the latter can be calculated with the help of the curve of frames or approximately estimated using block coefficients δ_i. Obviously, formulae involving structure surface areas S_i require more data.

Dimensions b and h of the end structures are dictated by the dimensions of the outer hull in way of PH end bulkheads while the length of end structures ℓ is counted from these planes.

Dimensions ℓ_{SAIL} and b_{SAIL} are measured conventionally at middle-height of the sail counting from the superstructure deck.

Table 3.3.

Weight Indices for Submarine Control Surfaces

Control surface	Index	Unit	Index values
Stabilisers	g'_{ST}	t/m²	0.30 ~ 0.45
Vertical rudder with actuators	g'_{VR}	t/m²	0.90 ~ 1.05
Fore hydroplanes with actuators	g'_{FPL}	t/m²	1.50 ~ 1.60
Sail hydroplanes with actuators	g'_{SPL}	t/m²	0.68 ~ 0.80
Aft hydroplanes with actuators	g'_{AFPL}	t/m²	0.80 ~ 1.05

In underwater shipbuilding, follow-on modifications of a certain submarine type very often mean increasing the length of the PH and of the whole submarine. Transverse dimensions usually remain unchanged, mostly due to construction restrictions. In this case the weight of light between-hull structures and of the superstructure changes proportionally with the PH length. Provided their outfitting remains unchanged the weight of end structures can be approximately considered unchanged.

Updating estimation for compartment bulkheads weight at the second level of detail elaboration includes finding the weight of each bulkhead in accordance with its diameter, design pressure and structural type:

$$P_{BHD} = g_{BHD}(P'_D; d_{BHD}) \frac{\pi}{4} d^2_{BHD} \qquad (3.34)$$

where d_{BHD} – bulkhead diameter.

The g_{BHD} index is derived from data on existing similar bulkheads of prototype submarines re-calculated to the subject design:

$$g_{BHD} = g_{BHD_0} \left(\frac{d_{BHD}}{d_{BHD_0}}\right)^{\ell} \frac{P'_F}{P'_{F_0}} \cdot \frac{\sigma_{T_0}}{\sigma_T} \qquad (3.35)$$

where $\ell = 1/2 \sim 2/3$ – power exponent accounting for the effect of the diameter;

P'_D and P'_{D_0} – design pressure for the bulkhead.

One may use results of systematic bulkhead strength calculations made varying d_{BHD} and P'_D, materials and other parameters like those shown in Fig.3.9 (for plane bulkheads).

The design weight of the bulkhead should include additions due to reinforcement and cutouts calculated from prototype data.

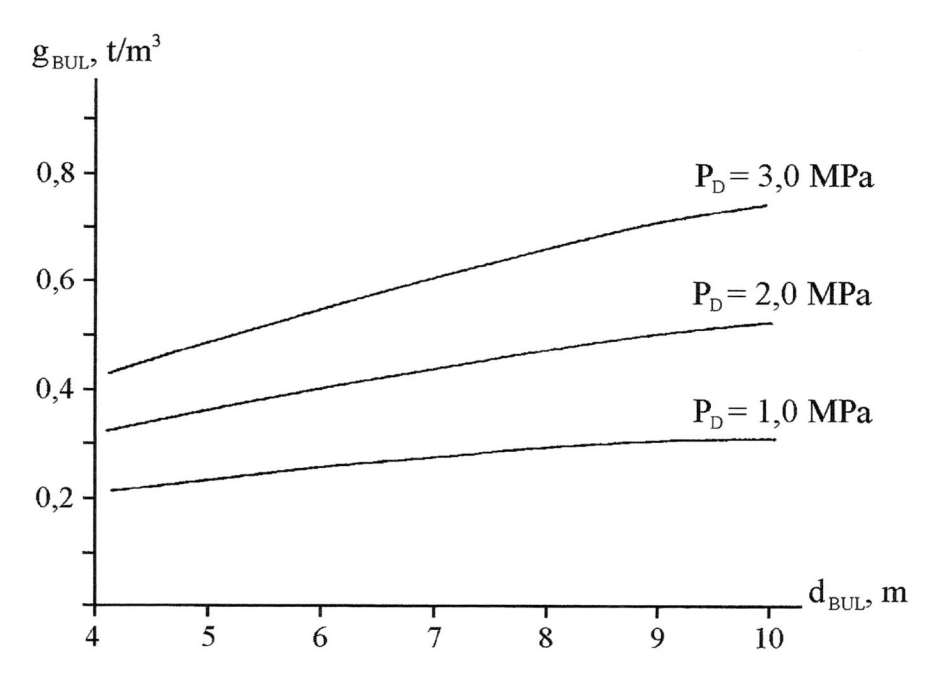

Fig. 3.9. Versus Design Load and Bulkhead Diameter

Weights of internal light structures can be refined by approximate estimation of the weight of larger internal tanks and decks based on their known volumes and areas. The weight of plating and stiffeners of light tanks inside the pressure hull may be estimated with $g_{LT} = 0.18$ to 0.19 t/m³; the weight of decks is calculated with $g'_{DEC} = 0.04$ to 0.09 t/m².

3.2. Loads for «Hull Gears, Fittings» and «Hull Systems» Groups

The total weight of these load groups makes up approximately the same share of the normal displacement on different submarines of various designations.

$$P_{GS} = p_{GS} D_0 = (0.08 - 0.9) D_0 \qquad (3.36)$$

As may be seen in Fig. 3.10, the p_{GS} index only depends slightly on the displacement.

Since prototypes chosen as sources of data for weight indices do not differ much in terms of the displacement from what is expected in the new design, we may ignore the influence of the displacement upon P_{GS} and take this index from the prototype without any calculations.

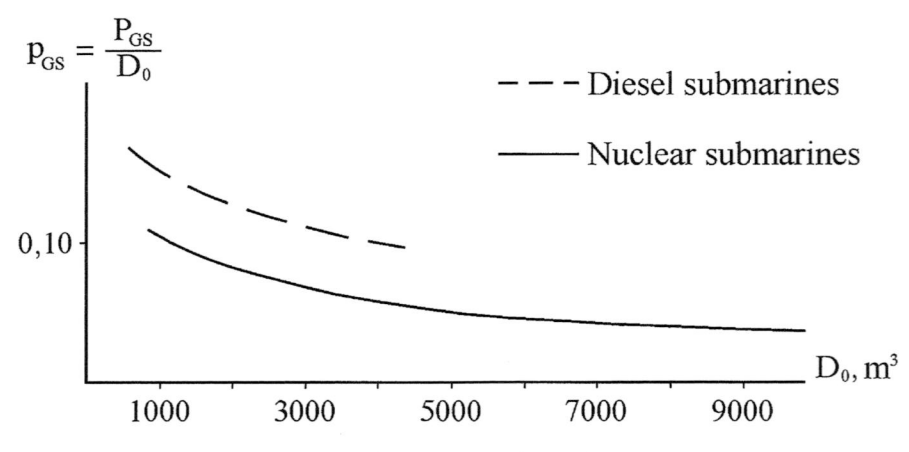

Fig. 3.10. The p_{GS} Index Versus D_0

Detail elaboration of loads due to gears and systems means finding individual weights of comparatively large gears and systems or groups of gears and systems with similar functions. Let us discuss some examples of such an elaboration.

Steering Gears (SG). For these devices it is advisable to apply scaling module $D_0^{2/3}\vartheta_{max}^2$, a value proportional to hydrodynamic forces on a plane/rudder, where ϑ_{max} is the full speed.

However, statistic data indicates that the most steady index is

$$P_{SG} = p_{SG}D_0 = (0.015 - 0.020)D_0 \qquad (3.37)$$

Compressed Air and Gas System (AGS). Submarines actually have three such general-purpose systems: high pressure air (HPA), medium pressure air (MPA) and low pressure air (LPA) systems, as well as special-purpose compressed air systems for launching torpedoes and missiles. The weight of these systems may be regarded as proportional to the displacement:

$$P_{AGS} = p_{AGS}D_0 = (0.022 - 0.027)D_0 \qquad (3.38)$$

The greater part of the AGS load is contributed by the high pressure air system. The basic purpose of the HPA system is to produce, store and deliver high pressure air for relevant consumers. To fulfil these functions, the HPA system has air bottles, compressors, air dryers, reducing and safety valves, filters, distributors and pipelines.

As long as the HPA storage capacity (Q_{HPA}) is known, we can easily derive an approximate formula for the AGS weight:

$$P_{HPA} = k_{HPA} \cdot P_{AB} = k_{HPA} \frac{Q_{HPA}}{V_{1AB}} (P_{1AB} + P_{1AIR}) \qquad (3.39)$$

or

$$P_{AGS} = k_{AGS} k_{HPA} \frac{Q_{HPA}}{V_{1AB}} (P_{1AB} + P_{1AIR}) \qquad (3.39a)$$

where k_{AGS} and k_{HPA} – coefficients for the weight of other system components;

P_{1AB} and V_{1AB} – weight of a standard air bottle and its payload volume;

P_{1AIR} – weight of the air in the bottle.

For the first approximation we assume that

$$Q_{HPA} = q_{HPA} D_0 \qquad (3.40)$$

where q_{HPA} – a conventional standard for HPA storage capacity per ton of the displacement. Then substituting (3.40) to (3.39a) we obtain a formula that more accurately describes the relationship between P_{AGS} and D_0:

$$P_{AGS} = k_{AGS} k_{HPA} \frac{q_{HPA}}{V_{1AB}} (P_{1AB} + P_{1AIR}) D_0 \qquad (3.41)$$

Life Support Systems – these are ventilation, air regeneration, air conditioning, air purification systems, as well as sanitary and refrigeration systems that collectively hold about the same share as AGS:

$$P_{HS} = p_{HS} D_0 = (0.020 - 0.025) D_0 \qquad (3.42)$$

Water Systems include, besides general-purpose ones, some weapon-support cooling and water systems. Altogether, their load accounts to about 1% of the displacement.

$$p_{WS} = (0.009 - 0.012)$$

The Diving and Surfacing System (DS) caters for the ultimate quality of the submarine: the ability to dive and surface. The system includes kingston valves and vent valves of MBT with their actuators and MBT LP blowing lines. The weight of this system depends on the number of MBTs (or indirectly on the reserve buoyancy) and on the number of tanks fitted with kingston valves. The latter factor has a significant impact, and therefore the weight index of the system is not very steady:

$$p_{DSS} = P_{DSS} / D_0 = 0.003 - 0.009$$

If the total number of MBTs and the contribution of tanks with kingston valves are known, we can approximately estimate the weight of this system as:

$$P_{DSS} = k_{DSS}\, n_{MBT} = (P_{1K} \cdot \overline{n}_K + P_{1VV}) \qquad (3.43)$$

where k_{DSS} – coefficient for other components included in the system;

$n_{MBT} = 2n_{MBT}^{side} + n_{MBT}^{end}$ – total number of MBTs on the submarine;

$\overline{n}_K = \dfrac{(n_{MBT})_K}{n_{MBT}}$ – relative amount of tanks fitted with kingston valves;

P_{1k} and P_{1VV} – weights of one kingston valve and one vent valve as taken from prototypes or earlier dedicated studies.

Usually they try to standardise these structures, at least within one submarine.

Hydraulic Systems. Due to everincreasing automation of control processes, hydraulic systems of modern submarines have become very extensive. Hydraulic actuators are used in control circuits of power plants, steering gears, diving and surfacing systems, for opening torpedo tube doors, caps of silos, etc. In the submarine load balance these systems take up approximately 0.4 to 0.6% of the displacement.

3.3. Loads for the «Mechanical Equipment, Pipelines and Systems of Power Plants» Group

The weight of the power plant (PP) is related to its power as

$$P_{PP} = p_{PP}\, N \qquad (3.44)$$

where p_{PP} – PP weight index (specific weight);

N – plant output, kW.

Fig.3.11 shows the trend of surface-to-submerged power ratio variation over the past 50 years [40], [50].

The specific weight depends on many factors including the power N, the growth of which results in the decrease in the specific weight of similar PPs. This is explained by the fact that any plant contains many components which show little response to changes in the power output [19], [73].

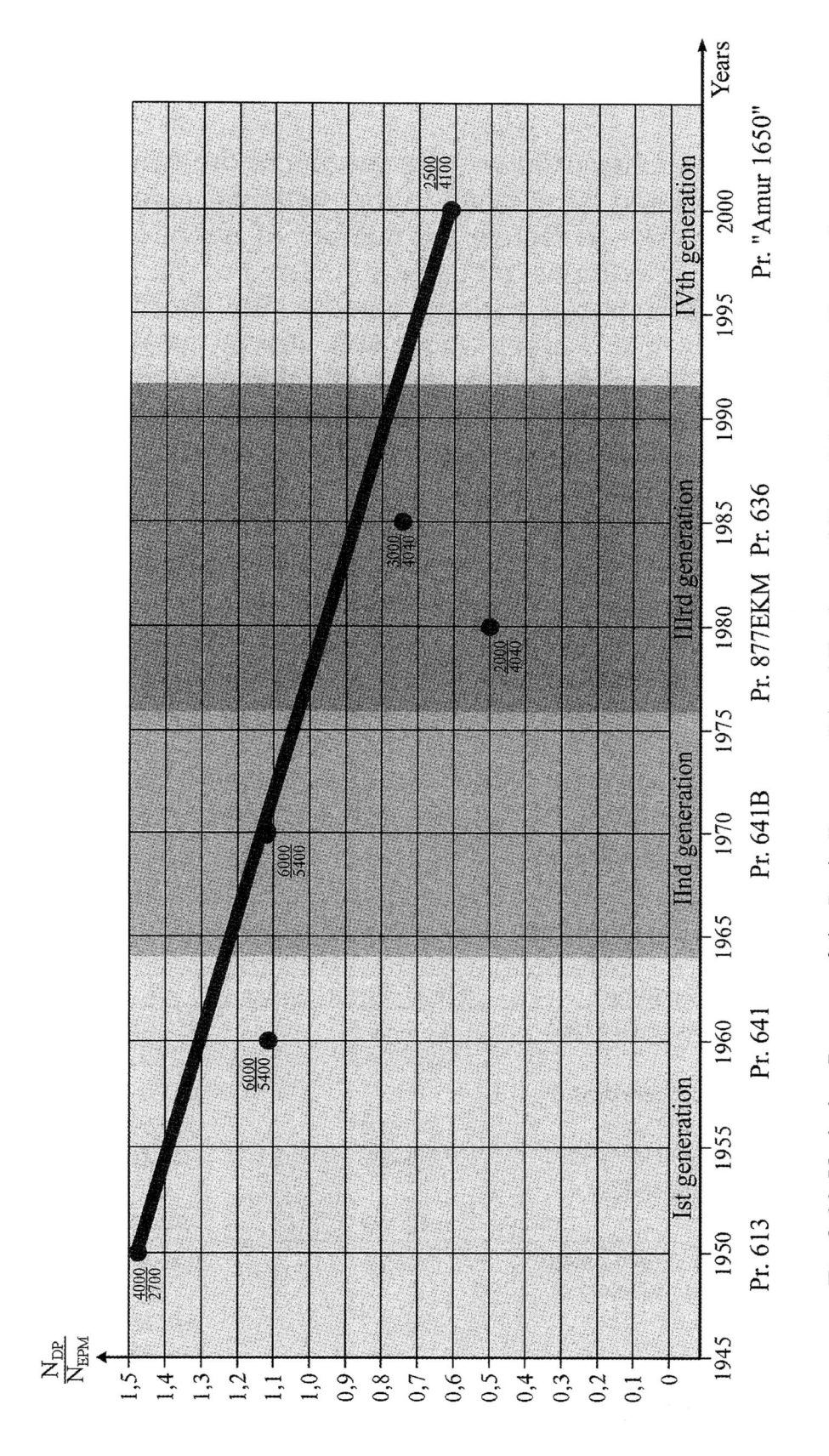

Fig.3.11. Variation Pattern of the Ratio Between Diesel Engine and Propulsion Motor Power Outputs for Diesel-Electric Submarines of Different Generations

Note. From the plot it is evident that diesel-generator capacity on Project 877EKM submarines was insufficient and that resulted in increasing the battery charging time and restricting submarine speed during charging. That design decision was forced by the lack of a diesel-generator of the required capacity. On Project 636, which is a follow-on of Project 877EKM they have installed a more powerful DG.

At early design stages the PP output is as a rule calculated from the formula of Admiralty coefficients

$$N_i = \frac{\vartheta_i^3 D_i^{2/3}}{C_i} \qquad (3.45)$$

Actual values of the displacement, the speed and the Admiralty coefficient in (3.45) should correspond to the particular considered service condition. In order to reduce the number of unknowns, all powers for all service conditions are usually referred to the normal displacement, though it is not exactly correct. We should also remember that this formula is approximate and, strictly speaking, it is true for geometrically similar bodies under all other assumptions.

At the same time the accuracy of power estimations depends on the correct choice of the Admiralty coefficient C_i which is a function of many parameters and characteristics.

Hence, the Admiralty coefficient should be selected very cautiously. Even when there is a prototype of very similar architecture, it should be scaled to the newly designed submarine with extreme care.

To give an idea of the order of magnitude of the submarine Admiralty coefficient under different service conditions, Table 3.4 demonstrates C_i statistics. At the same time for submerged sailing the Admiralty coefficient can be regarded as virtually independent of the speed because in this case the total drag coefficient (as will be shown later) is almost independent of the Reynolds number (Re) and, hence, of the submarine speed [31], [50].

Table 3.4

Admiralty Coefficient C_i Statistics

Hullform	Condition	Speed, knots	Ci
Stem type	Full submerged speed	12.0~16.0	85~100
		16.0~19.0	150
	Economic submerged speed	2.0~3.0	40~60
	Snorkel	7.0~8.0	70~85
	Maximum surface speed	12~16	140~150
Body of revolution	Full submerged speed	16~21	250~360
	Economic submerged speed	3~4	40~60
	Snorkel	7~10	150~250
	Maximum surface speed	10~12	140~150

The Admiralty coefficient governs the power required for submarine propulsion. The power required to supply general-purpose onboard consumers is accounted for by introducing an absolute (in kW) or, as will be shown below, relative allowance to the power used for propulsion (3.45).

The Diesel (Diesel-Generator) Plant

When the power is known, we can make detailed calculations of the power plant weight based on its major components. Table 3.5 shows principal components and weights for a diesel-electric submarine PP in accordance with the design weight breakdown approach.

Table 3.5

PP Weight Breakdown for a Diesel-Electric Submarine

Description	Symbol	% of PP
Diesel engine plant	P_{DP}	14.3
Electric propulsion plant	P_{EPM}	8.2
Shafting	P_{SH}	4.5
Total main plants and shafting (Group 400)	$P_{DP} + P_{EPM} + P_{SH}$	27.0
Storage battery	P_{SB}	37.0
Fuel and oil (normal capacity)	P_{FO}	36.0
Total energy carriers	$P_{SB} + P_{FO}$	73.0
Total P	ΣD_i	100

The specific weight of the entire power plant of a large diesel-electric submarine, including all articles listed with normal fuel capacity, referred to the PP output for submerged propulsion (EPM) is $p_{PP} = 200$ kg/kW. Excluding fuel and oil, it is $p'_{PP} = 135$ kg/kW. Principal characteristics of power plants of Russian diesel-electric submarines may be found in Table 3.6 [3], [40], [41], [42], [44], [73].

On modern SSs one can find diesel engine plants (DP) with mechanical power transmission to the propeller shaft (directly or via a gear box), combined plants, which have both diesels with mechanical transmission to propeller shafts and auxiliary diesel-generators, and diesel-generator plants (DGP) with electric power transmission for exclusively electric propulsion (Fig.3.12).

Table 3.6

Principal Characteristics of Power Plants of Russian Diesel-Electric Submarines

Plant parameters	Ist generation 613	Ist generation 611	Ist generation 641	IInd generation 641B	IIIrd generation 877EKM	IIIrd generation 636	IIIrd generation 636 *	IVth generation «Amur 1650»
Type of diesel engines	37D	37D	37D	2D42	4-2DL42M	7-2D42	7-2D42	84 26/26
Number of diesels x power Speed, r.p.m.	2x2,000 h.p. / 500	3x2,000 h.p. / 500	3x2,000 h.p. / 500	3x1,900 h.p. / 500	2x1,000 kW / 700	2x1,500 kW / 750	2x1,750 kW / 750	2x1,250 kW / 1,000
Type of generators	–	–	–	–	PG-142	PG-167	PG167	SBDG-122-1000
Type of diesel-generators	–	–	–	–	DG-PT	30DG	3-DGM	28DG (AC)
Type of MPM	PG-101	PG-101 (on sides) PG-102 (middle)	PG-101 (on sides) PG-102 (middle)	PG-101 (on sides) PG-102 (middle)	PG-141	PG-165	PG-165	SED-1 («Permazin» type)
Number of MPMs x power Speed, r.p.m.	2x1,350 h.p. / 420	2x1,350 h.p. 440 (on sides) 1x2,700 h.p. 540 (middle)	2x1,350 h.p. 440 (on sides) 1x2,700 h.p. 540 (middle)	2x1,350 h.p. 440 (on sides) 1x2,700 h.p. 540 (middle)	1x4,040 kW / 500	1x4,040 kW / 250	1x4,040 kW / 250	1x4,100 kW / 200
Type of economic speed PM	PG-103	PG-104	PG-104	PG-104	PG-140	PG-166	PG-166	SED-1 multi-speed
Number of economic speed PMs x power Speed, r.p.m.	2 x 50 h.p. / 150	1 x140 h.p. / 185	1 x140 h.p. / 185	1 x140 h.p. / 185	1 x139 kW / 150	1 x95 kW / 70	1 x95 kW / 70	SED-1 multi-speed
Number of propeller shafts	2	3	3	3	1	1	1	1
Type of reserve	–	–	–	–	PG-168	PG-168	PG-168	submersible

72

Plant parameters	Ist generation			IInd generation	IIIrd generation			IVth generation «Amur 1650»
	613	611	641	641B	877EKM	636	636 *	
propulsion motor								electric motor
Number of reserve PMs x power Speed, r.p.m.	–	–	–	–	$\frac{2 \times 75\ kW}{650}$	$\frac{2 \times 75\ kW}{650}$	$\frac{2 \times 75\ kW}{650}$	$\frac{2 \times 35\ kW}{500}$
Number of reserve propulsors	–	–	–	–	2	2	2	2
Type of SB	46SU	46SU	46SU	48SU	446	446	476	476
Number of SB groups x number of cells in a group	2 x 112	4 x 112	4 x 112	4 x 112	2 x 120	2 x 120	2 x 120	2 x 126
Submerged range, miles speed, kts.	$\frac{350}{2}$	$\frac{440}{2}$	$\frac{400}{2}$	$\frac{400}{2}$	$\frac{400}{3}$	$\frac{400}{3}$	$\frac{450}{3}$	$\frac{650}{3}$
Snorkel range, miles speed, kts.	$\frac{4,500}{8}$	$\frac{13,000}{8}$	$\frac{18,000}{8}$	$\frac{9,500}{8}$	$\frac{6,000}{7}$	$\frac{7,500}{7}$	$\frac{8,500}{7}$	$\frac{6,000}{7}$
Surface range, miles speed, kts.	$\frac{8,550}{10}$	$\frac{22,000}{9}$	$\frac{30,000}{8}$	$\frac{16,000}{8}$	**	**	**	**

* Under study

** For SS of the IIIrd and IVth generations the surface range is not regulated.

A diesel (diesel-generator) plant should cater for the following service conditions:
- snorkel and surface propulsion at the specified speed;
- battery charging underway at snorkel and surface trims;
- MBT blowing with diesel exhaust gases when rising from the low-buoyancy trim;
- power supply to onboard consumers under all above-listed conditions.

Service conditions that dictate the required DP (DGP) power are economic speed in snorkel or surface trim with simultaneous battery charging (preferably by the 1st stage current).

In both cases onboard consumers should continue to receive their power supply.

The required DP (DGP) power for propulsion plus charging may be formulated as:

$$N = N_{PR} + N_{BCH} + N_{GPS} + N_{LOS} \qquad (3.46)$$

where: N_{PR} – power required for propulsion;
N_{BCH} – power required for SB charging;
N_{GPS} – power for onboard consumers;
N_{LOS} – power for losses in cables and cable runs (plant efficiency).

The power required for SB charging may be formulated as:

$$N_{BCH} = I_{BCH} U_{BCH} 10^{-3} n_{SB} \qquad (3.47)$$

where: U_{BCH} – voltage supplied during charging to one cell (U = 2.4 V for lead-acid cells and U = 2.05 V for silver-zinc cells);
I_{BCH} – charging current amperage;
n_{SB} – total number of cells (112 or more for lead-acid cells, 152 for silver-zinc cells).

At the initial stage of design work direct calculations of N_{GPS} and N_{LOS} are impossible because they require detailed design information. Therefore, let us describe N_{LOS} as a ratio:

$$\alpha_{LOS} = \frac{N_{LOS}}{\dfrac{N_{EPM}}{\eta_{EPM}} + N_{GPS}} \qquad (3.48)$$

where α_{LOS} – coefficient for losses in cables and cable runs;
N_{EPM} – power of electric propulsion motors;

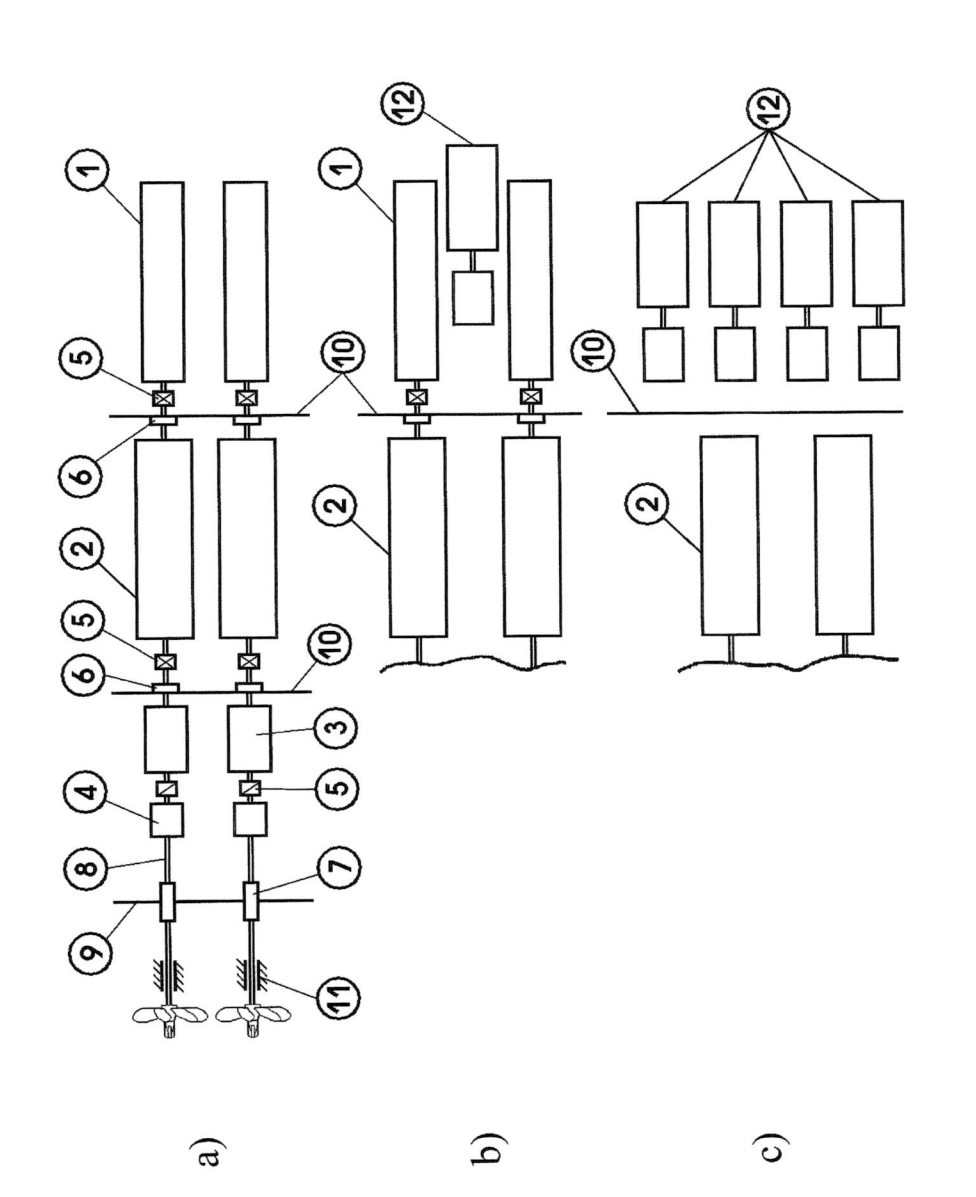

Fig.3.12. Diesel-Electric Submarine Power Plants

a) diesel-engine plant; b) combined plant; c) diesel-generator plant

1 - diesel engine, 2 - main propulsion motor, 3 - economic speed motor, 4 - thrust bearing; 5 - couplings, 6 - bulkhead glands, 7 - stern tube gland, 8 - shafting, 9 - pressure hull (or end bulkhead), 10 - compartment bulkheads, 11 - aft bossing, 12 - diesel-generator

75

η_{EPM} — efficiency of electric motors and assume that N_{GPS} for submarines of one type varies as:

$$N_{GPS} = a_{GPS} D_0^{2/3} = a_{GPS} \frac{1}{k_{FS}^{2/3}} D_{FS}^{2/3} \qquad (3.49)$$

where a_{GPS} — coefficient for the power required for general-purpose needs;

k_{FS} — coefficient relating the normal and the total submerged displacements.

Assuming that propulsion and charging tasks are distributed among individual motors in such a way that their power is utilised completely, we can obtain formulae for the total rated power of the DP (DGP) for the propulsion plus charging condition.

1. In snorkel (periscope) trim:

a) diesel plant, propulsion under diesel power:

$$N_{DE} = \alpha_{SNORT} \left[\frac{\vartheta_{SNORT}^3}{C_{SNORT}} + (J_3 U_3 \cdot 10^{-3} n_{SB} + \frac{a_{GPS}}{\eta_{EPM} k_{FS}^{2/3}})(1 + \alpha_{LOS}) \right] D_{FS}^{2/3} = \beta_i D_{FS}^{2/3} \qquad (3.50)$$

b) diesel-generator plant, propulsion under economic speed electric power:

$$N_{DE} = \alpha_{SNORT} \left[\frac{\vartheta_{SNORT}^3}{C_{SNORT} \eta_{EMES}} + (J_3 U_3 \cdot 10^{-3} n_{SB} + \frac{a_{GPS}}{k_{FS}^{2/3}})(1 + \alpha_{LOS}) \right] D_{FS}^{2/3} = \beta_i D_{FS}^{2/3} \qquad (3.51)$$

where N_{DG} — power output of the diesel-generator;

α_{SNORT} — numerical coefficient taken from the prototype;

β_i — coefficient relating the plant power with the displacement in formulae (3.50) through (3.52).

2. In surface trim:

diesel plant, propulsion under diesel power

$$N_{DE} = \frac{\vartheta_{SFB}^3}{C_{SFB}} D_{FSB}^{2/3} + (J_{BCH} U_{BCH} \cdot 10^{-3} n_{SB} D_{FS}^{2/3} + \frac{a_{GPS}}{\eta_{GEN}} D_{FS}^{2/3})(1 + \alpha_{LOS}) =$$

$$= \left[\frac{\vartheta_{SFB}^3}{C_{SFB}} (\frac{k_{FSB}}{k_{FS}})^{2/3} + (J_{BCH} U_{BCH} \cdot 10^{-3} n_{SB} + \frac{a_{GPS}}{\eta_{GEN} k_{FS}^{2/3}})(1 + \alpha_{LOS}) \right] D_{FS}^{2/3} = \beta_i D_{FS}^{2/3} \qquad (3.52)$$

In formulae (3.50) through (3.52) η_{GEN} — efficiency of propulsion motors in the generator mode.

It should be kept in mind that coefficients a_{GPS} and α_{LOS} for different conditions described by formulae (3.50) through (3.52) may be different.

The power required for charging batteries designed to ensure specified parameters of submerged propulsion, for general-purpose needs, as well as for losses in cables and cable runs, determines the power of diesel plant propulsion electric motors in the generator mode.

$$(N_{BCH}+N_{GPS})(1+\alpha_{LOS})=(J_{BCH}U_{BCH}\cdot 10^{-3}n_{SB}+\frac{a_{GPS}}{\eta_{GEN}k_{FS}^{2/3}})(1+\alpha_{LOS})D_{FS}^{2/3} \quad (3.53)$$

This power should be matched with the motor mode power. It should be remembered that in the generator mode electric motors allow about 40% higher outputs than in the motor mode, i.e.

$$N_{GEN} \leq 1.4 N_{MOT} \quad (3.54)$$

If $N_{GEN} > 1.4 N_{MOT}$, one should accordingly increase the motor mode power (to choose a motor with a higher output) or to install an auxiliary diesel-generator.

With the help of above-derived power expressions, we can obtain an approximate formula for PP (DG) weight estimation:

$$P_{DP(DGP)} = p_{DP(DGP)}N_{DE(DG)} = k_{DP(DGP)}\beta_i g_{DE(DG)}D_{FS}^{2/3} \quad (3.55)$$

where $k_{DP(DGP)}$ – coefficient of the plant weight (for plants with diesel engines directly running the shaft $k_{DP(DGP)} = 1.8$ to 2.0);

$g_{DE(DG)}$ – mean specific weight of diesel engines (diesel-generators) proper in the assembled plant.

An Electric Propulsion Plant should cater for all submerged propulsion conditions and, if diesel engines are mechanically connected to shafting, and battery charging in the surface trim. In the latter case electric motors are fully or partially run in the generator mode. Where the submarine has fully electric propulsion, these tasks are covered by the diesel-generator.

Assuming that the electric propulsion motor power is dictated by the specified full submerged speed and using the general formula (3.44), we can derive a formula for approximate estimations of the electric propulsion plant weight:

$$P_{EPM} = p_{EPM}\Sigma N_{EM} = k_{EPM}g_{EM}\frac{\vartheta_{\downarrow max}^3}{C_{\downarrow max}}D_{FS}^{2/3} \quad (3.56)$$

where g_{EPM} – mean specific weight of the propulsion motor proper, which depends on the type, speed and aggregate power of electric motors included in the plant;

k_{EPM} = 1.25 to 1.32 – coefficient for weight of the economic speed propulsion motors and auxiliary equipment of the plant (air coolers, propulsion motor fans, etc.).

The Weight of the Shafting is calculated from the power corresponding to the full submerged speed:

$$P_{SH} = p_{SH} N_{EPM} = p_{SH} \frac{\vartheta^3_{\downarrow max}}{C_{\downarrow max}} D_{FS}^{2/3} \qquad (3.57)$$

where p_{SH} = 4.0 to 4.5 kg/kW – specific weight of the shafting referred to the highest transmitted power.

If the speed of the propeller shaft and the shafting length are known, the weight can be calculated as:

$$P_{SH} = p_{SH_0} \left(\frac{N}{N_0}\right)^{2/3} \cdot \left(\frac{n_0}{n}\right)^{2/3} \cdot \left(\frac{\ell_{SH}}{\ell_{SH_0}}\right) \qquad (3.58)$$

where n_0; n – propeller shaft speed of the prototype and of the subject submarine;

ℓ_{SH_0}; ℓ_{SH} – respective lengths of propeller shafts.

The Storage Battery System (SBS) includes a storage battery (SB), wedging and connections of battery cells, loading and maintenance devices, mechanical systems for electrolyte agitation, ventilation and air conditioning in battery wells, storage battery water cooling.

For the first approximation to SB weight estimations the specific energy pick-up ΔW (kWh/t) is normally used, showing how many kilowatt-hours of electric power can be picked-up from one ton of the cell weight at the specified discharging rate. Fig. 3.13 shows the pattern of $\Delta W(t)$ curves for lead-acid (curve 1) and silver-zinc (curve 2) cells [24], [106].

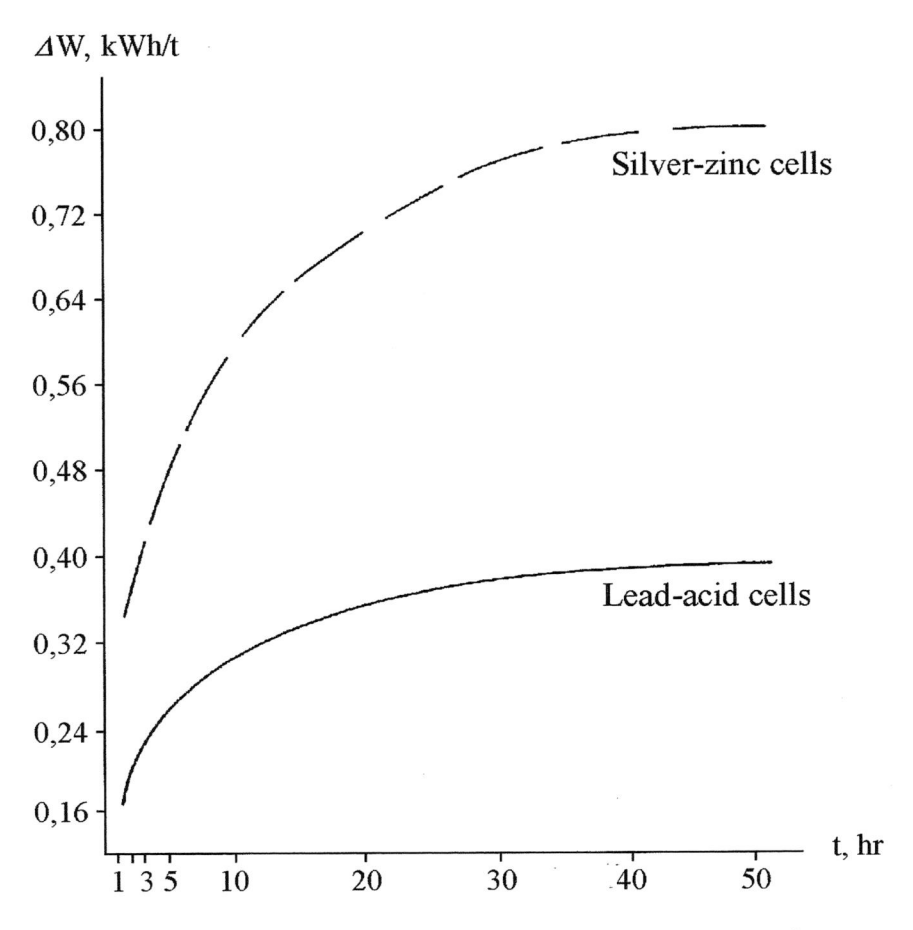

Fig. 3.13. Specific Power Pick-up As a Function
of the Battery Discharge Time

If discharge parameters of the cell – current I(t) and mean voltage $U_{MEAN}(t)$ – are known, the specific energy pick-up is found as:

$$\Delta W(t) = \frac{J(t) \cdot U_{MEAN}(t) \cdot t \cdot 10^{-3}}{P_{CELL}} \qquad (3.59)$$

where t — discharge time, hr;
P_{CELL} — weight of one cell found from delivery specifications, t.

A general formula for the storage battery system weight can be written as:

$$P_{SB} = k_{SB}k_{DIST}P_{SB} = k_{SB}k_{DIST}\frac{W}{\Delta W} \qquad (3.60)$$

where: $k_{SB} = 1.15$ to 1.20 – coefficient accounting for the system weight;
$k_{DIST} = 1.02$ to 1.03 – coefficient accounting for the weight of the distillate for topping the cells.

For SSDEs, the storage battery is the only source of power when submerged (in this case air independent plants are not considered) that provides speeds and ranges specified in SDS, as well as power supply to all other consumers.

The general expression for the required SB power may be presented as:

$$W = (N_{PR} + N_{GPS} + N_{LOS})t \quad (3.61)$$

where: $N_{PR} = \dfrac{N_{EPM}}{\eta_{EPM}}$ – SB power for propulsion;

N_{GPS} and N_{LOS} – power for general-purpose needs and losses found from (3.48) and (3.49);

$t = \dfrac{R_i}{\vartheta_i}$ – submerged time under subject service conditions.

Taking into account these functions and relationship (3.56), the required power expression becomes:

$$W = \left(\dfrac{\vartheta_i^3}{C_i \eta_{GEN}} + \dfrac{a_{GPS}}{k_{FS}^{2/3}}\right)(1+\alpha_{LOS})t\,D_{FS}^{2/3} \quad (3.62)$$

Substituting (3.62) to (3.60) and taking into account that the SB should ensure the cruising range under all service conditions specified for the design (usually it is the full submerged speed for 1 hour and long-term economic speed cruising at 2 to 3 knots), we obtain the formula for the storage battery system weight.

$$P_{SB} = k_{SB}k_{DIST}(P_{SB})_{max} =$$

$$= k_{SB}k_{DIST}\left[\left(\dfrac{\vartheta_i^3}{C_i \eta_{GEN}} + \dfrac{a_{GPS_i}}{k_{FS}^{2/3}}\right)(1+\alpha_{LOS})\dfrac{t_i}{\Delta W_{t=t_i}}\right]_{max} D_{FS}^{2/3} \quad (3.63)$$

where i is the designator of the subject service condition.

Having determined the SB weight for the specified service conditions, for subsequent calculations we take its maximum value. Value $\Delta W_{t=t_i}$ in formula (3.63) is found from delivery specifications for the cells planned for installation on the designed submarine.

The Weight of Fuel and Oil is calculated with the formula:

$$P_{FO} = g_F k_{OIL} k_F \frac{\vartheta_{SNORT}^2}{C_{SNORT}} R_{SNORT} D_{FS}^{2/3} \qquad (3.64)$$

where g_F = 0.20 to 0.22 kg/kWhr – specific fuel consumption in the snorkel mode;
k_{OIL} = 1.04 to 1.07 – coefficient accounting for the fuel stock weight
k_F = 1.15 to 1.02 – coefficient accounting for the fuel left in tanks.

The Nuclear Power Plant

With a chosen type of the steam generating plant (SGP) and MPP configuration, the specific weight of the main power plant P_{MPP} depends on:
- main power plant output;
- MPP configuration, in particular, the plant formula (number of reactors – number of main geared turbines – number of propeller shafts);
- arrangement of the plant in the submarine hull;
- propeller shaft speed;
- composition of auxiliary equipment.

In submarine shipbuilding they use, as a rule, water-cooled and water-moderated SGPs. Therefore, the following description is applicable to this type of plant.

Fig.3.14 shows basic configuration of the main power plant of a nuclear submarine [10].

Weight ratios of nuclear power plant (NPP) components are given in Table 3.7 [10], [18], [26], [71].

Specific weights of MPP components in Table 3.7 refer to the shaft power.

The relationship between the NPP total weight and submarine displacement can be presented as:

$$P_{MPP} = p_{MPP} N = p_{MPP} \frac{\vartheta^3}{C} D_{FS}^{2/3} = k_{HF} k_{MPP} p_{MPP} \frac{\vartheta^3}{C} D_{FS}^{2/3} \qquad (3.65)$$

where values N, ϑ, C correspond to the full submerged speed.

Available information on weights and dimensions is very often limited to the MPP only and, as a rule, does not cover foundations and auxiliary components of the plant.

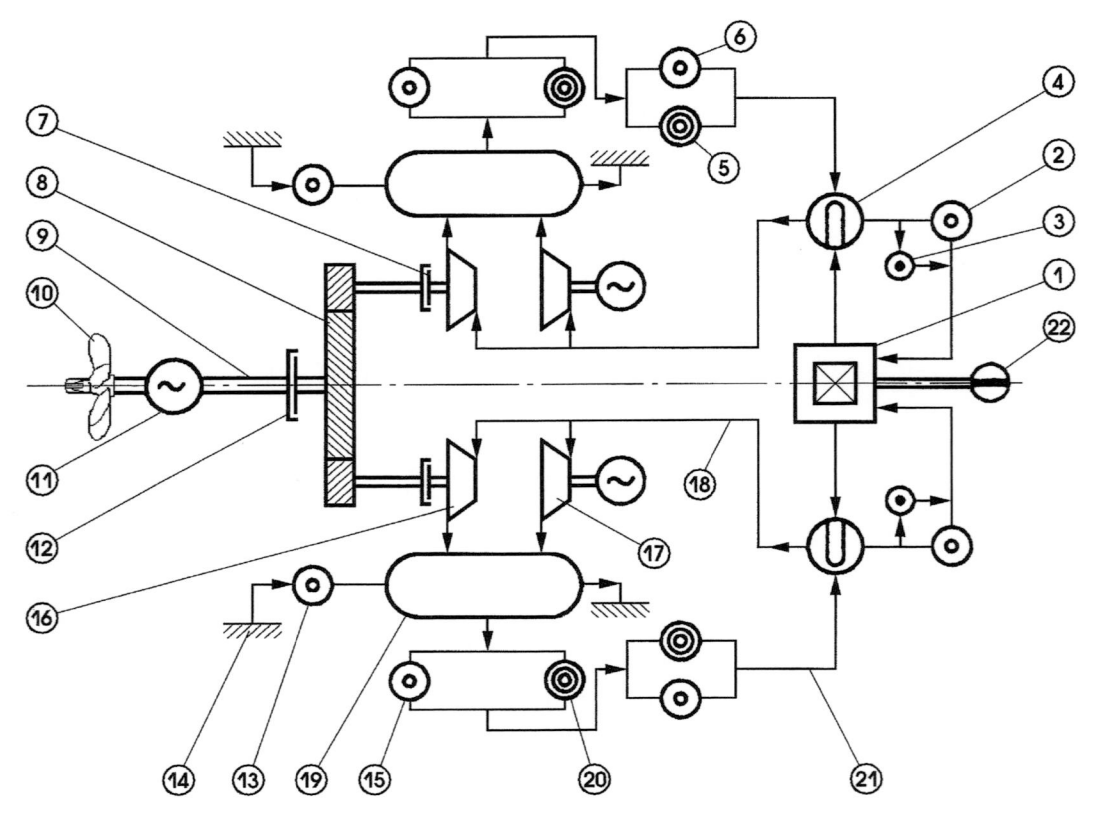

Fig. 3.14. Basic Configuration of a SSN PP

1 – nuclear reactor; 2 – main circulating pump of the 1st circuit; 3 – filter; 4 – steam generator; 5 – auxiliary feed pump; 6 – main feed pump; 7 – disengaging clutch: 8 – reduction gear; 9 – shaft line; 10 – propeller; 11 – propulsion electric motor; 12 – disengaging clutch of the propeller shaft; 13 – circulating pump of the main condenser; 14 – pressure hull; 15 – main condensate pump; 16 – main turbine; 17 – stand-alone turbo-alternator; 18 – main steam line; 19 – main condenser; 20 – auxiliary condensate pump; 21 – feed water pipeline; 22 – pressuriser.

Table 3.7

Nuclear Power Plant Weight Breakdown

Description	Specific weight, kg/kW	% of
Steam generating plant (SGP) without shield tank (SHT)	21.0 to 24.0	44.0 to 45.0
Steam turbine plant + turbo-alternator	11.0 to 10.0	22.0
Shield tank and steam-turbine plant foundations	6.0 to 7.0	13.0
Reserve propulsion motor	1.0	—
Shaft lines with propulsors	2.0	8.0
Diesel-generator plant	1.0	—
Total group 400	41.0 to 46.0	88.0
Storage battery system	4.5	9.0
Margins for power plant	1.5	3.0
NPP total	45.0 to 52.0	100

In formula (3.65) we introduce the following coefficients: $k_{HF} = 1.20$ taking into account the weight of foundations and $k_{MPP} = 1.20$ to 1.25 taking into account the weight of auxiliary components of the plant.

Fig. 3.15 shows the specific weight as a function of the power plant [10], [26], [71].

Weights of Auxiliary Components can be determined more accurately based on their own required power or energy values.

The Weight of Reserve Propulsion Motors (RPM) is found from the general formula (3.56):

$$P_{RPM} = p_{RPM} N_{RPM}$$

where p_{RPM} – specific weight referred to the power of the motor itself.

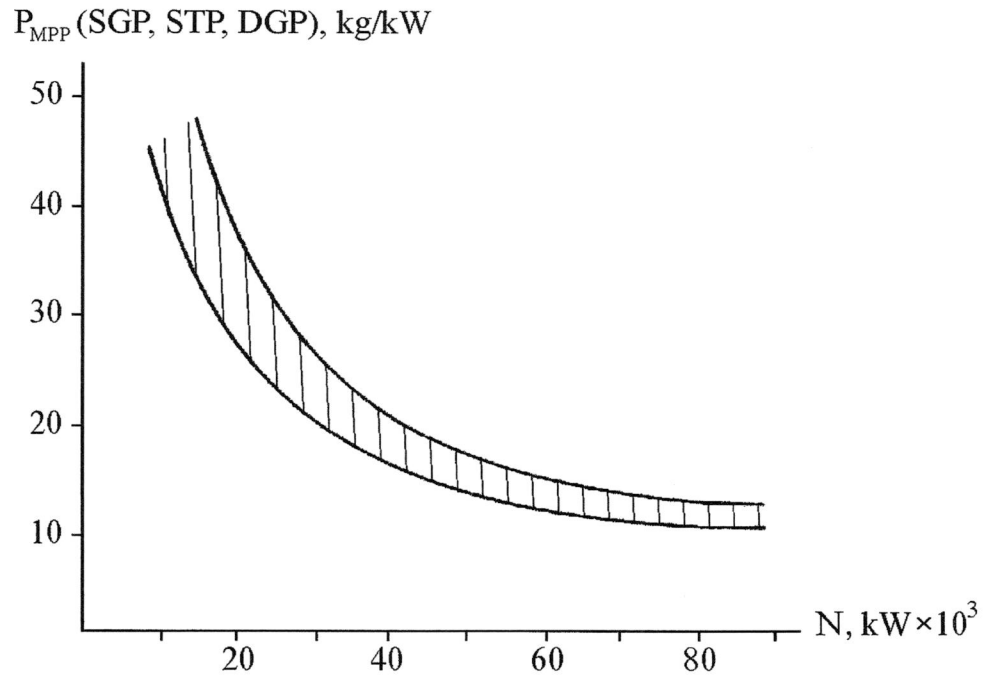

Fig.3.15. MPP Specific Weight to Power Ratio

The required SB capacity on a nuclear submarine is governed by the following requirements:
- SGP shut-down cooling in the case of shutdown at sea and subsequent restart;
- power supply to the minimum required number of general-purpose consumers during the interruption in SGP operation;
- propulsion at the minimum speed allowable due to submarine steerability considerations.

Taking into account the losses and assuming that the energy for the first of the above requirements depends on N while that for the second and the third requirements is a function of the displacement, we can write

$$W_{SB} \approx [(a'_{SB}N^e)t' + (a''_{SB}D_0^{2/3})t''](1 + \alpha_{LOS}) \qquad (3.66)$$

where a'_{SB} and a''_{SB} – coefficients relating to the required SB capacity to the MPP power and the submarine displacement;

$t'^{(")}$ – anticipated duration of SB operation during reactor shutdown cooling, start and repair interval taking into account submarine propulsion and power supply to consumers.

The nuclear submarine storage battery should be configured of groups with a standard number of cells. The number of groups is dictated by the required SB capacity. In the view of survivability considerations, it is desirable to have at least two groups. When the SB required capacity is known, the weight of the storage battery system is found with formula (3.63), very approximately: $P_{SB} = (0.10 - 0.15)P_{MPP}$.

The Weight of Feed Water and Oil Stocks for the MPP should ensure one complete replacement for a single-shaft plant or for one independent propulsion group of a twin-shaft plant. If capacities and types of plants of the prototype and of the new submarine are very similar, the weight of these stocks can be taken from the prototype as an absolute value.

We should note that the obtained relationships $P_i(D_{FS})$ can be used for power plant weight estimations based on calculated displacement value only for very early approximations as they take into account just the larger components of the plant. The power plant weight determined in this way should be updated taking into account a more refined configuration of the plant (particular chosen motors, number of propeller shafts and power distribution between them, as well as the SB compiled of standard cell groups).

Air-Independent Propulsion Plants

And finally, let us consider one more type of power plant that is nowadays developed both in Russia and in many other countries: air-independent propulsion plants (operating without access to atmospheric air) for conventional submarines of displacements from 1,000 to 3,000 tons. It appears that at larger displacements the nuclear power plant is always more optimal. The idea of such an engine is not new. In Russia, in 1912, sub-lieutenant Nikolsky suggested a concept of a closed-cycle internal combustion engine operating with oxygen supply. The opportunity to implement this concept was made possible at the same time, at a test stand at the Baltiysky Shipyard. Practical work on a submarine with a single propulsion motor for both surface and submerged operation was placed on a broad footing in our country in 1930s under the leadership of S.A.Bazilevsky [38] and immediately after the Great Patriotic war a large series of Project 615 «QUEBEC» submarines (Fig. 3.16) were built.

In 1943 German designers were working on the design of a submarine (XXVI series) equipped with a Walter turbine as a booster for submerged propulsion, and a similar plant was installed on Project 617 submarine in the USSR in 1951 (Fig.3.17).

Sea trials of the Project 613E «WHISKEY-AIP» submarine with fuel cells (Fig.3.18) were performed in 1988 [43], [46], [78].

The present boom in work on air-independent plants, which are used on submarines as auxiliary (additional) power plants, takes place at a qualitatively new level of science and engineering. There are evident successes in the creation of closed-cycle diesel engines (Russia, Germany, Holland, Great Britain, Italy), Stirling engines (Sweden), gas-dynamic turbines (France), and plants with fuel cells (Russia, Germany). Fig.3.19 schematically shows the fuel cell power plant compartment of a Russian submarine.

In a number of foreign publications [100] and national design studies [57], it has been shown that the arrangement of such auxiliary plants in separate compartments causes minor increases in the displacement and some reduction of the full submerged speed, but allows for several times greater submerged economic speed range in the search mode without using up the power of storage batteries (Fig.3.20).

At the same time, any engine type with all its advantages raises problems in submarine design and inevitably reveals some weak points. Therefore, new types of engines should be analysed in terms of major criteria for submarine power plants: specific power, explosion and fire safety, heat dissipation into the submarine environment, noise, operating costs, etc. Only after such an analysis can we make conclusions about the niche air-independent plants will find between classic diesel-electric and classic nuclear power plants, and on what kinds of submarines their installation is advisable.

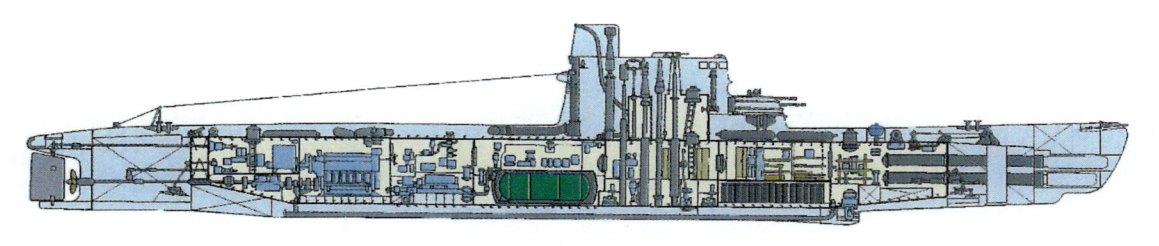

Principal Tactical and Technical Characteristics

Submarine construction period, years	1953-1959
Number of constructed submarines, pcs	30
Normal displacement, m³	405

Main dimensions, m:
  length
  beam
	
	56,8
	4,5

Maximum diving depth, m	120
Power of diesel plant, h.p.	3×900
Full surface speed, knots	16
Full submerged speed, knots	15
Cruising range at full submerged speed, miles	56
Submerged range at economic speed, miles	360
Submerged endurance, days	4
Endurance, days	10
Liquid oxygen, tons	8,5
Chemical absorber, tons	15

Closed Cycle Plant

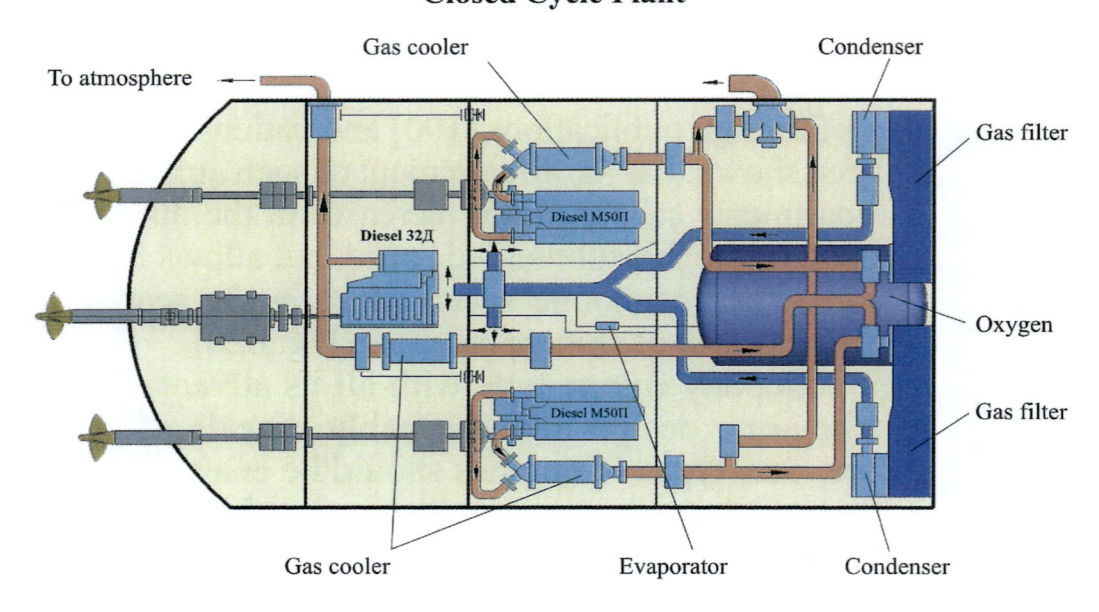

Fig.3.16. Project A615 «QUEBEC»
Submarine with a Single Motor for Surface and Submerged Propulsion

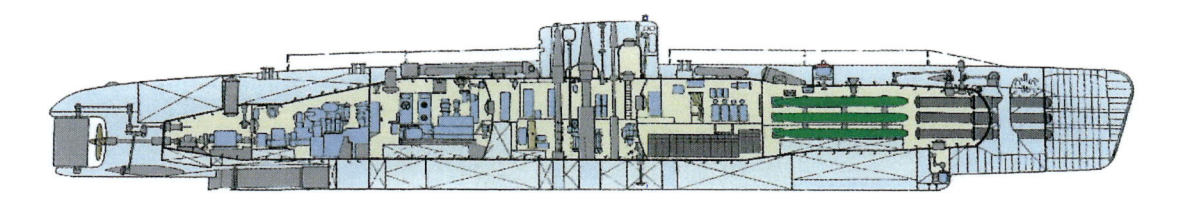

Principal Tactical and Technical Characteristics

Submarine construction period, years	1951-1952
Normal displacement, m³	950

Main dimensions, m:
    length
    beam
	
	62,2
	6,08

Maximum diving depth, m	200
Steam-gas turbine power, h.p.	7250
Full surface speed, knots	11
Full submerged speed, knots	20
Cruising range at full submerged speed, miles	120
Submerged range at economic speed, miles	132
Submerged endurance, days	8
Endurance, days	45
Hydrogen peroxide, tons	103,4
Light fuel, tons	13,9

Basic Configuration of a Steam-Gas Plant

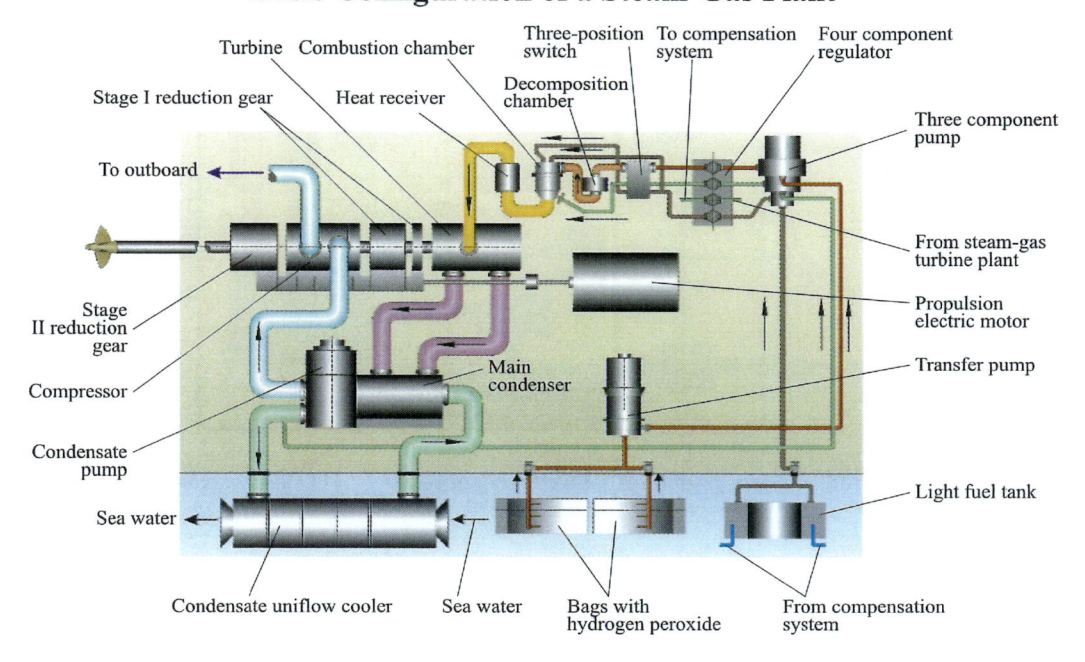

Fig.3.17. Project 617 «WHALE»
Submarine with a Steam-Gas Plant

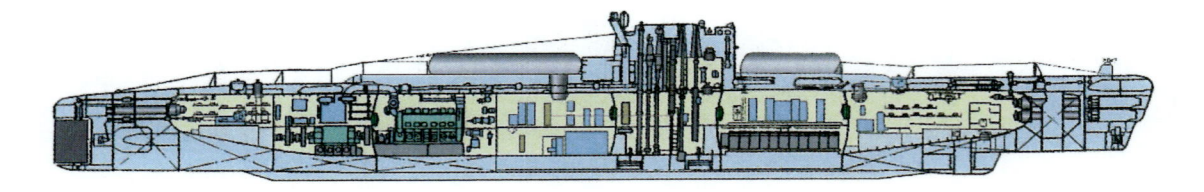

Principal Tactical and Technical Characteristics

Period of sea trials, years	1988-1989
Normal displacement, m³	1265

Main dimensions, m:
    length
    beam
	
	76,0
	7,3

Fuel cell power, kW	280
Full surface speed, knots	9
Full submerged speed, knots	5
Submerged range at economic speed, miles	1700
Endurance, days	30
Oxygen, tons	32
Hydrogen, tons	4

A Power Plant Based on Fuel Cells

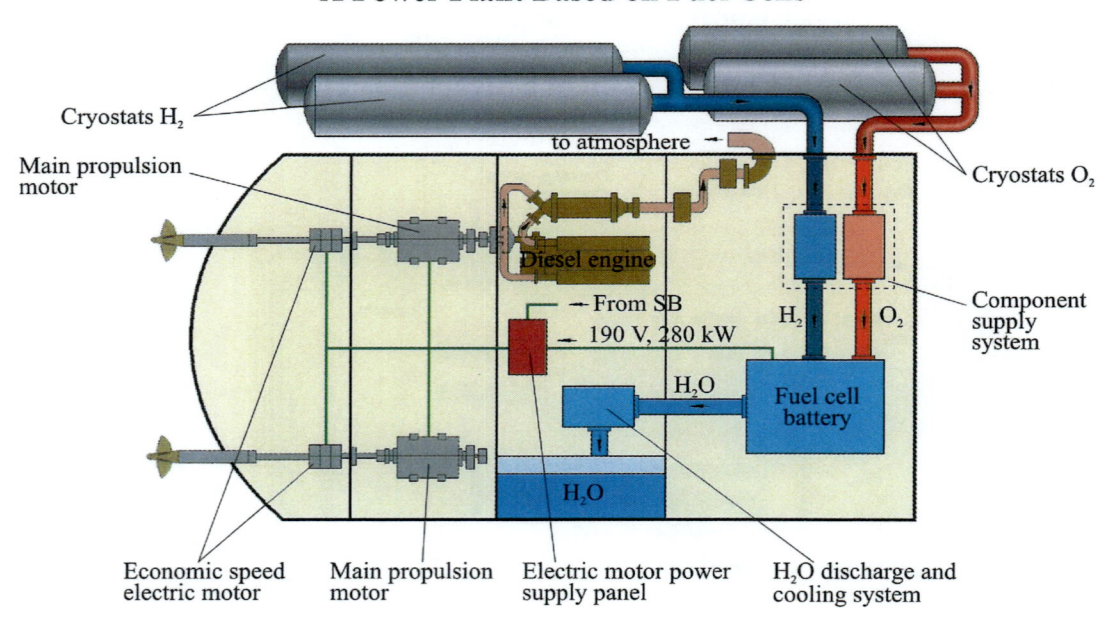

Fig.3.18. Project 613E «WHISKEY-AIP» Submarine with Fuel Cells

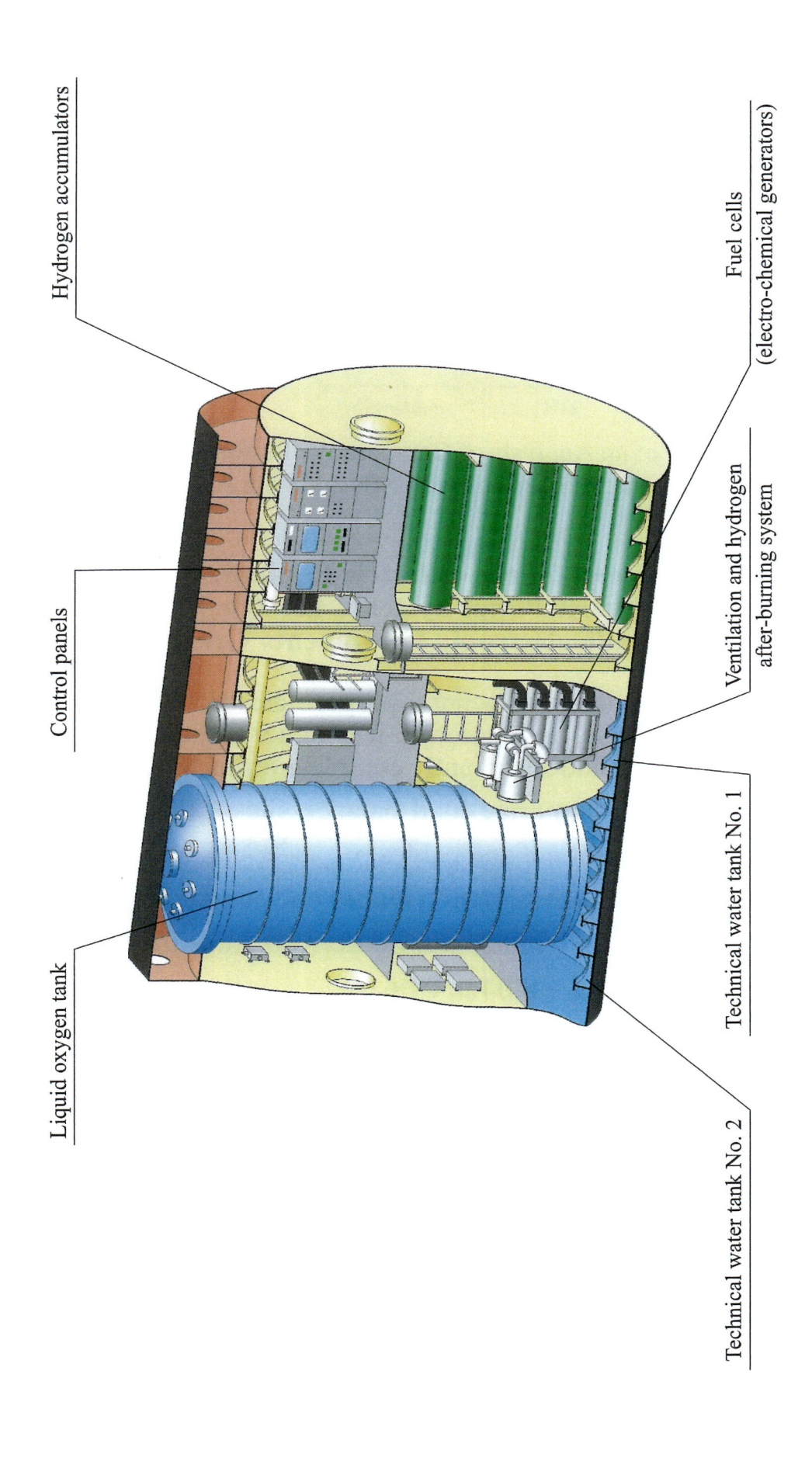

Fig.3.19. A Fuel Cell Power Plant Compartment

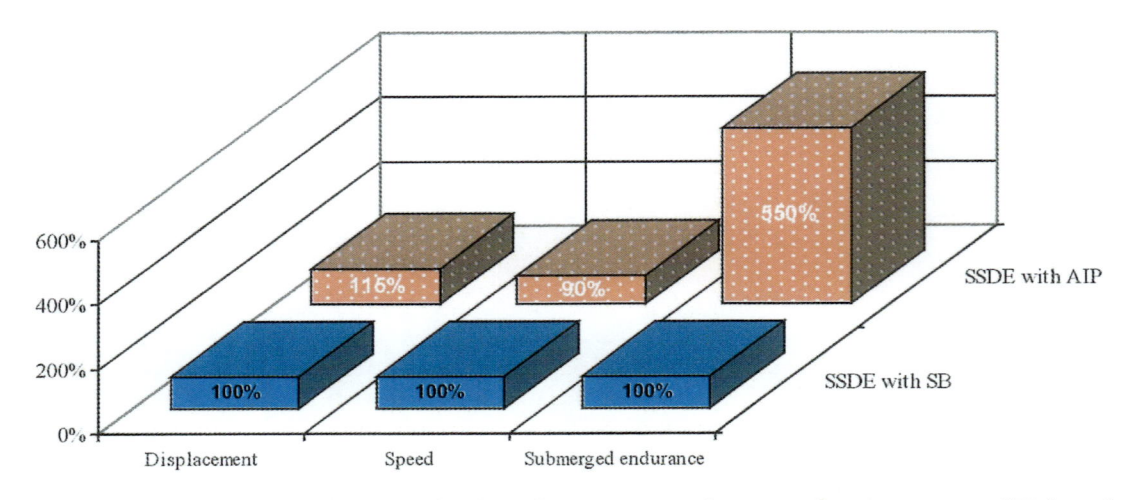

Fig.3.20. Results of an Analysis of Comparative Performances of Diesel-Electric Submarines with SBs and with AIPs.

3.4. Load for the «Electric Equipment, Cables of Electric Power Systems, Electric Networks and Radioelectronic Equipment» Group

Lists and designations of general-purpose electric equipment and cables are very diverse, and therefore it is rather difficult to obtain a physically justifiable and sufficiently complete formula for its weight.

Analysis of the submarine load balance shows that the weight index of electric equipment and cables $p_{EEG} = \dfrac{P_{EEG}}{D_0}$ is fairly steady for submarines of the same designation within one generation. Hence, at initial design stages this index can be derived from the prototype. Approximately it is $p_{EEG} = 0.04$ to 0.06.

The list of radioelectronic aids (REA), types of stations, and their main characteristics are indicated in the Submarine Design Specifications. The weight of equipment constituting radioelectronic aids is taken from delivery specifications or tentative data from the subcontractors-suppliers. In design studies one may use weights and dimensions of REA prototypes corrected for relevant design requirements and new technology developments.

The weight of REA does not depend on the design features of the submarine and therefore should belong to the group of independent («specified») weights. At the same time, the weight of outboard equipment, including antennas/arrays, depends on the diving depth (external pressure on the equipment). This should be taken into account in calculations. The weight of cables for all purposes (trunk and local cables) is about 2 to 3% of the normal displacement of the submarine.

3.5. Load for the «Weapons and Their Supporting Systems» Group

For a submarine, this group includes weights of tools for engaging the adversary, as well as the weight of systems and units supporting their operation (fire control systems, etc.).

These weights are estimated based on the Submarine Design Specifications. Additionally, they also use data received from subcontractors and statistics from constructed submarines.

The weight of torpedo weapons can be estimated as:

$$P_{TS} = k_{TT}n_{TT}P_{TT} + (P_{WPT} + P_{SHUT} + P_{TOR})n_{TT} + k_{RAC}n_{RELT}P_{TOR} + P_{TLG} \quad (3.67)$$

where $k_{TT} = 1.2$ to 1.3 – coefficient accounting for control devices;
n_{TT} – number of torpedo tubes (TT);
P_{TT} – weight of one assembled TT;
P_{TOR} – weight of one torpedo;
$P_{WPT} = 0.5$ to 1.2 t – weight of water in the round-torpedo annulus depending on the type of TT per one torpedo tube;
$P_{SHUT} = 0.3$ to 0.6 t – weight of the tube muzzle shutter;
$k_{RAC} = 1.4$ to 1.5 – coefficient accounting for the weight of storage racks;
n_{RELT} – number of reload torpedoes on the racks;
P_{TLG} – weight of the torpedo-loading gear.

Strictly speaking, this group can be considered as a constant weight that can be calculated with a high accuracy at the earliest stages of design work.

With a fixed calibre, the tube weight P_{TT} depends on the TT type (pneumatic, hydraulic, etc.), ejection depth, maximum diving depth of the submarine and TT material characteristics. At initial stages of the design, P_{TT} should be derived from the prototype. In case the maximum diving depth is changed, TT parts exposed to the full outboard pressure are re-calculated.

If the TT calibre is fixed, the weight of the shutter is determined by its length which depends on the outer hull shape in way of the torpedo exit point, as well as on the firing cone which is usually 1/18 [35].

Even with the same calibre, different types of torpedoes can be considerably different in weight (up to 1.5 times) [32], [52], [68]. This affects the weight of the water in the round-torpedo annulus. In conceptual design, when the weapon package is not yet specified in details, it is advisable to assume one type of torpedo for each calibre.

For approximate estimation of the missile weapon package weight one may use a formula similar to (3.67):

$$P_{MS} = n(P_{SILO} + P_{SIO} + k_{CAP}P_{CAP} + P_{MIS} + P_{WRT}) \quad (3.68)$$

where n_{SILO} – number of silos for ballistic missiles or containers for cruise missiles;

P_{SILO} – weight of one silo/container;
P_{SIO} – weight of internal outfitting of one silo/container;
$k_{CAP} = 1.25$ to 1.30 – coefficient accounting for the weight of the silo fairing and cap actuator;
P_{CAP} – weight of one silo cap;
P_{MIS} – weight of one missile;
P_{WRT} – weight of the water in the round-missile annulus in one silo.

The silo body weight depends on its dimensions which are dictated by overall dimensions of the missile, by the maximum diving depth of the submarine and by characteristics of the material. When scaling P_{SILO} from a prototype, variations in these factors may be approximately accounted for with formulae for the PH weight. However, it should be kept in mind that g_{SILO} may be considerably larger:

$$P_{SILO} \approx P_{SILO_0} \frac{g^T_{SILO}(P_{MIS}, \sigma_T)}{g^T_{SILO_0}(P_{MIS_0}, \sigma_{T_0})} \cdot \frac{V_{SILO}}{V_{SILO_0}} \quad (3.69)$$

where $g^T_{SILO} = \frac{P_{CYL}}{V_{CYL}}$ – weight index of the cylinder simulating the silo (similar to the g^T_{PH} index but including the bottom plate).

For scaling the cap weight we may use formula:

$$P_{CAP} = P_{CAP_0} \left[\frac{d_{SILO}}{d_{SILO_0}}\right]^{2,5} \left[\frac{P_{MIS}}{P_{MIS_0}}\right] \left[\frac{\sigma_{T_0}}{\sigma_T}\right] \quad (3.70)$$

where d_{SILO_0} and d_{SILO} – diameters of the prototype submarine and the new submarine silos, respectively.

Other values in (3.68) are derived from delivery specifications for weapons, subcontractors' information and prototypes with similar weapons [34], [68].

3.6. Load for «Stocks and Complement», «Displacement Margin» and «Solid Ballast» Groups Stocks and Complement

Stocks and Complement

After the subtraction of stores for power plants, which has been done above, the index of the remaining part is $P_{SCR} = 0.02$ to 0.03, and the smaller value corresponds to submarines of greater displacements.

For a more accurate load estimation for the «Stocks and Complement» group, it is necessary to resolve a very important problem: to determine the number of crew members. On one hand, the submarine crew is required to maintain combat and general-purpose hardware but, on the other hand, personnel need food, water, air and berthing. If the number of personnel is too high, there are difficulties with accommodation spaces. For nuclear submarines of large displacements this issue is not very acute, but for diesel submarines with their limited displacements it becomes a problem and solving this problem is a crucial design task.

As a rule, the submarine complement can be approximately established at initial stages of the design. For this purpose they take Tables of Complement of the prototype submarine and make necessary corrections, including the account for two- or three-shift watch keeping, the assumed damage control and post-accident repair scenario, the extent of submarine control automation envisaged in the design.

When the number of personnel $n_{äë}$ is established, we can calculate the weight of the crew and associated stores as:

$$P_{SCR} = p_{PER} n_{PER} + (P_{PROV} + P_{DRW}) A n_{PER} \qquad (3.71)$$

and regard the result as a constant (independent) weight. Here:

$p_{PER} = 100$ to 125 kg – weight of one man with personal belongings;

$P_{PROV} = 3.5$ kg – food allowance (including packing) per man per day;

$P_{DRW} = 6$ litres – fresh drinking water allowance (received at the base) per man per day;

A – endurance, days.

An additional amount of fresh water (in addition to 6.0 litres of drinking water) is provided by distilling plants.

In addition, these calculations should include an emergency subsistence of provisions and fresh water.

After we have calculated the weight of the complement and associated stocks, the share of trimming water, trapped water in tanks and air in the PH volume amounts to $p_{TTW} = P_{TTW} / D_0 = 0.005$ to 0.010.

The Displacement Margin

The displacement margin (DM) is divided into two parts: the design and construction margin, which is at the disposal of the design bureau and the shipyard, and the upgrading margin, which is at the disposal of the customer – the Navy.

The value of the design and construction margin depends on the design stage, the novelty of the project, and the availability of a close prototype. At the Technical Proposal development stage it can be $p'_{DM} = P_{DM} / D_0 = 0.030$ to 0.050, decreasing at the Detail Design phase to $p'_{DM} = 0.005$ to 0.010.

The unused portion of the DM is balanced by solid ballast.

The upgrading DM is specified in SDS and usually amounts to $p''_{DM} = 0.005$ to 0.020.

The issue of the advisable DM for future upgrading is rather complicated. From the military–and–economic analysis point of view, to restore submarine combat capabilities when military equipment (especially weapon and sensor packages) progresses very rapidly, it was reasonable, until recently, to keep a higher than above-indicated margin. However, since the mid 1980s, rapid advances in microelectronics and the introduction of integrated platform-sensor-weapon automatic control systems has resulted in drastic reductions of weights and dimensions of such equipment: by 1.5 to 2.0 times on submarines of the IVth generation. Therefore, this problem yet needs additional studies [93].

All types of displacement margins should be taken into account in the load balance table in such a way as to ensure the reserve stability. Usually, for submarines it is considered sufficient if the ordinate of the centre of gravity of the displacement margin is not lower than the total CG of groups «Gears» and «Systems», i.e. $Z_{DM} = Z_{GS}$. At early design stages this offset can be taken at the level of the pressure hull axis.

The abscissa of the displacement margin X_{DM} should be in the area of its anticipated utilisation or, when that is uncertain, at midsection.

The Solid Ballast

Solid ballast is an obligatory accessory of any submarine. It plays the role of a kind of governor necessary to achieve the vital equality $D_0 = \rho g V_0$, to trim the submerged submarine and to maintain required stability values.

The variety of solid ballast functions makes it difficult to offer a relationship for estimating its weight P_{BAL} based on any physical considerations.

When the particulars of a submarine are initially determined, it is conventionally assumed that $P_{BAL} = p_{BAL} D_0$. For most submarines of traditional types $p_{BAL} = 0.02$ to 0.04.

The required amount of the solid ballast and co-ordinates of its CG are more accurately determined after compiling the load balance and the constant buoyant volume tables.

The final amount of the solid ballast and the position of its CG are determined after the constructed submarine has been re-ballasted and heeled.

FEDERAL SCIENTIFIC & PRODUCTION CENTER
STATE UNITARY ENTERPRISE "AVRORA"
CORPORATION SCIENCE & PRODUCTION

Control systems for technical facilities of Naval ships.
Automated combat control systems.
Control systems for technological processes of the industrial power generation objects.
Electric transport automation systems.
Automated processes of oil and gas extraction and refining.
Special simulators and training centers.

НПО аврора

15, Karbyshev St., St. Petersburg, 194021, Russia
Tel.: (812) 247-2250, 247-6179 Fax: (812) 247-8061
E-mail: aurora@peterlink.ru

4. THE LOAD EQUATION AS A FUNCTION OF THE DISPLACEMENT. DETERMINATION OF THE DISPLACEMENT

4.1. Load Equation Derivation and Solution for the First Approximation

As stated in Chapter 2, the displacement, otherwise the weight of a buoyant body, D_0 is equal to the weight of water ρV displaced by this body. The displacement D_0 is a sum of weights of structures, machinery, equipment, weapons fuel, etc. constituting the submarine load balance, i.e.:

$$D_0 = \sum_{i=1}^{i=n} P_i \qquad (4.1)$$

Expression (4.1) is called the equation of load or the weight equation.

Some weights in formula (4.1) can be expressed as functions of the displacement $P(D_0)$. Other weights, let us designate them P_{IMD}, are specified and assumed to be independent of D. Therefore, we can write:

$$D_0 = P(D_0) + P_{IMD} \qquad (4.2)$$

The solution of the (4.2) equation gives the sought displacement corresponding to the specified independent weights P_{IMD}, as well as to parameters included in the $P(D_0)$ function, in particular, the sea range, relative weights, etc. Weights $P(D_0)$ can be calculated for any given D_0, but the sought displacement will correspond only to the solution of (4.2).

The graphic solution of this equation is shown on Fig.4.1. The solution of equation (4.2), i.e. $D_{0_{SOL}}$ is the intersection point of the weight curve $P(D_0) + P_{IMD}$ and the buoyancy straight line $D = D_0$. This is also the point where the law of Archimedes is satisfied. As we may see from Fig.4.1, the equation has the only solution.

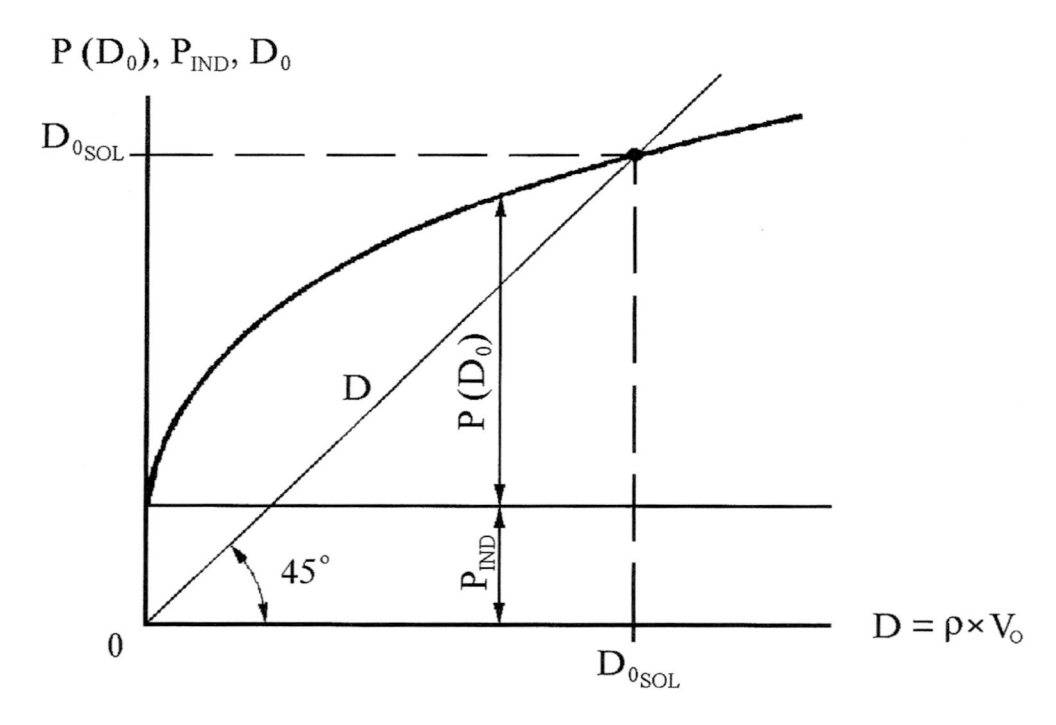

Fig. 4.1. Graphic Solution of the Equation of Load

It has been already mentioned in Chapter 3 that some weights depend on the displacement as a linear function while some are functions of $D_0^{2/3}$. Then, in a general form, function $P(D)$ can be written as:

$$P(D_0) = AD_0 + ED_0^{2/3} \qquad (4.3)$$

where coefficient A is the sum of relative weights directly proportional to D_0 while E is the sum of those depending on $D_0^{2/3}$.

The load equation corresponding to equation (4.2) is:

$$D_0 = AD_0 + ED_0^{2/3} + P_{IND} \qquad (4.4)$$

Substituting $x = D_0^{1/3}$, formula (4.4) is reduced to a cubic equation and solved analytically [1].

Solving equation (4.4) by any method, we obtain the normal displacement in the first approximation.

4.2. Load Equation for the Second Approximation

Updating the displacement and the load balance of the designed submarine, as well as other design particulars, for the second approximation, is made by selecting the power plant configuration and updating other components of the normal displacement.

As nuclear submarines are usually designed for a given or assumed-in-advance power plant, the output and weight/dimension parameters of which are known, there is no need to select a plant. Let us consider setting up an equation in the second approximation for a SS, though it should be mentioned that the general approach to the equation of load is as true for SSNs.

The selection of the power plant components, i.e. motors, SB and other PP parts, should be preceded by the decision on the number of shafts and on the distribution of the total power among them. When this issue is settled taking into account all relevant factors, we can start the selection of actual motors from those available (by Manufacturer's catalogues) or under development.

Selection of Diesel Engines and Propulsion Motors

When selecting diesel engines for submarines, one should consider the following factors:
1) adequate power;
2) engine speed, especially if it drives the shaft line directly;
3) acceptable weight and dimensions allowing the arrangement of diesel engines in the submarine;
4) acoustic characteristics of the diesel engine;
5) ability of the chosen diesel engine to operate at increased counterpressures in the snorkel mode;
6) fuel consumption rate.

When selecting diesel engines by the required power N_i estimated from first-approximation data (see Chapter 3), it should be kept in mind that in real life the required power never coincides with the output of available engines. However, this is quite acceptable considering that any power requirement estimated with the formula of Admiralty coefficients is approximate. Thus, when diesel engines are selected and their weight is known, it is possible to find out the weight of the diesel plant:

$$P_{DP} = k_{DP}\sum_{1}^{n} P_{DE} \qquad (4.5)$$

When selecting electric propulsion motors, the requirements are similar to that mentioned above about the selection of diesel engines. Additionally, it is necessary to take into account the fact that if the power plant of the submarine is not single-shaft, the power of electric motors obtained from first-approximation data has to be divided by shaft lines and this division may not always agree with the division of the diesel engine power. On the other hand, when diesel engines operate for the shaft directly, their revolutions should be matched in a certain way with the electric motor speed. Otherwise the propeller will be overloaded in one regime and underloaded in another. To match revolutions of the diesel engine and of the electric motor operating for the same propeller, it is necessary to have a power-propeller revolutions curve.

Thus, electric propulsion motors should be selected both by the power N_i and by the number of revolutions n. This condition makes the task more difficult and not always solvable. For completely electric propulsion designs this problem is non-existent, which was one of the major reasons for choosing electric-only propulsion plants for the IVth generation submarines.

When electric motors have been selected, similarly to the calculation of the diesel engine weight, we can estimate the weight of the propulsion plant with a formula like (4.5). $P_{EPM} = k_{EPM} \sum_{1}^{n} P_{EM}$

When the weights of above-discussed plants are established, they are transferred to the category of constant weights.

Selection of the Storage Battery System

With the help of the functions we have used when setting up the equation of load for the first approximation and of the obtained displacement D01 it is possible to calculate the weight of the storage battery system and the weight of the battery as such.

As already mentioned (see 3.3), the storage battery is made up of groups and the number of cells in a group is by no means arbitrary as it is determined by the required voltage delivered by the group and by the type of the storage battery.

The battery type is selected when the equation of load is set up for the first approximation because it largely dictates the value of the specific energy pickup ΔW and, hence, the battery weight (see Fig.3.13).

Thus, for the battery type established at the first approximation we have several alternative cells of different dimensions and weights P^i_{CELL}.

The required number of cells of weight P^i_{CELL} is determined as:

$$n^1_{CELL} = \frac{P_{SB}}{P^1_{CELL}} \tag{4.6}$$

Knowing n^1_{CELL}, we find out whether it is possible to set groups with the given number of cells per group. After that we determine the number of groups and the number of cells n_{CELL}. In a general case n^1_{CELL} and n_{CELL} may differ one or other way. If this difference is not significant (let us assume it is within 5%), the storage battery is assembled of cells of weight P^1_{CELL}. If the difference is considerable, we repeat the procedure to find the number of cells of another type of weight P^2_{CELL} etc.

Having configured the storage battery, we can calculate the weight of the system and transfer it to the category of constant weights:

$$P_{SB} = k_{SB} \cdot k_{DIST} P^i_{CELL} n_{CELL} \tag{4.7}$$

where k_{SB} – coefficient accounting for the weight of the system;
k_{DIST} – coefficient accounting for the weight of distillate for topping the cells.

In addition to the power plant components we can update some other weights included into the normal displacement of the submarine.

Based on updated results, the equation of load for the second approximation is set up as:

$$A'D_0 + E'D_0^{2/3} + P'_{IND} = D_0 \tag{4.8}$$

Equation (4.8) will be considerably different from the equation of load for the first approximation (4.4).

Constant weights are now considerably larger as $P'_{IND} = P_{IND} + P_{DP} P_{EMP} + P_{SB} + P_i$ and coefficient E' is smaller. Coefficient A' also may change.

Having obtained the displacement value from the second-approximation equation of load, we can compile the load balance table of the designed submarine for those major parts which were taken into account in setting up this equation. In this table it is helpful to enter data on submarine prototypes which have served as the basis for deriving respective weight indices and other parameters of the designed submarine.

Such a table enables us to check the correctness of the load equation solution and to compare the load balance of the subject project with those of other submarines. Deviations in values of weights per groups, particularly relative deviations (in percents of D_0), should be explainable from the point of view of Design Specifications.

4.3. Effects of Variations in Submarine Particulars and Independent Weights Upon the Load Balance. The Differential Form of the Load Equation As a Function of the Displacement. The Normand's Number

In various types of design estimations it is necessary to find out how the submarine displacement would change due to variations in different parameters (indices, tactical, technical or economical features) [6], [27]. In such tasks it may be convenient to use the differential form of the load equation describing the pattern of displacement variations due to small increments of relevant parameters:

$$\Delta D = \eta_{NOR} [\Delta P(D_0) + \Delta P_{IND}] \qquad (4.9)$$

Coefficient η_{NOR} is called the displacement variation factor or the Normand's number. From (4.9) it follows that

$$\eta_{NOR} = \frac{\Delta D}{\Delta P(D_0) + \Delta P_{IND}} \qquad (4.10)$$

i.e. the Normand's number is the ratio of the displacement variation due to a change in submarine parameters and independent weights. The limits of formula (4.9) applicability have not yet been studied adequately but for practical purposes we may expect rather accurate results at weight variations within 20% of D_0.

To calculate the variation of displacement ΔD by formula (4.9), it is necessary to find out both η_{NOR} and the weight variation $\Delta P(D_0)$ due to changes in submarine parameters assuming the displacement remains unchanged.

The variation of independent weights is calculated quite easily:

$$\Delta P_{IND} = P_{IND_1} - P_{IND_0} \qquad (4.11)$$

Variations of weights which depend on the displacement and a number of submarine parameters can be found by finite difference calculations. E.g., when it becomes $P_{PH} = P_{PH_1}$ the PH load changes as:

$$\Delta P_{PH} = P_{PH_1} - P_{EST_0} = (p_{PH_1} - p_{EST_0})D_0 = \Delta p_{PH} D_0 \qquad (4.12)$$

Similarly, it is possible to formulate load variations for all other groups. The result is:

$$\Delta P(D_0) = \sum_1^n \Delta P_i \qquad (4.13)$$

The simplest expression for the Normand's number is written when the subject change $P(D_0) + P_{IND}$ occurs only due to the change in P_{IND}, i.e. $\Delta P(D_0) = 0$, and $\Delta P_{IND} \neq 0$.

$$\eta_{NOR} = \frac{1}{1 - \dfrac{dP_0(D_0)}{dD}} = \frac{1}{1 - A^{2/3}E\,\dfrac{D_0^{2/3}}{D}} \qquad (4.14)$$

where $\dfrac{dP_0(D_0)}{dD} = \dfrac{P_{PH_0}}{D_0} + \dfrac{P_{LH_0}}{D_0} + \dfrac{P_{GS_0}}{D_0} + \ldots + \dfrac{2}{3}\left(\dfrac{P_{PH_0} + P_F + P_{SB}}{D_0}\right)$

With these formulae we can calculate η_{NOR}, e.g., when there is a change in weapon or ammunition weights. Quite often formula (4.13) is used even for $\Delta P(D_0) \neq 0$ (the very condition it was originally derived for by Jacques-Augustin Normand). The accuracy of calculations is slightly reduced, but at the same time calculations are simplified and, besides, it becomes possible to use one constant η_{NOR} for several options of parameter and independent weight variations.

Using relationship (4.13) it is possible to formulate displacement variation for A (when, e.g., the diving depth is changed and, accordingly, the weight of pressure structures becomes different):

$$\Delta D = \Delta A D_0 \eta_{NOR} \qquad (4.15)$$

displacement variation for E (when PP power, speed and, accordingly, the weight are changed):

$$\Delta D = \Delta E D_0 \eta_{NOR} \qquad (4.16)$$

displacement variations for ê, i.e. when the constant weights are changed (weapons, sensors):

$$\Delta D = \Delta P \eta_{NOR} \qquad (4.17)$$

The value of the Normand's number for submarines is within $\eta_{NOR} = 3$ to 4 and it shows by how much the displacement can increase when the load grows. E.g., when the load increases by 100 t, the displacement increases by 300 to 400 t. The reason for this growth is a large contribution of displacement-dependent weights.

As follows from the above-said, the application of the differential form of the load equation is very closely related to the weight breakdown approach. The latter is governed by the considered task and, depending on the aim of the study, the composition of weight groups may change.

It should be mentioned that the larger the number of submarine equipment weight classed as constant weights, the more reliable the result obtained with the load equation and the Normand's number, and the less the value of this number.

Thus, the process of submarine load balance calculations at different design stages can be presented as a continuous growth of the constant weight share. At the Detail Design stage, when all weights are determined rather accurately and the load balance can be changed only at the expense of the displacement margin, the displacement increment becomes zero.

4.4. Load Control

By «load control» we understand a set of administrative and technical measures directed toward ensuring that the boat can be trimmed at all stages of her design, construction, trials and service.

The essence of submarine load control efforts boils down to the following procedures:
- prediction of the submarine load balance variations at all stages of design, construction, trials and service life;
- load control for equipment installed in the submarine according to the design documents issued in the process of design;
- displacement margin management;
- formulation and enforcement of load limits;
- weight control during construction (control by weighing);
- reballasting and heeling the submarine upon the completion of construction;
- load control for equipment installed for the period of submarine sea trials;
- load control for equipment installed during submarine upgrading;

Prediction of Submarine Load Balance Variations. The prediction of changes in the submarine load balance starts at the very early stages of the submarine design. Submarine load calculations made based on design formulae and statistic indices serve as initial information for work at subsequent design stages.

Having such a calculation available, the designer applies the method of expert evaluations to determine the most probable directions of submarine load balance changes based on information about innovations made in submarine design. Predictions, made jointly by specialists of the design bureau and companies developing the equipment, of possible changes in loads of individual types of submarine equipment serve as a basis for making decisions on assigning the value of the displacement margin for later design stages. In addition to the prediction of changes in the submarine equipment weight, they also make a prediction of the centre of gravity position for this additional weight.

Simultaneously with the prediction of the design-associated submarine load variations, a similar prediction is made for submarine load changes during construction. The results of this prediction are used for selecting the value of the construction displacement margin. As a rule, this value depends on the experience and equipment (machinery and appliances) of the shipyard selected for the submarine construction, as well as on non-traditional technologies required for the subject submarine (when the shipyard has to introduce the block and module method or different construction materials, e.g., when the chosen main structural material is a high-tensile steel which is better for strength but more labour consuming in terms of welding).

In parallel with the above-described work, they evaluate prospects of upgrading the boat within her life cycle and consider how such modifications may affect the submarine load balance.

All components of the displacement margin (for design, construction and upgrading of the submarine) are closely related to each other. Thus, if the displacement margin is not fully expended for design purposes by the completion of the Detail Design documents (the Detail Design stage), the remaining portion is usually included into the upgrading displacement margin. Similarly, the upgrading displacement margin increases if the construction displacement margin is not spent completely. However, in case of an unexpected increase in the submarine load at any design stage the «overweight» can be cleared at the expense of reducing the construction displacement margin. Then they have to apply more «stringent» load control measures during construction and to reconcile some design decisions.

Load Control by Detail Design Documents (drawings, diagrams, and lists in accordance with which the submarine is constructed) aims to obtain complete and reliable information about the submarine load balance. The load control is a component of the general Quality Assurance philosophy for documents issued by the design bureau. It is guaranteed by a system under which all generated documents must be approved by the Load Department. Such a form of control enables any design documents with incomplete or unreliable information in terms of the load balance to be spotted in time.

In the process of load control by the Detail Design documents they check all documents for the following information:
1) number of the Detail Design document;
2) description of equipment installed in the submarine according to this document;
3) subject equipment station (numbers of frames between which it is installed);
4) weight in tons (with the accuracy of up to 0.001 t);
5) coordinates of the centre of gravity in the submarine reference grid (from the midship station – X, from the base plane – Z, from the centre line – Y) – with the accuracy of up to 0.01 m;
6) moments of weights (with the accuracy of up to 0.01 tm);

If the design document is developed for the installation of some equipment containing various weights that may be removed from it in the process of operation, the document should contain information per items 3 through 6 of the above list, for each type of such removable weights. E.g., pipes, valves, pumps, etc. make up the load of the so-called «basic structure», and water (or other liquids) in this system is regarded as the «filler». The load data indicated in the design document should specify accordingly the type of such a «filler» (water, lub oil, diesel fuel, etc.). To simplify the subsequent information handling, the «fillers» are assigned numerical codes.

After the completion of this step of the design document check (for the completeness of information about the load), the document is checked for the reliability of information contained in it. E.g., the weight specified in a document can be checked by a simplified calculation as a function of any structural parameter (thus, to check the weight of deck plates one may use its function of the area). Co-ordinates of the centre of gravity of the entire structure (or installed equipment) can be approximately determined as co-ordinates of the centre of gravity of a uniformly filled geometrical body of a constant density.

The co-ordinates of centres of gravity of individual parts (or groups of parts) are checked selectively. When the calculation of the equipment load installed according to this design document is made «manually», they may check the correctness of arithmetic operations (e.g., summation of weights of individual parts or their weight moments).

After the completion of such (trivial) checks, the load as per the given design document can be included into the database on the submarine load balance of this project. The new load item is added to already checked equipment loads of the given structural group and the resulting value is compared against the specified limit. When all documents of the given structural group are completed, the final result is compared to the assigned limit to establish deviations of the weight and the weight moment. If the deviation exceeds the permissible value, the load for this structural group is revised. The aim of this revision is to find out reasons for such a deviation (whether it is the result of a design mistake or a consequence of some decisions directed toward equipment improvement in this structural group). The same procedure is applied if the document is not completed because the load limit has already been exceeded. In the detailed revision of structural documents developed for the given structural group, the loads per each document (or group of documents) are compared with similar loads estimated at previous design stages (for example, comparison of loads of the submarine light hull structures in the fore block with the similar loads obtained at the Engineering Design stage).

Construction Weight Control (Control by «Weighing»). In the process of submarine construction at a shipyard the load of equipment designated for installation is checked for its compliance with Detail Design documents. Practically, all equipment is weighed. The aim of this exercise is to guarantee trimming of the completed submarine and to meet requirements the submarine in terms of the value of the initial transverse metacentric height, as well as to be able to calculate the amount and the arrangement of the trimming ballast before launching the submarine.

Submarine construction weight control procedures include several types of weighing:
- part-wise (weighing of individual hull parts, equipment components, devices, etc.);
- assembly-wise (weighing of individual structural assemblies);
- section-wise (weighing of sections).

At each stage of weighing they have to make a decision whether revealed deviations in weights of parts, assemblies, or sections due to technological reasons are allowable. If the difference between weights established by weighing and those obtained by calculations is significant, the latter are revised in order to find out possible mistakes. If they find any mistakes in Detail Design documents, the load limits are corrected as described above.

In case it is impossible to compensate the load increase during construction by any corrections of design documents, the Chief Designer of the project can take a decision to reduce the displacement margin left for the submarine construction.

When the excessive weight of a part, assembly or section is a result of poor workmanship, this item is rejected and a new one manufactured.

The completing stage of construction load control is submarine ballasting and heeling performed during sea trials. The aim of this operation is to check whether the load matches the buoyancy and to get values of the initial transverse metacentric height for main service (usually in submerged, full surface and, sometimes, low-buoyancy trims).

In the process of re-ballasting and heeling they establish the need to lay (remove) additional solid ballast in order to match the buoyancy and the loading of the submarine, as well as the need to move some amount of solid ballast from the bow to the stern (or from the stern to the bow) and from side to side, in order to eliminate any angles of trim and heel.

Based on load balance calculation results they plot the load distribution along the submarine length. This profile is an input for trimming calculation (Fig.4.2).

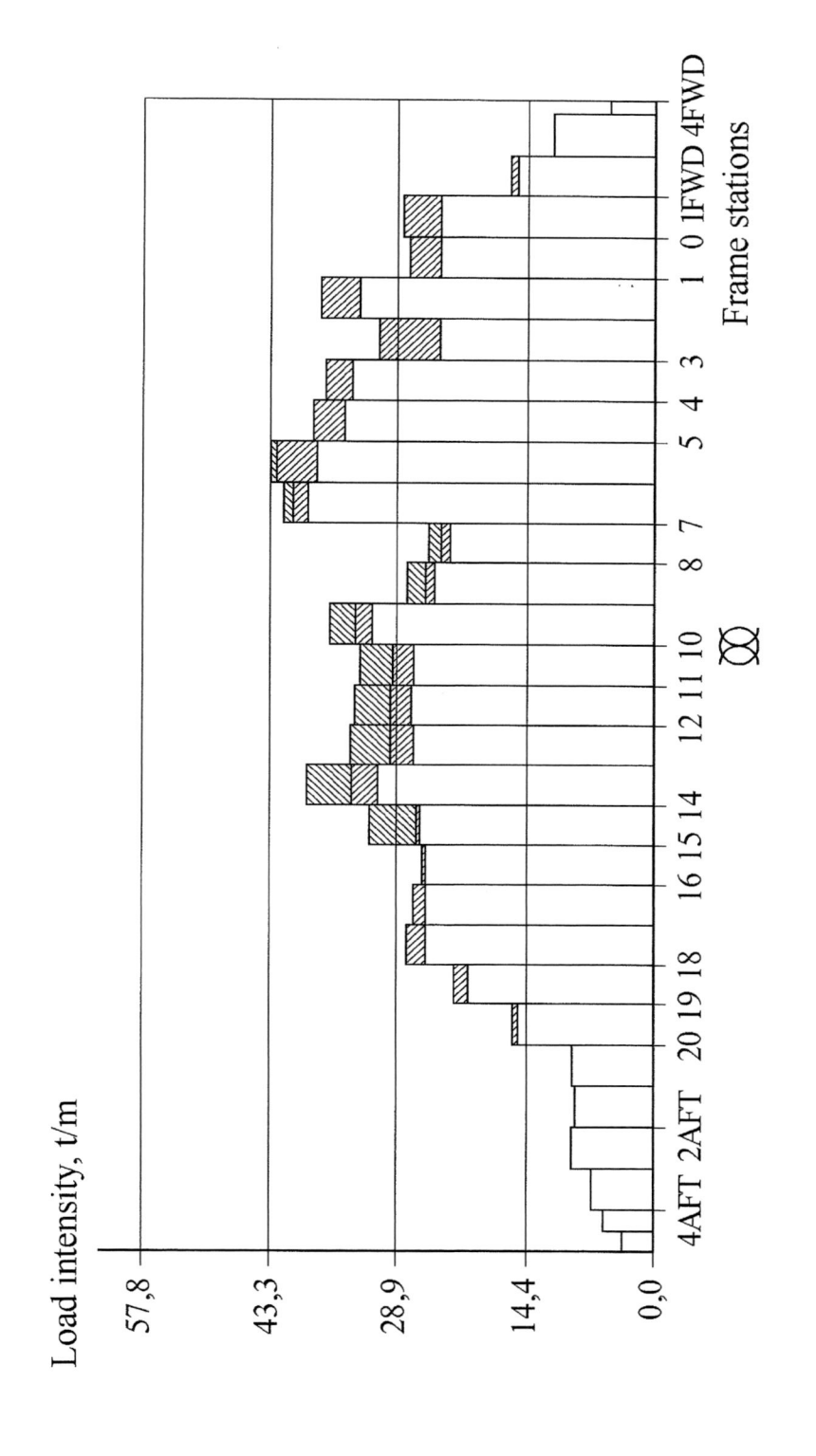

Fig.4.2. Load Profile by Station Spacings

CRI GRaniT

State Unitary Enterprise Central Scientific & Research Institute GRANIT is a traditional designer and supplier of complexes and systems for submarines, such as:

■ Radar stations for submarines of practically all classes beginning with legendary radar stations called Flag installed on hundreds of submarines being in service with national and foreign Navies, and up to the up-to-date automated jam-proof electronic digital integrated radar complexes for the Naval Forces of Russian Federation and foreign countries;

■ Fire control systems of anti-ship strike cruise missiles of submarines and surface combatants. Several generations of Soviet-design submarines are equipped with such systems. At present they form the basis of the Russian Naval Forces and Navies of a number of other nations.

3, Gospitalnaya str., Saint-Petersburg, 191014, RUSSIA
Tel.: (812) 271 6756, fax: (812) 274 6339
E-mail: cri-granit@peterlink.ru

5. SUBMARINE CONSTANT BUOYANT VOLUME (CBV)

5.1. The Constant Buoyant Volume and Its Relationship with the Submarine Normal Load

As we now know, the normal load of the submerged submarine corresponds to the constant buoyant volume (CBV), which includes all volumes displacing in water when the submarine is submerged (Fig.5.1).

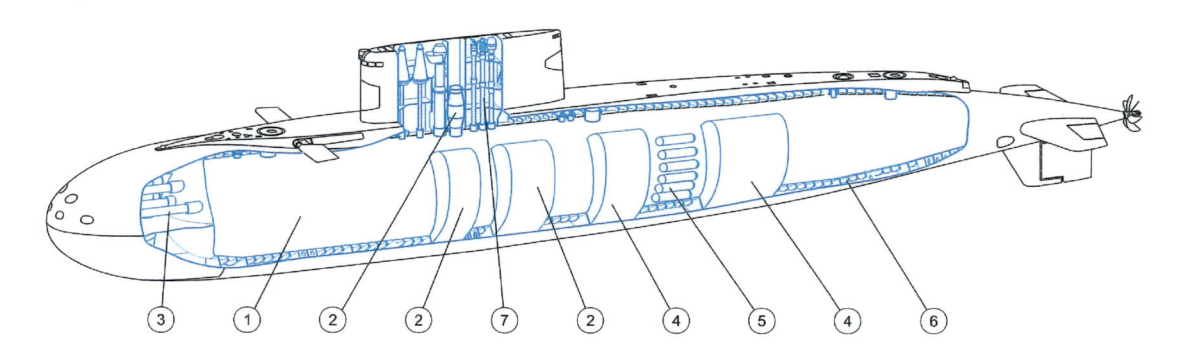

Fig.5.1. Basic CBV Components of a Diesel-Electric Submarine.
1 – pressure hull; 2 – external pressure-proof hatches, houses and tanks; 3 – torpedo tubes; 4 – external fuel tanks; 5 – systems, gears; 6 – plating, framing, coating of the pressure and the outer hulls; 7 – other

They include, in the first instance, volumes of the pressure hull and equistrong structures placed outside the PH; the volume of the metal of the outer hull and of structures located in the between-hull space; the volume of diesel fuel in external fuel tanks, etc. (See Chapter 3).

The results of CBV calculations are presented in the form of tabularised summaries (CBV Tables). Such a Table usually includes 30 to 50 volumes with their levers and moments like in the normal load table (Table 5.1).

Table 5.1

Constant Buoyant Volume Table

No	Description	Volume V m^3	Vertical lever, Z m	Vertical-moment, M_Z m^4	Longitudinal lever, X m	Longitudinal moment, M_X m^4
1	Pressure hull volume					
2	Regulating tank					
3	Quick diving tank					
4	Trim tanks *					
5	External fuel tanks					
6	Plating and framing of the pressure hull					
7	Plating and framing of pressure tanks					
8	Plating and framing of the light hull					
9	Plating and framing of the fore end					
10	Plating and framing of the aft end					
11	Plating and framing of the superstructure and the sail					
12	Projecting parts of torpedo tubes					
13	Pressure houses and hatches					
14	HPA bottles with pipelines					
15	Steering gears					
16	Masts					
17	Kingston and vent valves					
18	Miscellaneous, elsewhere non-accounted items					
	TOTAL constant buoyant volume					

* If arranged outside the pressure hull.

The CBV Table scope is more or less the same regardless of the design stage, but calculations, detailing and accuracy increase as the design is elaborated.

If general arrangement drawings or schematics, as well as the normal load balance, are available, calculations for the CBV Table pose no special difficulties. Basic buoyant volumes and co-ordinates of their centers are determined by formulae for elementary geometrical bodies (cylinders, cones, parts of spheres, etc.).

The submarine should be designed in such a way as to satisfy the static equilibrium condition without trim and heel:

$$D_0 = \rho V_0 \text{ at } X_g = X_C, Y_g = Y_C \qquad (5.1)$$

It is also necessary to satisfy the submerged stability condition:

$$Z_C - Z_g = h_0 \qquad (5.2)$$

where h_0 – submerged initial metacentric height.

Conditions (5.1) and (5.2) express the relation between the CBV and the normal load of the submarine.

Beside these conditions, there is a certain relation between some weights of the external outfitting of the submarine and their respective buoyant volumes. Volumes of such structures are dictated exclusively by their weights and by densities of their materials while co-ordinates of their centres of gravity and volume coincide [36]:

$$V_i = \frac{P_i}{\rho_{M_i}} \qquad (5.3)$$

$$X_{g_i} = X_{C_i}; Y_{g_i} = Y_{C_i}; Z_{g_i} = Z_{C_i} \qquad (5.4)$$

where V_i and P_i – buoyant volumes and respective weights;
ρ_{M_i} – material density.

Formula (5.3) is applicable to structures like:
– plating and framing of the pressure hull (except the part of PH included into the external pressure tanks) because calculations are usually made with moulded volumes;
– hull structures that are located outside PH moulded surfaces and have no voids (foundations in the light hull, outer hull plating);
– special-purpose external coatings.

Equality (5.3) may be considered true for any external equipment if we understand density ρ_i as the mean value for the subject type of equipment.

Equalities (5.4) are completely or partially applicable to any equipment and structures which have planes or axis of symmetry in terms of shapes and weight profiles. For equipment and structures with minor deviations from the axis of symmetry, they can be used at the initial stage in approximated design formulae.

Thus, when generating Load Balance and CBV Tables for any project, it is necessary:
- to cross-check totals of both tables, i.e. to make sure that conditions (5.1) and (5.2) are satisfied;
- to cross-check individual components that should satisfy conditions (5.3) and (5.4).

Such a check allows the degree of stability and trimming of the submarine to be assessed.

5.2. Relationships among CBV Components Considered at Early Design Stages

Volumes listed in the CBV Table can be grouped into the following larger articles:
- pressure hull;
- large volumes outside the PH (various external pressure tanks V_{EPT}, external parts of missile silos and containers, the sail, the sonar bay V_{SC}, external fuel tanks V_{EFT});
- small volumes outside the PH (HPA bottles, external parts of TTs, pressure-proof external pipelines);
- various minor unaccounted volumes.

Table 5.2 shows relative values of the largest CBV components for double-hull and single-hull submarines of different designations and types, with different weapons and power plants. To obtain steadier values, volumes V_i are referred to the constant buoyant volume, ignoring the volumes of external coatings ($V_0' = V_0 - V_{COAT}$) [36].

Table 5.2

Relative Shares of CBV Table Components (in % of V_0')

Srl. No.	Description	\multicolumn{4}{c}{CBV components}	$\sum V_i$
		V_{PH}	V_{EPT}	V_{SC}	V_{EFT}	
1	Diesel submarine (large)	76~77	3~4	-	14~15	93~96
2	Diesel submarine (middle)	78~80	3		13~14	94~97
3	Nuclear submarine (torpedo)	92~94	2-3	-	-	94~97
4	Nuclear submarine (cruise missiles)	87~89	1.5~4	4	-	92.5~97
5	Nuclear submarine (ballistic missiles)	86~87	7~8	2~3	-	95~98
6	Diesel-electric submarine (single-hull)	96	0.5	-	-	96.5

Analysing Table 5.2 we can make the following conclusions:

1. The relative volume of the pressure hull χ_{PH}, generally speaking, is not very stable for submarines considered in the Table and amounts to

$$\chi_{PH} = \frac{V_{PH}}{V_0'} \approx 0.76 - 0.96 \qquad (5.5)$$

The PH volume depends on other CBV volumes, mainly: external fuel and pressure tanks, missile containers and silos and their location with respect to the PH (inside or outside). The architecture of the submarine also affects the χ_{PH} value.

At the same time, for submarines with similar types of weapons and power plants, as well as for boats of well-established architectural types, the χ_{PH} value is rather stable and can be used in design, e.g., for approximate estimations of the CBV from the already determined PH volume.

2. The ratio of the sum of the largest volumes to V_0' is remarkable for its high stability.

$$\chi = \frac{V_{PH} + V_{EPT} + V_{SC} + V_{EFT}}{V_0'} \approx 0.73 - 0.98 \qquad (5.6)$$

Volumes in the numerator of (5.6), except V_{PH}, can be approximately found from statistic data. They are expressed through the displacement or respective weights included into the load balance, as well as derived from the data of design specifications, taking into account general arrangement considerations.

Special attention should be paid to the calculation of the main CBV component: the pressure hull volume V_{PH} and the position of its centre of buoyancy.

5.3. Determination of the Pressure Hull Volume. Volume Indices.

The scope of information about the equipment and facilities arrangement in the pressure hull available to the designer at early design stages may be different.
   a) Dimensions and maintenance access requirements of the basic, largest equipment items in different compartments are known. In this case compartment volumes are determined by means of a graphic study of the equipment arrangement. This may be when motors were selected by delivery specifications based on the first approximation (solution of the load equation).
   b) Only main motor outputs, weights of equipment and facilities are known. Then the required compartment volumes and rooms are estimated with the help of volume indices.
   c) Some room volumes are assumed as absolute values from the prototype submarine.

Thus, at early design stages there are basically two methods to determine the PH volume: graphical and analytical.

The advantages of the graphical method of establishing the required PH volume may include: firstly, sufficient accuracy when determining the sought value; and secondly, the possibility to find not only the volume of the hull but also its main dimensions, as well as overall dimensions of individual compartments.

To check decisions made by the designer, they build rather large-scale models of those compartments that are especially difficult from the point of view of equipment layout.

The main drawback of the graphical method, as has been mentioned earlier, is that it is time-consuming, and therefore does not allow any large number of options to be considered.

For the purposes of approximate estimations of required PH compartment volumes at early design stages, when the scope of graphic data is minimum (general arrangement schematic), they apply volume indices that, when logically grouped, indicate:

1) Volumes required for the arrangement of a unit of weaponry, larger facilities, the complement (m^3/1TT, m^3/man, etc.). Volume indices of the first group are applicable to those compartments that do not depend on the submarine dimensions because they are dictated by design specifications. Their volumes, like in the load breakdown of a submarine at early design stages, can be regarded as specified or independent volumes.
2) Volumes required for the arrangement of a unit of weight of actual equipment (m^3/t). Such indices are as a rule used when dealing with equipment of one and the same type, e.g., storage batteries.
3) Volumes per unit of power of steam or refrigerating capacity (m^3/kW, m^3/kg of steam) for power plants.

The above-listed groups of indices are used to find volumes of rooms and compartments with identifiable major pieces of equipment. It should be emphasised that in all cases we understand «volume» not just as overall dimensions of major equipment items but as including access volumes required for their maintenance and repair and associated unused volumes. The latter include, e.g., portions of spaces between internal frames because it is not always possible to fully utilise them.

Let us consider two key values required for V_{PH} estimations.

The space fill factor is the ratio of the equipment volume itself to the volume of the room where it is installed:

$$\chi_{DENS} = \frac{V_{EQP}}{V_{ROOM}} < 1.0 \qquad (5.7)$$

This factor takes into account the additional access volume required for maintenance and repair of the equipment. It should be noted that value χ_{DENS} is to a large degree determined by the type of equipment and by the chosen installation procedure.

The second crucial value used for V_{PH} estimations is the unit weight υ_{EQP} either calculated for the actual equipment or taken as a mean value for a group of equipment items of one designation:

$$\upsilon_{EQP} = \frac{P_i}{V_i} \qquad (5.8)$$

With the help of these values it is possible to approximately estimate the volume of a compartment, from the specified weight of equipment to be placed in this compartment taking into account access volumes for maintenance and repair:

$$V_{ROOM} = \frac{P_i}{\chi_i \upsilon_i} \qquad (5.9)$$

These χ_i and υ_i values are especially helpful when estimating room volumes required for smaller equipment items that are so numerous in any submarine.

As follows from the above-said, the analytical method for required PH volume estimations is based on extensive use of statistics from earlier constructed ships.

The application of indices and statistic data has already been discussed in Chapter 2. Nevertheless, it may be appropriate to reiterate here that it is in PH volume estimations that inapt use of statistic data leads to the greatest errors.

Let us for the sake of an example consider the arrangement of torpedo tubes.

Fig.5.2 shows torpedo tube arrangements on different submarines [16], [65].

Obviously, all these TT arrangements and their reloads have different characteristic volumes. It is evident that the designer who has prototypes with only one TT arrangement option, but chooses a different configuration, will never get valid prototype-based data about the compartment volume, and therefore will miscalculate the CBV.

Thus, when using numerical volume indices and coefficients available in different publications, one should be fully aware of what sort of equipment arrangement these values have been originally derived from.

Pressure hulls of modern Russian submarines are shaped as cylindrical and conical shells divided by flat transverse bulkheads into a number of compartments with internal rooms and bays. The subdivision into compartments is made based on the minimum required length dictated by the equipment to be installed, by habitability considerations and by the need to satisfy surface floodability standards. With the now available hull diameters, every compartment has one to two decks.

The compartments are numbered from the bow aftwards and named after the major pieces of equipment installed in them (in other words, by their primary designations). Both the number of compartments and their functions may differ among various submarines, as

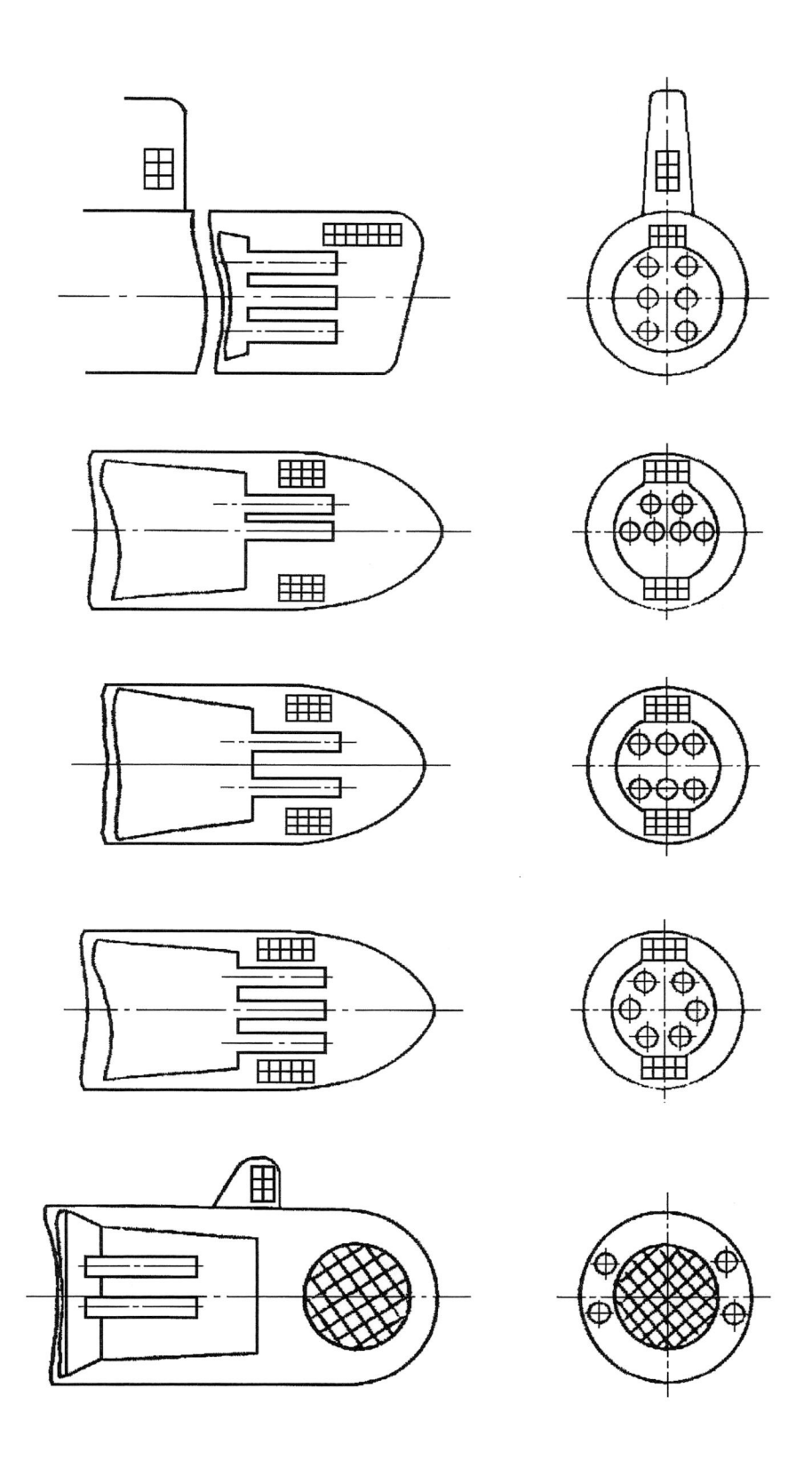

Fig. 5.2. Submarine Torpedo Tube Arrangements

they depend on class and designation of the submarine. Therefore, the designer has to arrange compartments in a new way every time. Still, there are certain general regularities (requirements) that should be followed in compartment-wise equipment arrangement.

So, the submarine layout of combat and general-purpose equipment should ensure:
- uniform distribution of loads along the hull for the purpose of trimming;
- convenient and fast control, monitoring and onboard maintenance;
- easy access to all equipment for inspection, adjustment and troubleshooting.

Presently available hull diameters allow each compartment to be divided into several decks. Heavy equipment should be placed for the sake of submerged stability on lower decks and in the bilge.

Every piece of equipment or a dedicated space (bay) should be, irrespective of its significance, tentatively placed in the most suitable area. When it is impossible to arrange all equipment of a system in a certain compartment or area, the priority should be given to the articles that are more crucial for the overall efficiency of the submarine. For the less critical items the designer should find trade-off solutions and alternative places.

Although each compartment has a certain function (reflected in its name), that does not mean that it accommodates only equipment associated with this primary function. E.g., the bow compartment is the torpedo compartment but it also contains other equipment that has no relation to the torpedo package.

Each submarine compartment is self-contained. The crew should be able to perform their functions of submarine and system control in their fully isolated compartment. This means that every compartment should be provided with independent life support and damage control facilities.

These comments partially reflect only the most general principles of compartment-wise equipment arrangement. It is only in the process of general arrangement development work on an actual submarine design that all numerous factors affecting the arrangement, including complicated relationships between different kinds of equipment, can be taken fully into account and all arising contradictions can be satisfactorily resolved.

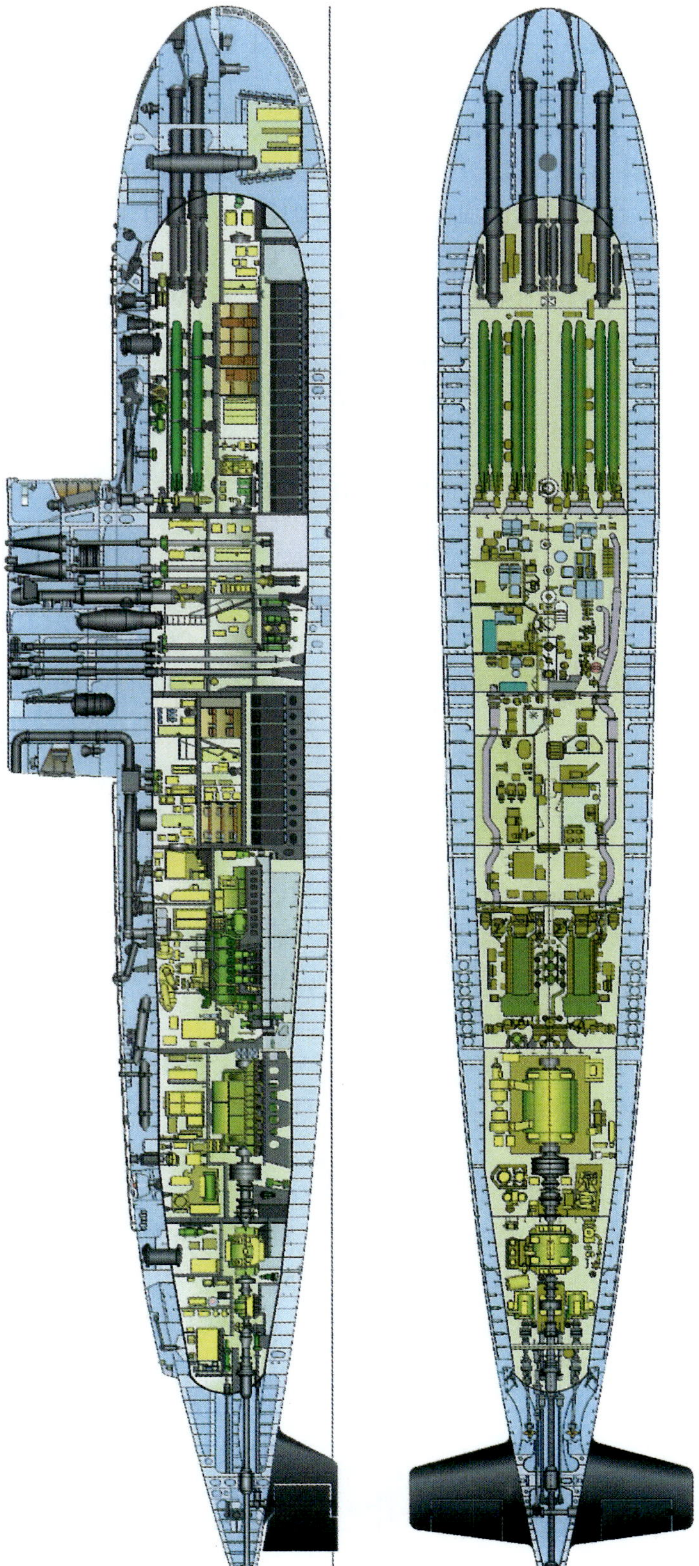

Fig. 5.3. Project 636 "KILO" Diesel-Electric Submarine

Let us take for an example compartment subdivision and major equipment layout of a modern submarine (Fig.5.3).

The first compartment - the torpedo and battery compartment is divided by two decks.

Torpedo tubes (TT) and racks for reload torpedoes are on the upper deck. The local torpedo fire control station and TT automatic control devices sit in the same area, as well as hydraulic actuators of hull gears placed in the fore portion of the conning tower.

The middle deck contains living spaces (crew cabins, wardroom, commanding officer's cabin and doctor's cabin), sanitary conveniences. The backup control station for midship hydroplanes, the storage battery control and monitoring station are located on the middle deck, and the air-foam fire-fighting system station is in its dedicated bay.

In the bilge there are: the storage battery, the fore trim tank, the torpedo tube flooding tank, as well as a dirty water tank.

The second compartment - central (control) compartment.

The first deck of the compartment is occupied by the main control centre (MCC).

Arranged on the second deck are: the sonar room, the gyro room, the air conditioning system equipment and the converter room with converters and distribution switchboards of electric mains.

The bilge accommodates provision stores and various bilge equipment (the main drain pump, hydraulic system pumps, the bilge pump).

The third compartment - the living and battery compartment is divided by two decks; on the upper and the middle decks there are: the galley, sanitary conveniences and crew cabins; the air-conditioning equipment is arranged in enclosed bays; the storage battery is located in the bilge.

The fourth compartment - the diesel-generator compartment. The equipment located in this compartment is: diesel-generators, the electric compressor, the power distribution board, the diesel-generator control station. The bilge accommodates diesel-generator auxiliary equipment, dirty water tanks, fuel tanks and the circulating oil tank.

The fifth compartment - the motor compartment. It accommodates the main propulsion motor, the control station for the power plant electric equipment, the silencing coupling, the plummer block.

The sixth compartment - the aft compartment. The economic speed motor, two reserve motors (starboard and portside), the reserve propulsion complex backup control station, plane/rudder actuators and the backup plane/rudder control station, chemical and air-foam fire-fighting system stations are located in this compartment. The bilge accommodates a hydraulic system pump, a fuel tank and the aft trim tank.

Thus, the volume of the submarine pressure hull, as we can see from the above descriptions, may be presented as a sum of volumes of compartments and spaces required for equipment, technical facilities and the crew:

$$V_{PH} = \sum_{1}^{n} V_i = V_{TS} + V_{MS} + V_{REA} + V_{PP} + V_{SYS} + V_{DEV} +$$
$$+ V_{EEG} + V_{LIV} + V_{MCC} + V_{ACC} + V_{MAST} + V_{ABT} ... + V_{PPT} \quad (5.10)$$

where V_{TS} – volume of the torpedo compartment;
V_{MS} – volume of the missile compartment;
V_{REA} – volume of spaces for radioelectronic aids;
V_{PP} – volume of the power plant compartment;
V_{SYS} – volumes of spaces for hull systems;
V_{DEV} – volume of spaces for hull gears;
V_{EEG} – volume of electric equipment spaces;
V_{LIV} – volume of living spaces occupied by the crew and associated stocks;
V_{MCC} – volume of the main control centre;
V_{ACC} – volume of the acoustic coatings inside the PH;
V_{MAST} – volume of spaces for masts;
V_{ABT} – volume of internal variable ballast tanks;
V_{PPT} – volume of tanks for the submarine power plant.

Let us consider how to estimate these components of the submarine volume.

The Torpedo Compartment Volume V_{TS}

Overall dimensions of the torpedo compartment and, hence, its volume are determined by the total length of the breech end of the torpedo tube, reload torpedoes and clear spaces for opening the TT breech door and placing the quick loading gear (QLG).

The compartment diameter can be defined by the number of reload torpedoes as well as by the arrangement, type, calibre and number of TTs. It is also necessary to account for volumes required for PH tanks of the torpedo package support systems.

Thus, the volume required for the arrangement of the torpedo package consists of the following components:

$$V_{TS} = V_{TTB} + V_{RELT} + V_{TSS} \qquad (5.11)$$

where V_{TTB} – volume of space occupied by breech ends of torpedo tubes;
V_{RELT} – volume of space occupied by reload torpedoes;
V_{TSS} – volume of tanks supporting the torpedo package.

The PH volume required for the arrangement of TT breech ends can be estimated as:

$$V_{TTB} = v_1 n_{TT} \qquad (5.12)$$

where v_1 – volume occupied by one TT;
n_{TT} – number of TTs.

On most submarines torpedo weapons are placed only in the bow following the traditional arrangement with TTs set parallel to the centre line of the submarine (Fig.5.4). In this case for the most common calibre (533 mm) we need $v_1 \approx 10.0$ m³/TT. [32], [54], [68].

The increase in overall dimensions of sonars and increasingly stringent requirements to platform noise levels in way of their stations, in some cases forces torpedo tubes to be moved from the first compartment to the aft and to set them at 10 to 14° to the CL. With this arrangement v_1 for the same calibre reaches $v_1 = 12.5$ to 14.5 m³/TT.

Submarines may have external torpedo tubes. This allows the torpedo salvo power to be increased without increasing the PH volume. When TTs are arranged in the between-hull space, their contribution to the CBV can be estimated with (5.12) but numeric values of the v_1 index should correspond to the entire torpedo tube volume.

The volume required for reload torpedoes and servicing access depends on the chosen type of storage racks. In case the reloads are placed in shock-mounted racks with the QLG set parallel to the CL, the volume of space per one reload is $v_2 = 12.5$ to 25 m³, and for torpedoes arranged at an angle to the CL $v_2 = 15$ to 30 m³. Then, the volume occupied by reload torpedoes is:

$$V_{RELT} = v_2 n_{RELT} \qquad (5.13)$$

a)

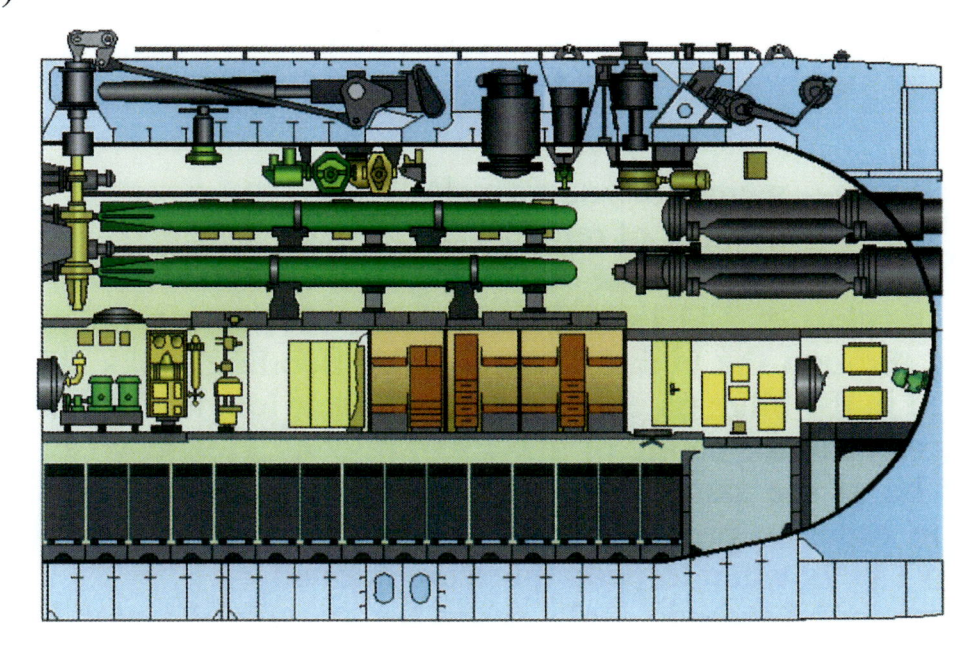

b)

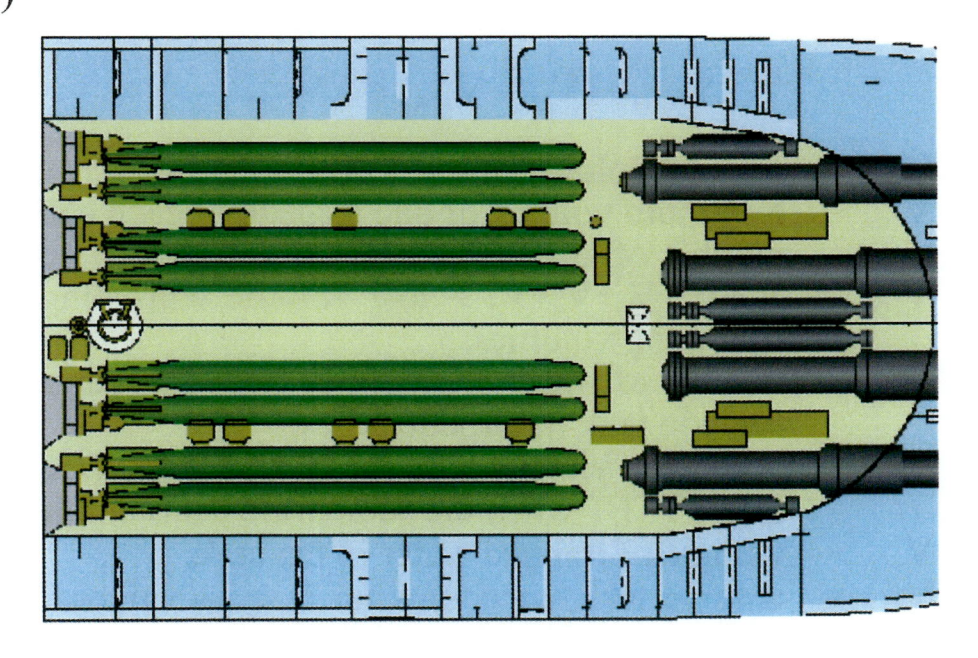

Fig.5.4. Torpedo Compartment of a Submarine

a) elevation b) deck plan

Supporting tanks for torpedo tubes include tube flooding tanks (TTFT), water-round-torpedo tanks (WRTT tanks) and poppet valve drain tanks (PVDT tanks). These tanks may be placed either inside the PH or outside it. Torpedo tube flooding tanks are designated to compensate for torpedoes that were not taken onboard or already used. The volume of torpedo tube flooding tanks is found as:

$$V_{TTFT} = \frac{1}{\rho} n_{RELT} P_{TOR} + V_{LOST} \qquad (5.14)$$

where P_{TOR} – weight of one torpedo;
n – number of reloads;
V_{LOST} – lost volume consisting of the trapped water volume («dead stock») in TTFTs and the volume accounting for pipelines passing through the tanks. Torpedo tube flooding tanks should be arranged in the bilge on either side, with their longitudinal centre of gravity close to the centre of gravity of reloads. Height-wise, TTFTs should be at the same level as the water-round-torpedo tanks or lower in order to provide natural overflow of water from WRTT tanks to TTFTs when torpedo tubes are drained.

Water-round-torpedo tanks serve for storage of water intended for flooding annular spaces around torpedoes in torpedo tubes. The volume of a water-round-torpedo tank is estimated as:

$$V_{WRTT} = (V_{TT} - V_{TOR}) n + V_{LOST} \qquad (5.15)$$

where V_{TT} – internal volume of one torpedo tube;
V_{TOR} – minimum volume of one torpedo of each calibre;
n – number of torpedo tubes;
$V_{LOST} = V_1 + V_2 + V_3 + V_4 + V_5$ – lost volume consisting of the following volumes:
V_1 – volume of water in pipelines from the water-round-torpedo tank to torpedo tubes;
V_2 – volume of water which escapes from the WRTT tank to the TTFT when the submarine trims and heels;
V_3 – volume of trapped water in the tank;
V_4 – volume which accounts for by-pass valves from the WRTT tank to the TTFT;
V_5 – volume of water overflowing from the WRTT tank to the TTFT.

At early design stages lost volumes may be ignored. Water-round-torpedo tanks should be located below torpedo tubes so that, if necessary, the tubes could be drained by the force of gravity.

The poppet valve drain tank is intended to release the air bubble from the tubes and to take water partially balancing the negative buoyancy of torpedoes. For small firing depths the PVDT tank is made light and placed inside the PH. In this case its volume is:

$$V_{PVDT} = (3.0 \div 4.0)\left[\frac{V_{TOR}}{\rho} - P_{TOR}\right]n_{TT} \qquad (5.16)$$

where $\left[\frac{V_{TOR}}{\rho} - P_{TOR}\right]$ – negative buoyancy of one torpedo.

When it is intended to fire from deeper waters, the PVDT tank is made pressure-proof and may be located either inside the pressure hull or in the between-hull space. In this case the tank volume is:

$$V_{PVDT} = (3.0 \text{ to } 4.0)n_{TT} \qquad (5.17)$$

If all torpedo package tanks are located inside the PH, the volume $V_{TST} = V_{TTFT} + V_{WRTT} + V_{PVDT}$ should be taken into account in the hull design.

The Missile Compartment Volume V_{MS} can be determined with formulae similar to (5.12) through (5.17), or from

$$V_{MS} = 0.785 d_{PH}^2 d_{SILO}(kn_{SILO} + b) \qquad (5.18)$$

where n_{SILO} – number of silos in one longitudinal row;

k and b – numerical coefficients taken from the prototype.

Statistic analysis data about foreign nuclear submarines with ballistic missiles allows one to conclude that for «Poseidon»-class missiles one silo together with the space for firing control equipment occupies a volume of 120 to 150 m³ [16].

At early design stages volumes V_{TS} and V_{MS} may be, like in the load balance, considered as specified because the weapon package is already known.

The Diesel (Diesel-Generator) Plant Compartment Volume V_{PP}.

Depending on the information available, the volume of the diesel compartment can be found in different ways.

The first way is to establish the minimum acceptable dimensions of the compartment by the least clearance method. In this case it is considered that the minimum acceptable diameter of the PH in way of diesel engines is determined by the following conditions:
- the compartment height should allow removal of covers, as well as of pistons and cylinder liners, of the engine taking into account the spaces for lifting gears;

- there should be a normal walkway provided between engines and a man-size crawl-way between the wall and the engine;
- it is necessary to allow for structurally acceptable dimensions of foundations and for the gravity drain of the oil from the crankcase to the tank.

The length of the diesel compartment is determined by the length of the chosen engine and by minimum permissible clearances between it and the compartment bulkheads (Fig.5.5).

The clearance a between the aft bulkhead and the flange of the engine crankshaft, which is required for mounting the isolating–and–silencing coupling and the bulkhead gland, is determined proportionally to the engine capacity. The length b, i.e. from components mounted on the engine to the fore bulkhead of the compartment, is related to the compartment layout. In case the engine operates directly to the shaft, these clearances between the engine and the compartment bulkheads are within: $a = 1.2$ to 1.5 m; $b = 1.5$ to 2 m. In electric propulsion designs these values are less: $a = b = 1.0$ to 1.2 m.

It should be kept in mind that above-indicated values a and b are minimum requirements. When the overall dimensions of the compartment are known, it is possible to find its volume. The ratio of the compartment length to the diesel engine length is usually $L_{DC}/L_{DE}=1.8$ to 2.0 [36].

The second method is applied when the engines to be installed are not yet known. Then, using volume indices, it is possible to estimate the volume of the compartment in the first approximation:

$$V_{DC} = v_{DC} \sum_{1}^{n} N_{DE} = v_{DC} \frac{\vartheta^3_{SNORT} (k_{FS} V_0)^{2/3}}{C_{\downarrow max}} \qquad (5.19)$$

where v_{DC} = 50 to 55 l/kW.

k_{FS} — conversion factor for going from the normal displacement to the full submerged displacement.

For electric propulsion submarines (Fig.5.6) the diesel-generator compartment length (two-side arrangement) can be estimated as:

$$L_{DGC} = 0.10 \, N_{DG}^{1/3} + b \qquad (5.20)$$

where b — numerical coefficient taken from the prototype.

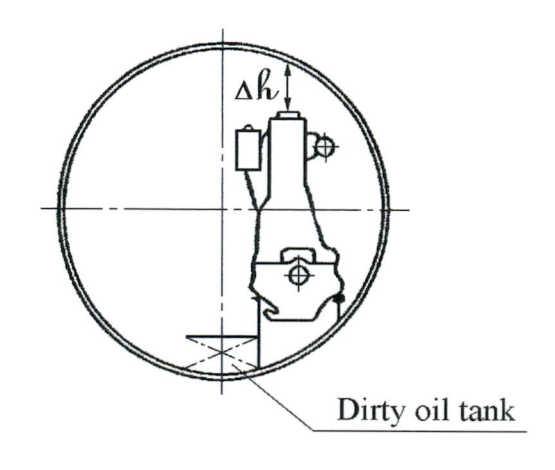

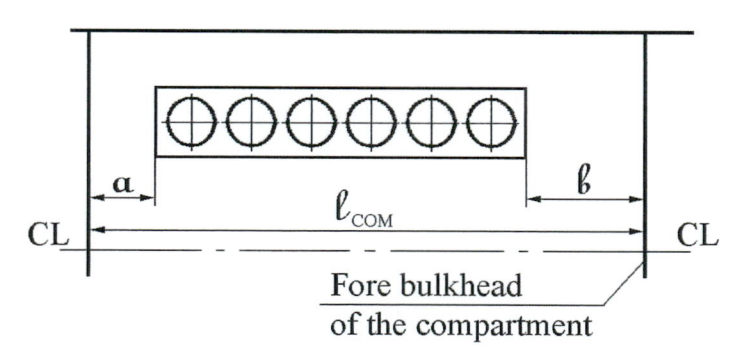

Fig. 5.5. Diesel Engine Layout in the Diesel Compartment

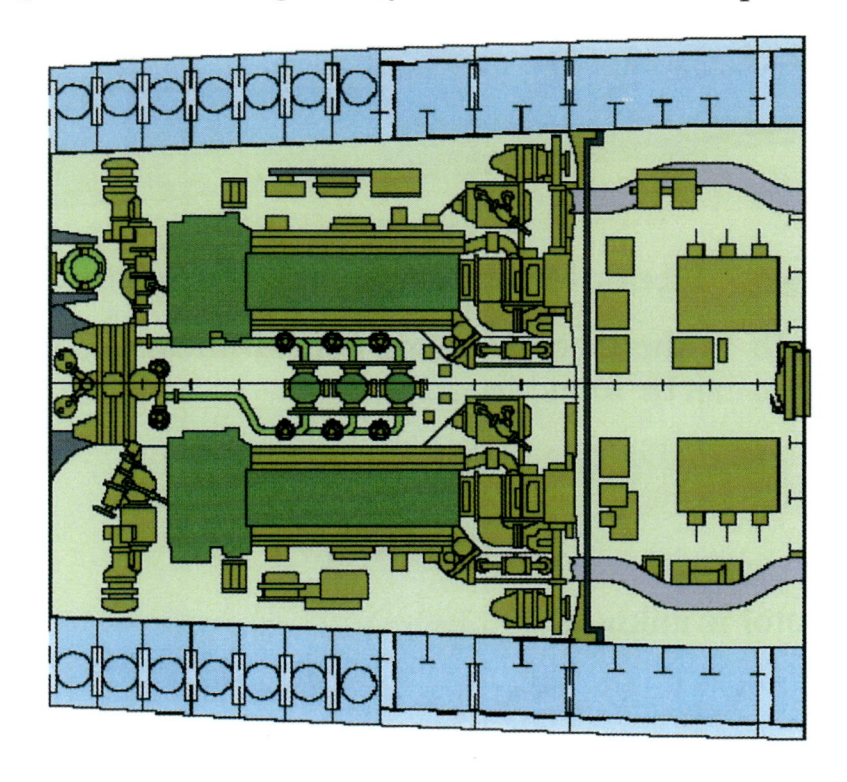

Fig.5.6. Diesel-Generator Compartment of a Submarine

The Electric Motor Compartment

The diameter of the electric motor compartment is determined by the diameter of the chosen electric motor and by the arrangement of control panels (Fig.5.7).

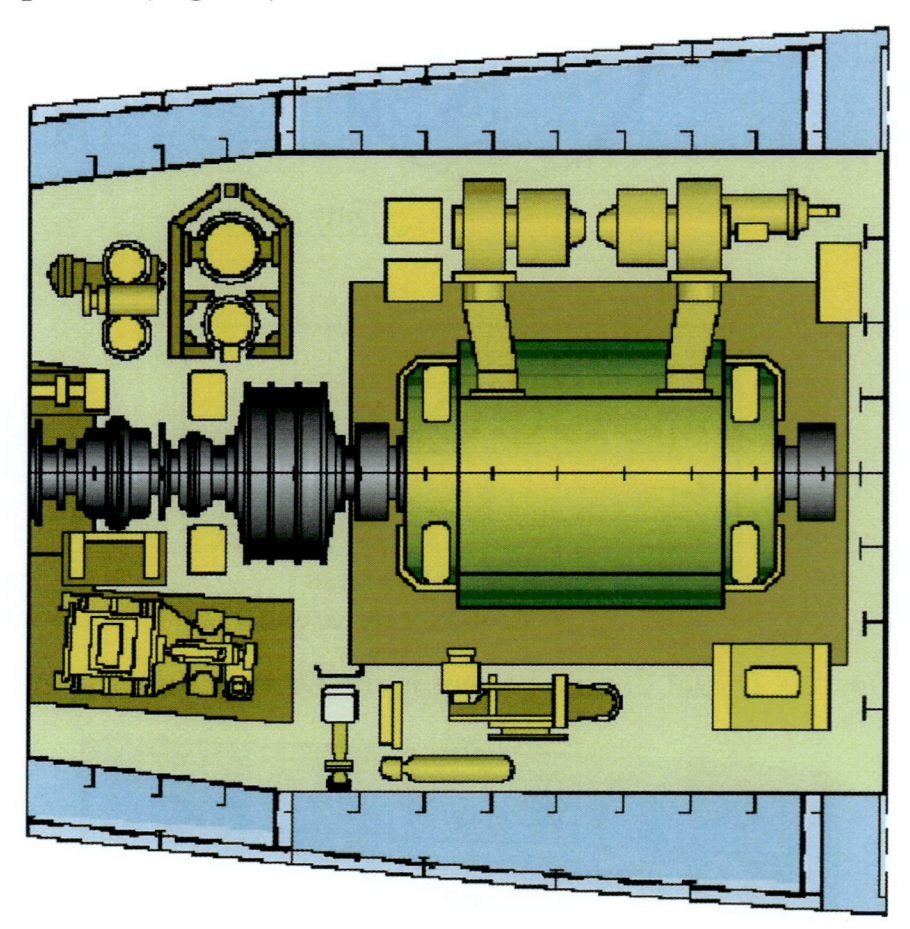

Fig.5.7. Electric Motor Compartment of a Submarine

The length of the electric motor compartment L_{EMC} in the first approximation can be found from:

$$L_{EMC} / L_{EPM} = 1.7\sim1.9 \qquad (5.21)$$

$$L_{EMC} = 0.35 N_{EPM}^{1/3} + b \qquad (5.21a)$$

If the motor is unknown:

$$V_{EMC} = v_{EMC} \sum_1^n N_{EMP} = v_{EMC} \frac{\vartheta^3_{\downarrow max}(k_{FS}V_0)^{2/3}}{C_{\downarrow max}} \qquad (5.22)$$

where $v_{EMC} = 40.5$ to 44.5 l/kW

The Battery Wells (BW) Volume should be selected based on the possibility of accommodating the chosen storage battery. Additional volumes should be provided for servicing the SB cells. Thus, overall dimensions of the battery well, in the first place, depend on the type of cells and their dimensions, as well as on servicing procedures for SB cells. Servicing can be performed either from walkways between the cells or from trolleys moving above them.

There are two methods to determine the required BW volume. The first one is applicable when the SB type and configuration are known. In this case we first establish the overall dimensions and then estimate the BW volume (Fig.5.8).

The width is determined by the number of cells n in a row and by clearances between them, as well as by the number and the width of servicing walkways. As a rule, they arrange one or two walkways. The walkway width is taken as the least dimension of the cell plus a clearance required for easy transportation of cells along the walkway when loading or unloading them.

One should also account for clearances between battery well walls and end cells.

The battery well width is found as:

$$b_{BW} = mb_w + n\beta_{CELL} + \Delta b_i \cdot (n-1) + 2\Delta b_2 \qquad (5.23)$$

where m   — number of walkways;
b_w — width of one walkway;
n   — number of cells in one row;
β_{CELL} — breadth of one cell;
Δb_i — clearance between cells across the BW;
Δb_2 — clearance between BW walls and end cells.

The battery well length is also determined by the number of cells, by clearances between cells, as well as by clearances between end cells and transverse bulkheads. The size of these clearances is taken the same as for BW width estimation. To prevent battery cells from being displaced, they are wedged after every 2 or 3 transverse rows. This increases between-cell clearances in those places.

The BW length formula is accordingly written as:

$$\ell_{BW} = n\lambda_{CELL} + mb_w + \Delta\ell_i + \frac{n}{3}\Delta\ell_{CL} \qquad (5.24)$$

where $\Delta\ell_i$ — clearance between cells along the BW;
$\Delta\ell_{CL}$ — clearance for wedging.

131

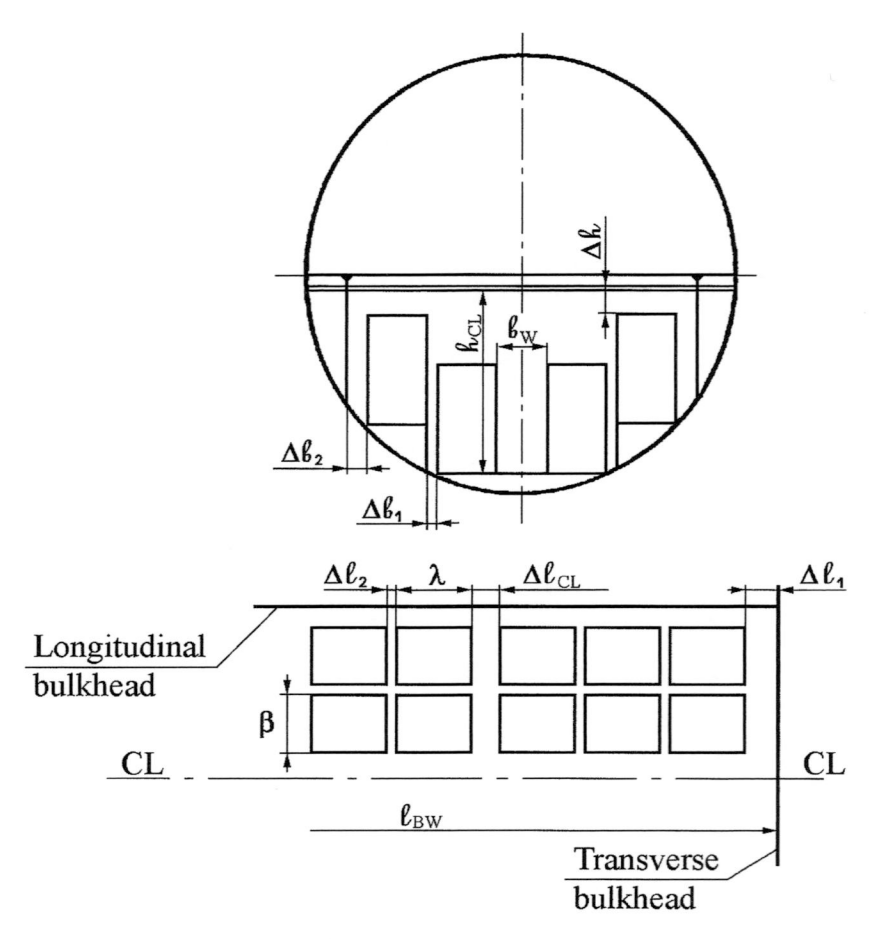

Fig. 5.8. A Schematic Layout of SB Cells in a Battery Well for Servicing from Walkways

The battery well height depends on the battery type and its servicing method (Fig.5.9).

If the cells are serviced from walkways, the distance from the upper plane of clamp bolts to deck beams should allow for cell servicing. The walkway height should allow the personnel to pass there. When cell servicing is performed from trolleys, the well height is as a rule a little larger, but the total BW volume is less thanks to the fact that there are no walkways.

Taking into account the height of wooden furrings or rubber shock mounts under each cell, the height of each tier of the battery well is found as:

$$h_{BW} = h_{CELL} + h_{SM} + \Delta h \quad (5.25)$$

where h_{CELL} — height of the cell;
h_{SM} — height of the shock mount;
Δh — height over SB cells for servicing access.

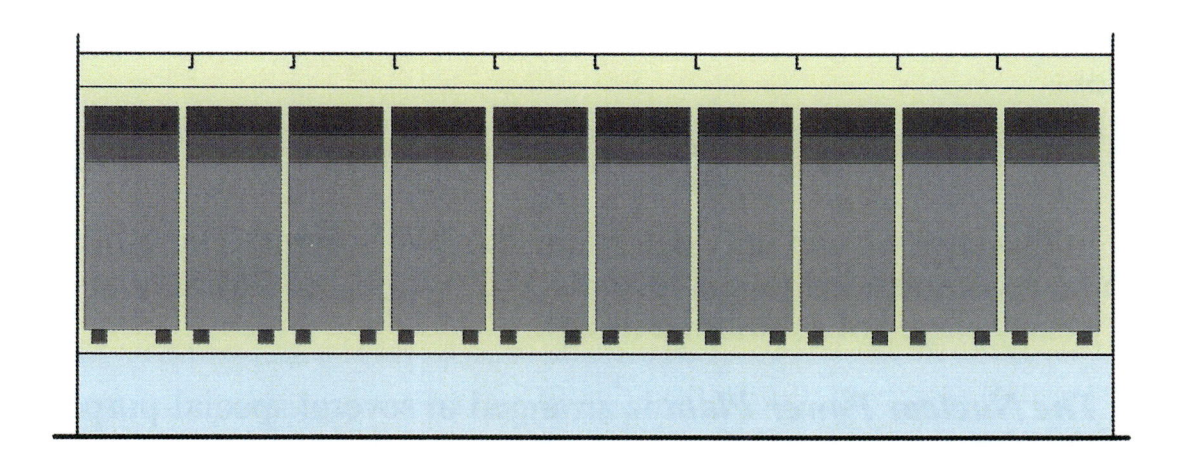

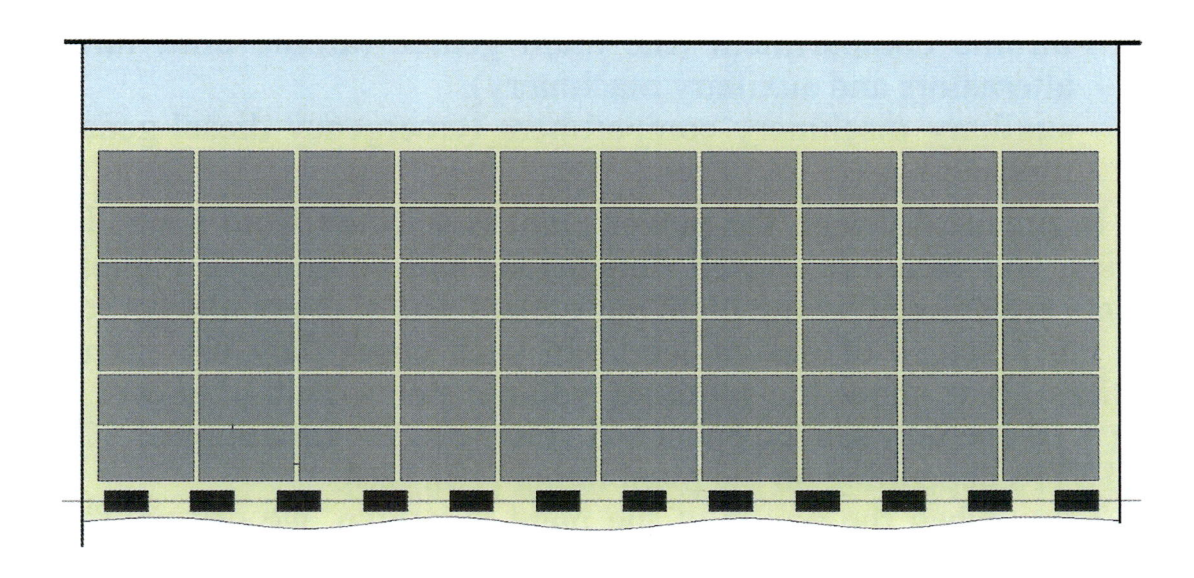

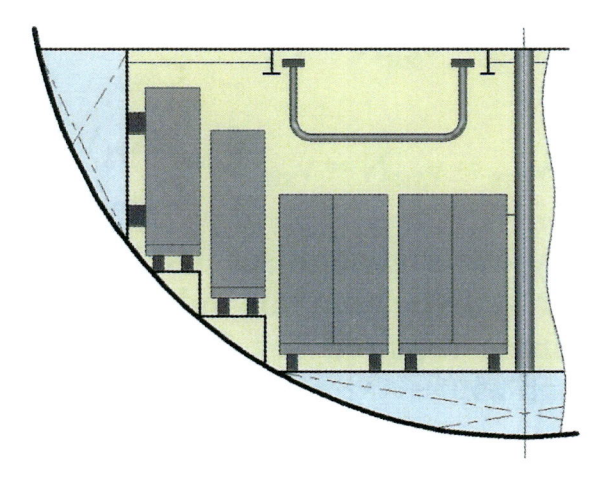

Fig.5.9. SB Cell Layout in the Battery Well
for Servicing from Trolleys

133

The second way of BW volume calculations involves a volume index:

$$v_{BW} = \frac{V_{BW}}{\rho_{SB}} \approx 0.95 - 1.20 \text{ m}^3/\text{t} \qquad (5.26)$$

In this case we can only determine the BW volume. Therefore, it can be recommended to use formula $V_{BW} = v_{BW}P_{SB}$ (5.26) only at the very early stages of submarine design.

The Nuclear Power Plant is arranged in several special-purpose compartments of the pressure hull:
- reactor compartment (including equipment of the steam generating plant),
- turbine compartment (the main geared-turbine unit, turbo-alternators and auxiliary machinery),
- auxiliary machinery compartment (emergency diesel-generators, etc.).

In practical design the power plant is selected from a standard series, and therefore overall dimensions and volumes of compartments are defined by the plant particulars. But it is not always so at the early stages of design work. In such cases they use volume indices that show the required volume per a unit of power. In Fig.5.10 one may see the pattern of volume index variations of SGPs and STPs depending on the plant rating [10], [26].

Thus, when the power of the plant is known, we can approximately estimate volumes of SGP and STP compartments.

$$V_{SGP} = v_{SGP}N = v_{GP} \frac{\vartheta^3_{\downarrow max}(k_{FS}V_0)^{2/3}}{C_{\downarrow max}} \qquad (5.27)$$

$$V_{STP} = v_{STP}N = v_{STP} \frac{\vartheta^3_{\downarrow max}(k_{FS}V_0)^{2/3}}{C_{\downarrow max}} \qquad (5.27a)$$

Sometimes they use a generalised nuclear power plant volume index. According to statistic data this index is $V_{NPP} = (43 \sim 48)$ l/kW. Then, the PH volume required for the arrangement of the entire power plant is found as:

$$V_{NPP} = v_{NPP}N = v_{NPP} \frac{\vartheta^3_{\downarrow max}(k_{FS}V_0)^{2/3}}{C_{\downarrow max}} \qquad (5.28)$$

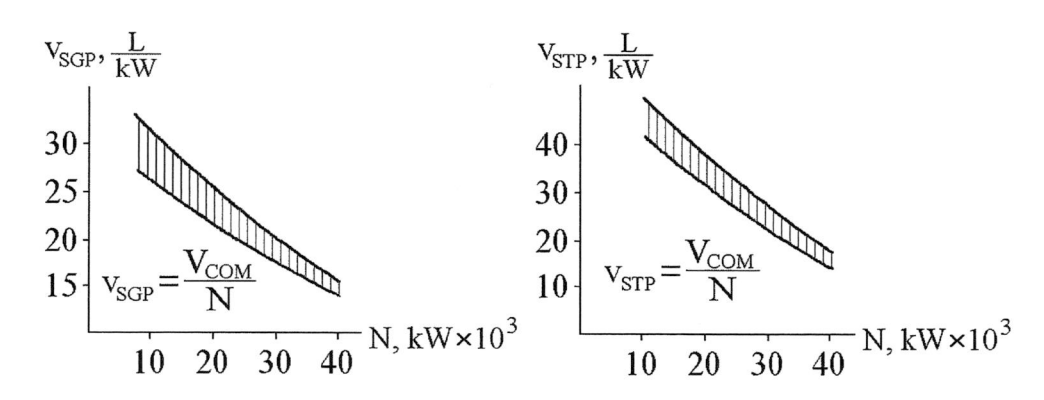

Fig.5.10. Indices v_{SGP} and v_{STP} Versus the Nuclear Plant Output

Volume of Spaces Occupied by the Complement and Stocks

On modern submarines the volume of spaces per one crew member is dictated by sanitary norms. When the PH diameter grows, the number of decks is increased and PH volumes can be used in a more rational way. Free volume standards per one crew member, depending on the number of decks, are given in Table 5.3.

Table 5.3

Number of decks	Volume per 1 crew member (m³)
2	6~7
3	10~15

Figs.5.11 through 5.15 provide examples of possible layouts of living spaces on submarines [40], [41].

Volumes of provision storerooms, as well as volumes of fresh (drinking and washing) and dirty water tanks, are directly related to the complement strength and to the design endurance of the submarine.

Provision storerooms on submarines are subdivided into low temperature (t = starting with −12°), cooled with t = −3° and non-cooled with t = t in the compartment. The volume (gross) of storerooms should be calculated for each room type as:

$$V_{PST}^{GR} = k_{PST} \cdot k_{SUR} \cdot V_i nA \qquad (5.29)$$

where k_{PST} − 1.8 to 6.5 − coefficient accounting for equipment and insulation of storerooms;

k_{SUR} = 1.0 to 1.2 − coefficient accounting for emergency (survival) stock of provisions;

V_i – numeric values of provision volumes per man per day (including packing) for each type of storerooms;

n – complement;

A – endurance.

The net volume of fresh water tanks is calculated as:

$$V_{FWT}^{NET} = V_{DWT}^{NET} + V_{WWT}^{NET} = (v_{DW} + v_{WW})nA = \sum_{1}^{n} v_i \cdot nA \qquad (5.30)$$

where V_{DWT}^{NET} – volume of drinking water tanks;

V_{WWT}^{NET} – volume of washing water tanks;

v_{DW} and v_{WW} – daily consumption of drinking and washing water per man.

On a diesel submarine with a distilling plant the total consumption allowance for fresh and distilled water should be at least 15 litres per man per day, including at least 6 litres for drinking purposes.

The volume of dirty water tanks is found as:

$$V_{DIRT} = v_{DIRW} \cdot n \cdot t \qquad (5.31)$$

where v_{DIRW} = 60 to 65 *l/man* per day – design standard of waste water;

t – possible duration of stay in an area where waste water overboard discharge is prohibited.

Thus the volume of pressure hull compartments occupied by the complement and stocks will be:

$$V_{SCR} = V_{PER} + V_{PST} + V_{FWT} + V_{DIRT} \qquad (5.32)$$

One of the present tasks of submarine design is to find radical solutions for environment protection issues associated with submarine operation. To prevent environmental pollution, modern Russian diesel-electric submarines are equipped with a complex array of systems for treatment and purification of human waste (Fig.5.16) [107]. In the first approximation, the required volume for the arrangement of this package can be assumed as an absolute value $V_{ENV} \cong 3.0$ to 5.0 m^3.

Fig.5.11. Commanding Officer's Cabin

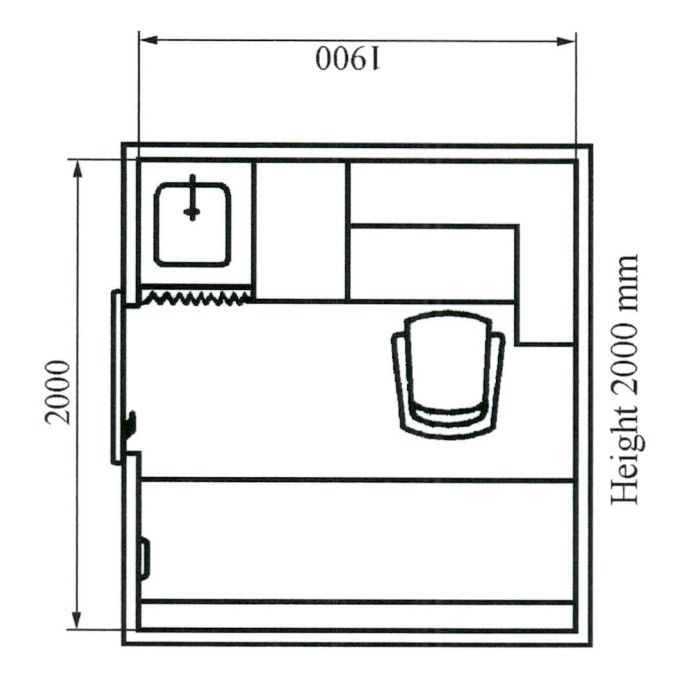

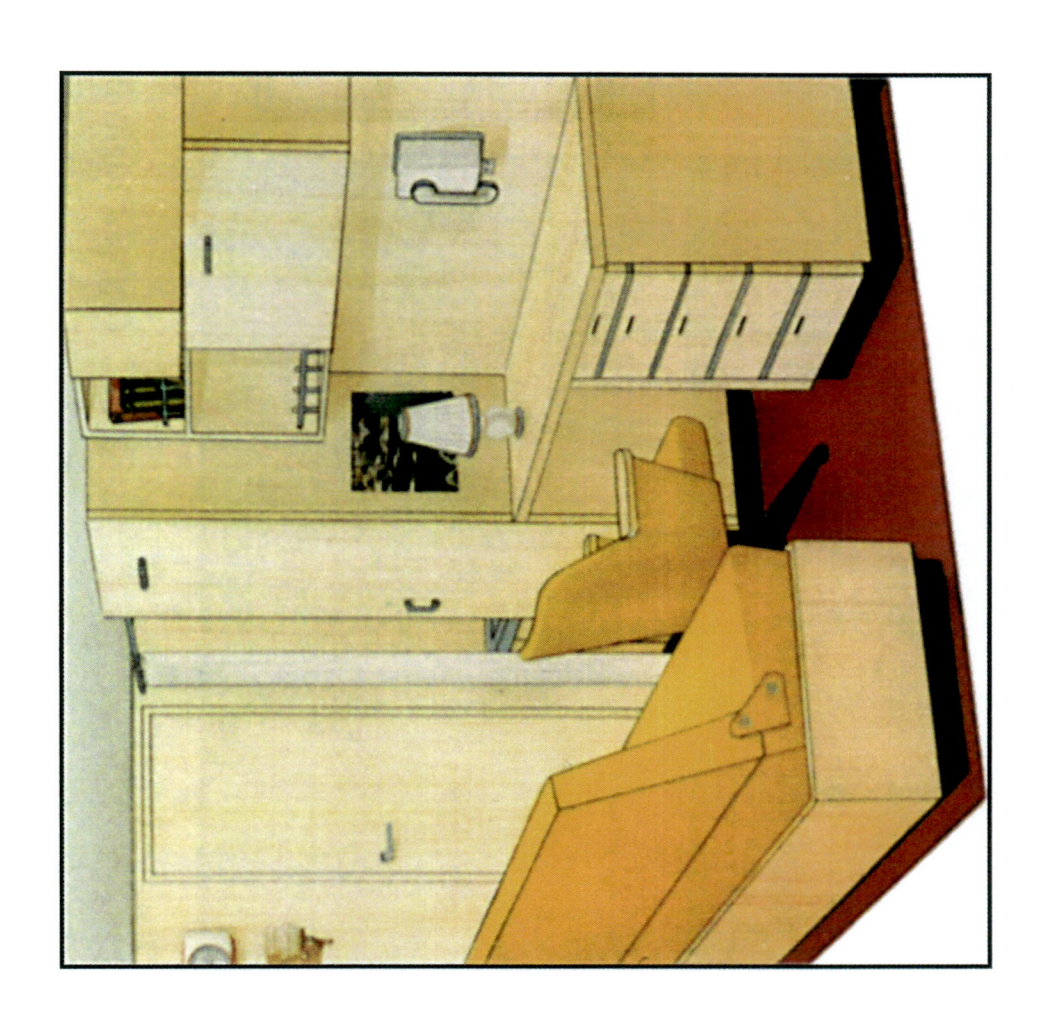

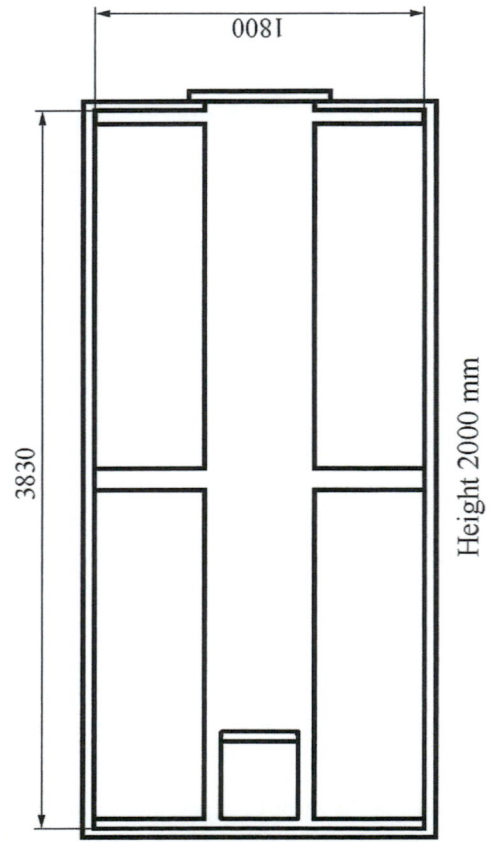

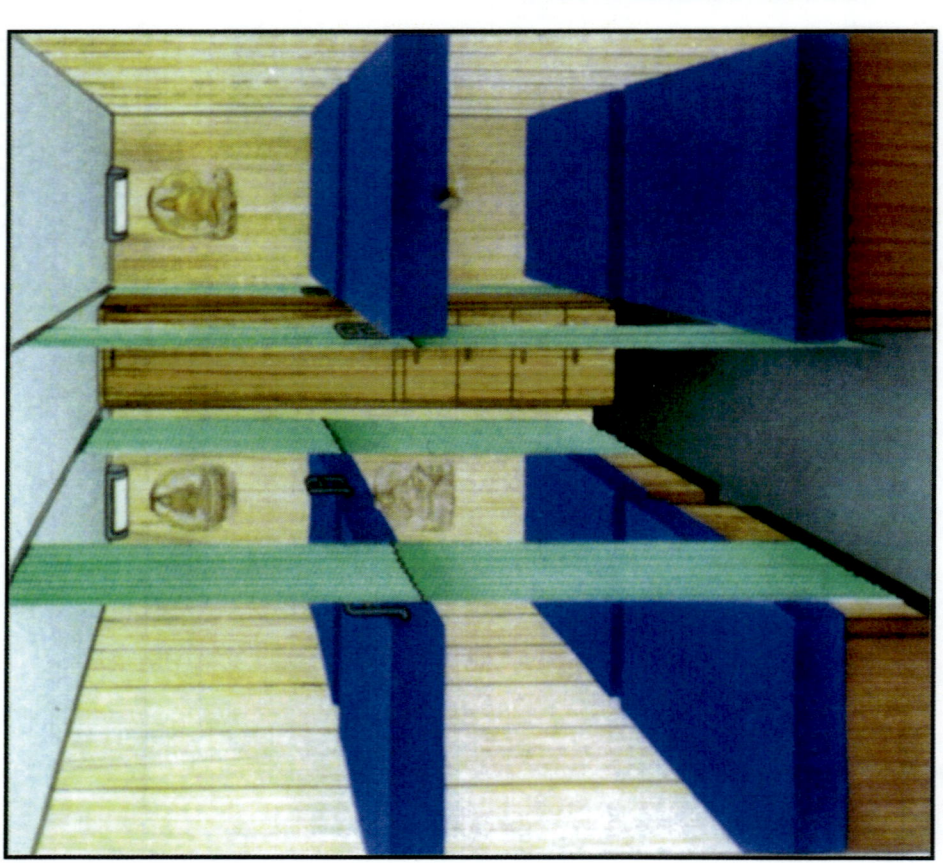

Fig.5.12. Eight-Berth Cabin

Fig.5.13. Wardroom

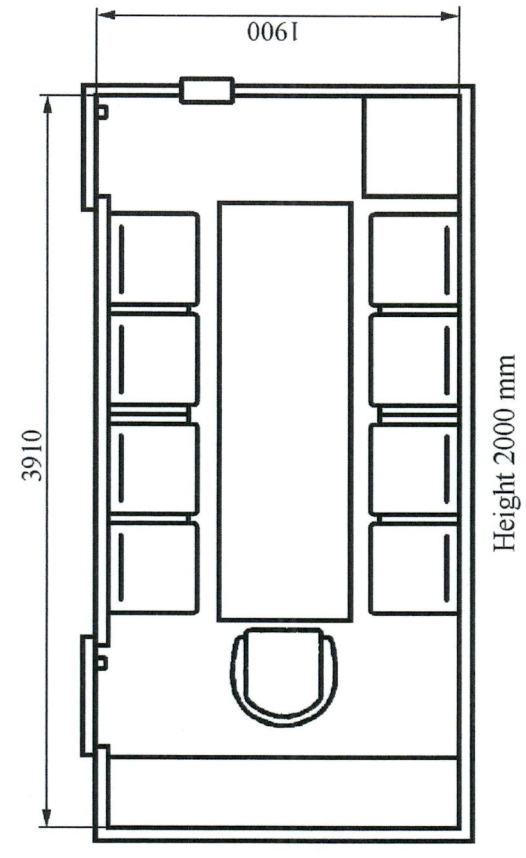

139

Fig.5.14. Living Quarters of «Amur 1650» Submarine. Deck 1

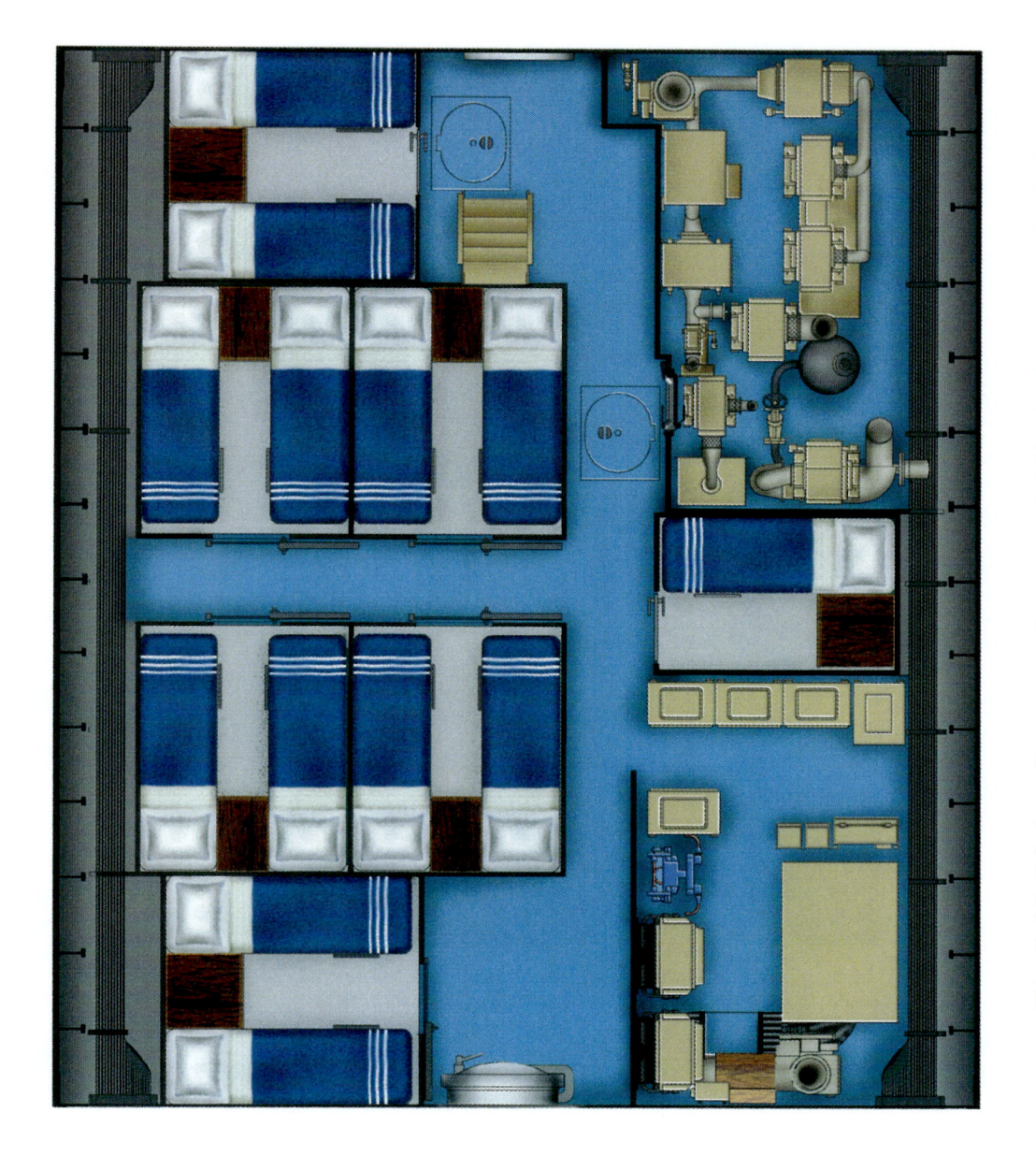

Fig.5.15. Living Quarters of «Amur 1650» Submarine. Deck 2

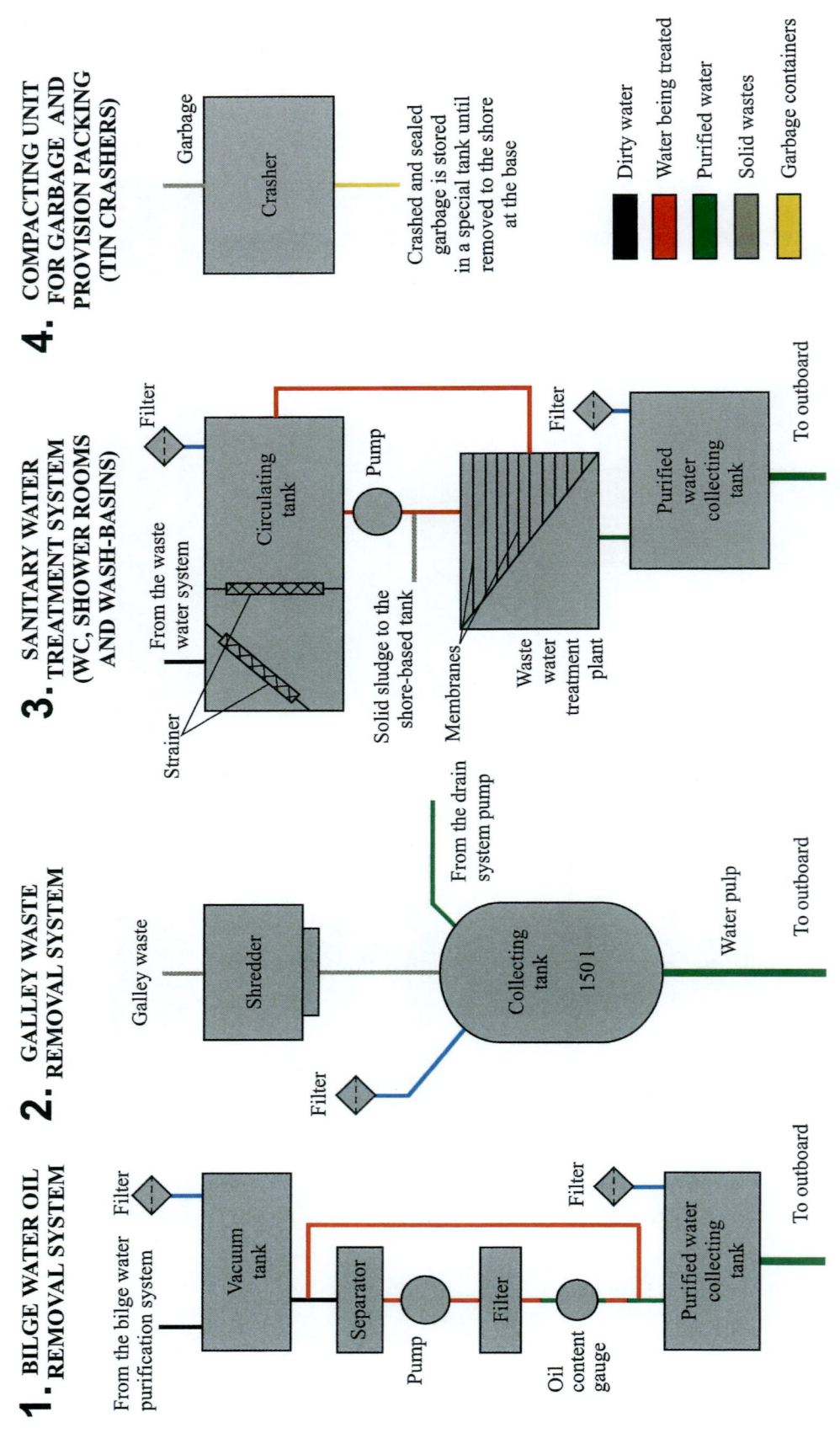

Fig.5.16. Systems for environment-friendly submarine operation

Other Volumes of the Pressure Hull

Volumes of many spaces of the pressure hull occupied by various equipment can be calculated with simplistic formulae relating volumes of these spaces either to the pressure hull volume or to the constant buoyant volume. Such spaces include the main control centre – V_{MCC} (Fig. 5.17–5.20), spaces for hull systems – V_{SYS}, spaces for hull gears – V_{DEV}, electric equipment spaces – V_{EEG}, spaces for sensors – V_{REA}. At initial design stages it is usually difficult to identify what particular equipment will dictate arrangements in these spaces. As a rule, such volumes are either determined by a graphic study or taken from the prototype. Statistic analysis has made it possible to obtain approximate ratios for calculating volumes for these spaces. They are given in Table 5.4.

Table 5.4

Other submarine pressure hull volumes

No.	Description	Design formula	Location
1	Volume of MCC space	$V_{MCC} = 0.03V_0 - 17.6$	CIC
2	Volume required for masts	$V_{MAST} = 0.008V_0 + 6.0$	CIC
3	Volume required for hull gears	$V_{DEV} = 0.007V_0 + 14$	All compartments
4	Volume required for hull systems	$V_{SYS} = 0.1 - 0.15 V_{PH}$	All compartments
5	Volume required for electric equipment	$V_{EEG} = 0.08 - 0.1 V_{PH}$	All compartments
6	Volume of acoustic coatings inside PH	$V_{ACC} = 0.0015 - 0.0025 V_{PH}$	All compartments
7	Volume of spaces of REA	$V_{REA} = 0.03V_0 + 1.08$	Compt.1 and 2, CIC

At present, masts penetrating the pressure hull are replaced by non-penetrating (telescopic) masts (Fig.5.21). On SSs of the IVth generation it has been possible to develop more rational arrangement of control panels in the space of the main control centre, with better respect to ergonomic considerations (see Figs.5.19 and 5.20).

Fig.5.17. Main Control Centre, Project 877EKM "KILO" submarine

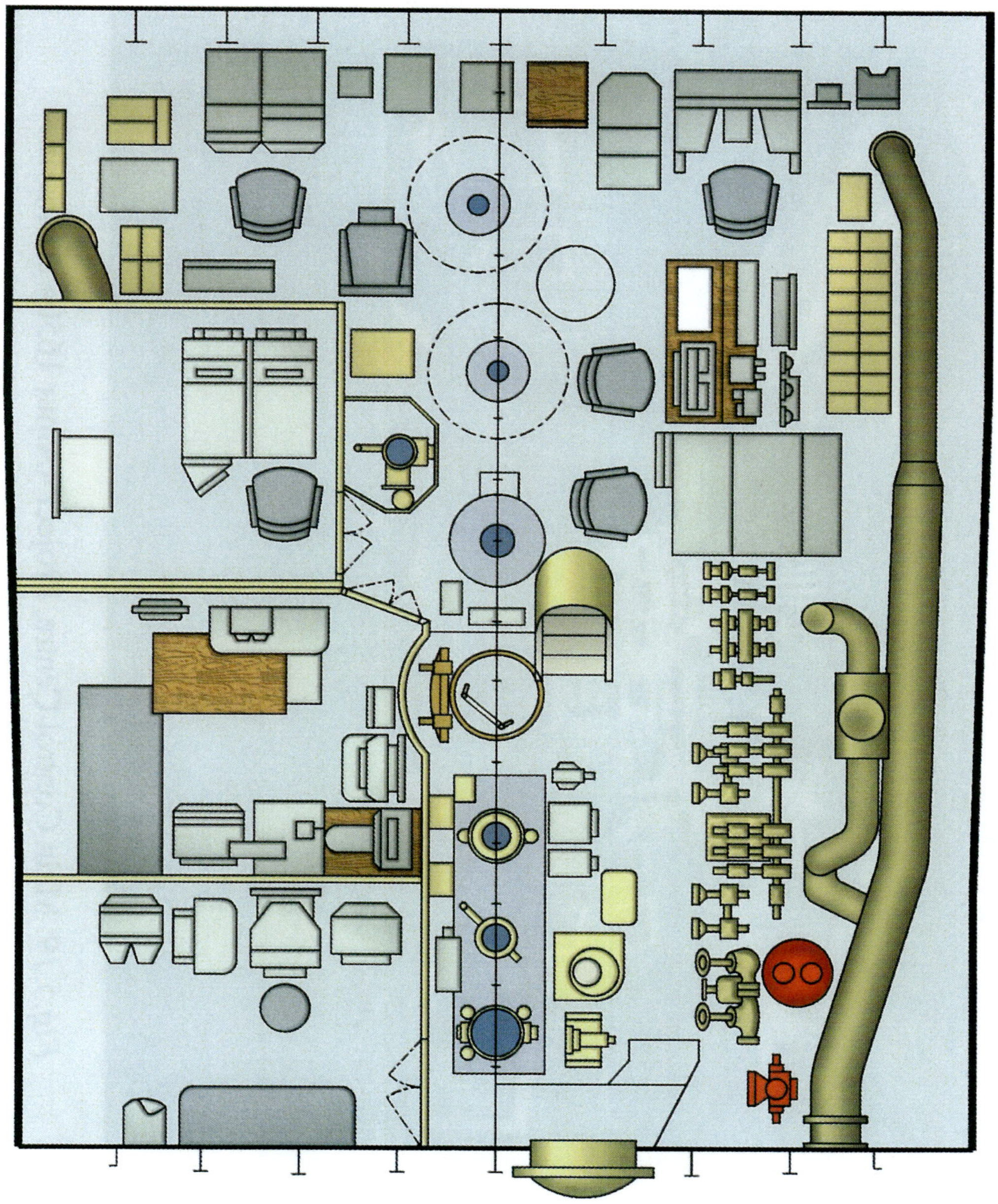

Fig.5.17. Main Control Centre, Project 877EKM "KILO" submarine

Fig.5.19. Main Control Centre, Project «Amur 1650» submarine

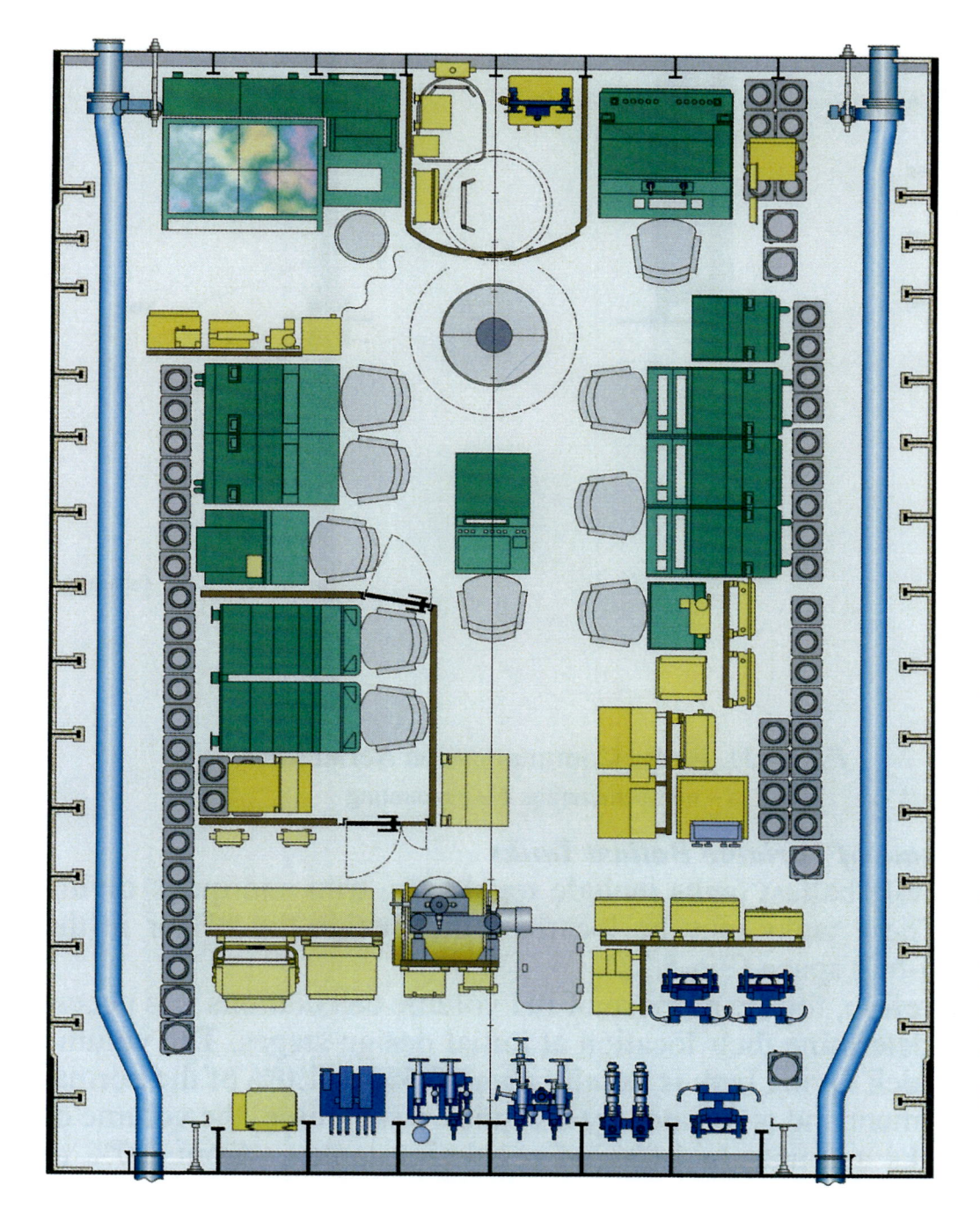

Fig.5.20. Main Control Centre, Project «Amur 1650» submarine. Plan View

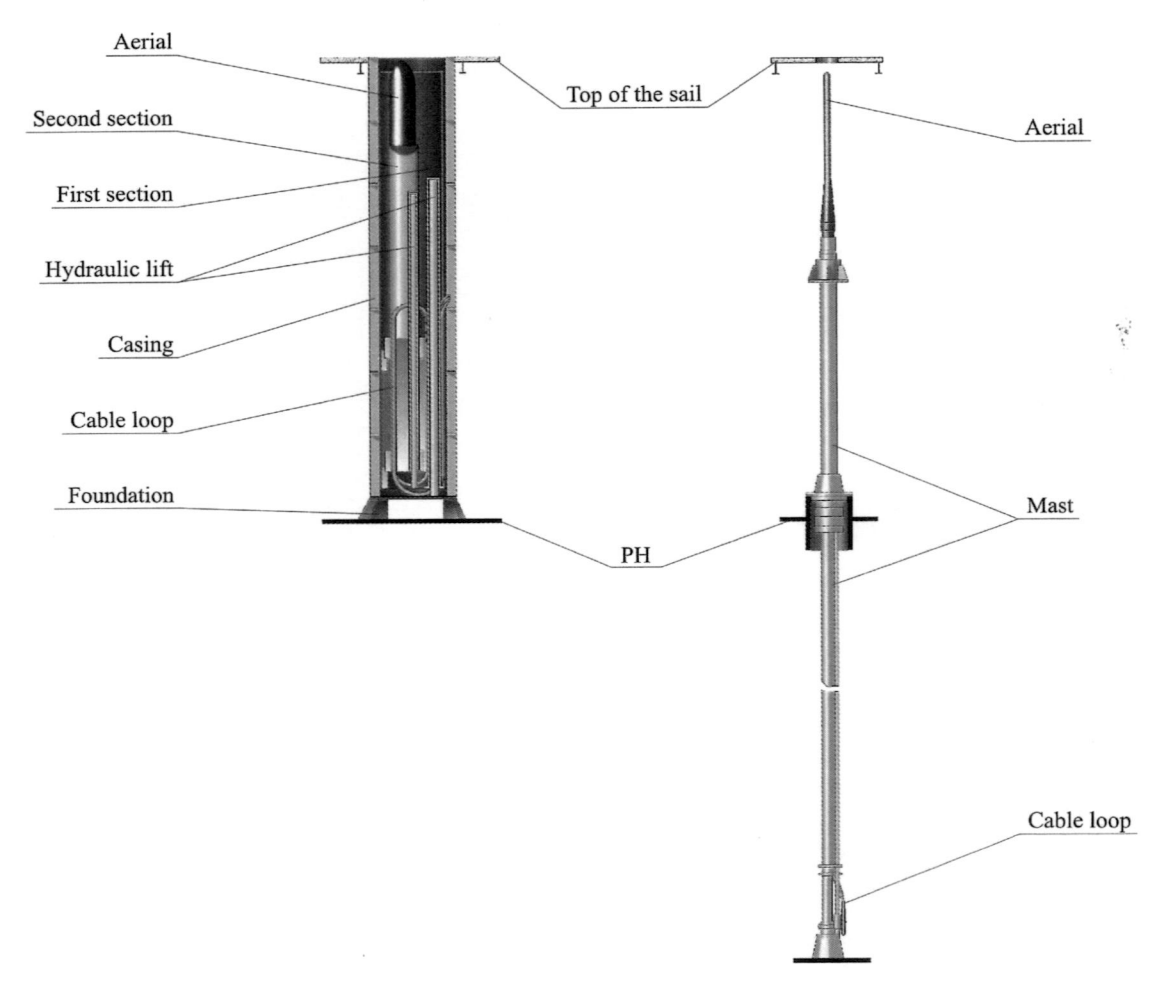

Fig. 5.21. Radio Communication Aerial Mast
a – non-penetrating, b – penetrating

Volume of Variable Ballast Tanks

Variable ballast tanks include regulating, trim and quick diving tanks. These tanks may be located either inside the PH or in the between-hull space (Fig.5.22).

Therefore, for the purpose of PH volume calculations it is necessary to determine their location at initial design stages. The volume of the quick diving tank is usually about 0.5% to 1.0% of the normal displacement and is dictated by the quick diving time, the volume of trim tanks amounts to 1.0%, of regulating tanks – from 2.0% to 4.0%. Thus, at the early design stages the total volume of variable ballast tanks can be calculated as:

$$V_{ABT} = k_{ABT} V_0 \qquad (5.33)$$

where $k_{ABT} = 0.055 \sim 0.060$.

During the subsequent development of the design the volume of these tanks is determined based on the calculation of the variable weight compensation.

Volume of Submarine Power Plant Support Tanks

Tanks that support the operation of diesel and diesel-generator plants, compressor stations and storage batteries include:
— fuel tanks;
— drain fuel tanks;
— clean oil tank;
— dirty oil tanks;
— compressor scavenging tanks;
— compressor oil tanks;
— distilled water tanks.

Fuel Tanks are divided by their location into internal and external (between-hull) tanks. The «gross» volume of external fuel tanks can be found by applying formula (3.64) without the k_{OIL} coefficient:

$$V_{EFT}^{GR} = k_{NFC} \frac{P_F}{\rho_F} = k_{NFC} \frac{1}{\rho_F} g_F k_F \frac{R_{SHORT} \vartheta_{SHORT}^2}{C_{SHORT}} (k_{FS} V_0)^{2/3} \qquad (5.34)$$

The rest of the fuel capacity is located in the pressure hull and should be taken into account when estimating the PH volume.

Drain Fuel Tanks are intended for draining of water and fuel discharged from all tanks through sampling and vent pipes, as well as for draining sediments from service tanks. At the early design stages the volume of drain fuel tanks is assumed to be approximately 1.0 to 2.0 m³. At Engineering and Detail Design stages the volume of these tanks is calculated to allow for collecting and storing all possible fuel leaks for the mission period because this tank is drained only when at the base. The drain fuel tank should be located in the bilge of the diesel compartment, below the points of drain.

The Clean Oil Tank serves for clean oil storage and is located in the compartment accommodating major oil consumers. On submarines the clean oil stock is, as a rule, within 4.0 to 7.0% of the total fuel capacity.

Dirty Oil Tanks are intended for collecting and storing dirty oil. At the initial design stages the volume of dirty oil tanks is taken as 10 to 20% of the clean oil tank volume. At the Engineering Design stage the volume of these tanks is calculated so as to allow for collecting and storing all possible oil leaks for the entire endurance period, as well as for complete oil replacement in one of the engines. In case of

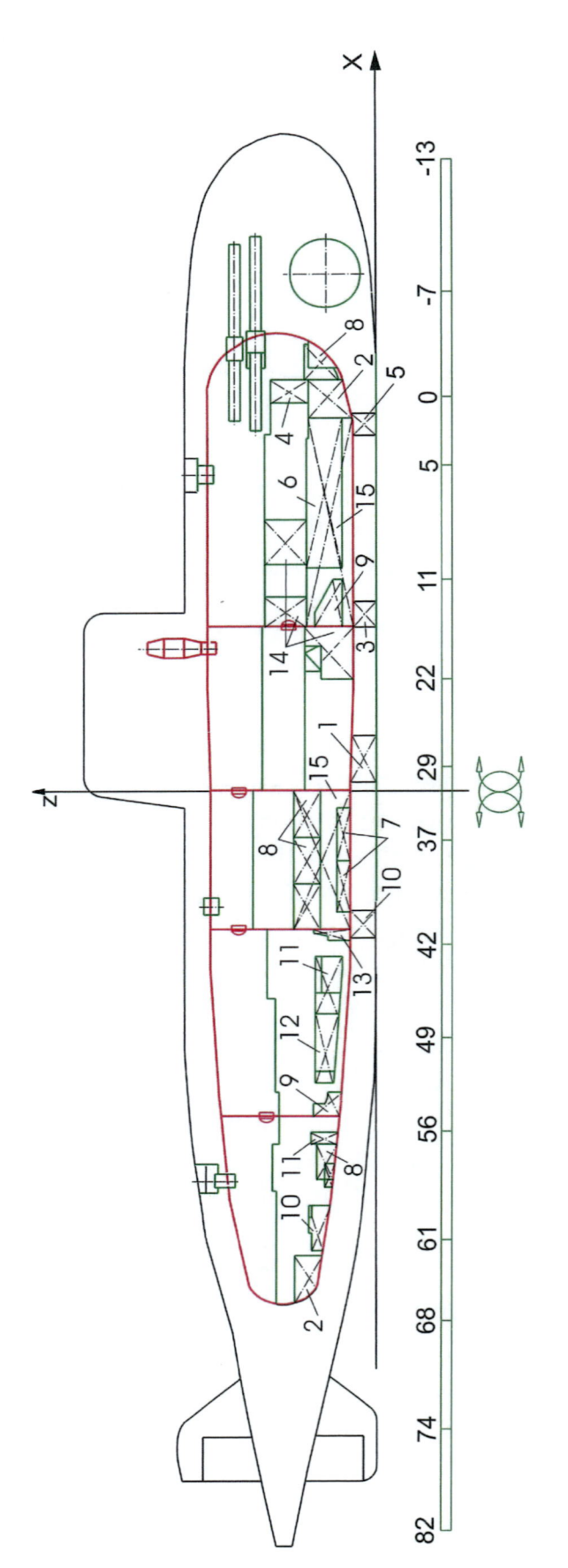

Fig.5.22. Tanks Inside the Pressure Hull

1. Regulating tank
2. Trim tanks
3. Quick diving tank
4. Water-round-torpedo tank
5. Poppet valve drain tank
6. Torpedo tube flooding tank
7. Distilled water tank
8. Drinking water tank
9. Dirty water tank
10. Fuel tanks
11. Clean oil tanks
12. Circulating oil tank
13. Dirty oil tank
14. Provision rooms
15. Battery wells

150

an accident these tanks should be located in bilges of power compartments, below the points of drain.

Compressor Scavenging Tanks are required for collecting and storing oil and water leaks from the operating compressor. The compressor is used for the replenishment of high pressure air stocks and for removing excessive air pressure in compartments during long submerged voyages. Approximately, the volume of such a tank for one compressor station is about 2.5 to 3.5 m^3.

Compressor Oil Tanks are required for storage of the compressor station lub oil. The compressor oil stock is calculated for the actual type of compressors installed on the submarine, based on the total number of the compressor operating hours (t) for the full endurance, per-hour oil consumption for the compressor lubrication (Q_{OIL}) and an emergency oil replacement for some of the installed compressors (V_{INC}). Thus, the compressor oil tank volume is found as:

$$V_{COT} = Q\,t + V_{INC} \qquad (5.35)$$

Distilled Water Tanks are intended for storage of water and its supply for topping storage battery cells and for replenishing the storage battery cooling system. The net volume of distilled water tanks is calculated with formula:

$$V_{DIST} = V_{TOP}\,n + V_{COOL} \qquad (5.36)$$

where V_{TOP} – volume of water required for topping one cell;
  n – number of cells;
  V_{COOL} – volume of water required for the replenishment of the storage battery cooling system.

The number of distilled water tanks on the submarine is not regulated but it is advisable to have at least two such tanks located as close as possible to battery wells.

Volumes of Containers and Silos can be determined from the data of Submarine Design Specifications in terms of types and numbers of missiles. Containers for cruise missiles are located either inside the PH or outside it, and ballistic missile silos as a rule pierce the pressure hull, and therefore only external portions of their volumes are taken into account in CBV estimations. On double-hull submarines with twin pressure hulls, silos are located in permeable structures between the twin hulls. In this case the CBV includes their total volume.

5.4. The Equation of Volumes. Submarine Volume Ratios

The need to ensure submarine submerged zero buoyancy equilibrium from the very beginning of the design forces to balance the load and the constant buoyant volume has already been emphasised several times.

The value of the constant buoyant volume can be determined in different ways.

One option is to use equation:

$$V_0 = \sum_1^n V_i \qquad (5.37)$$

where $\sum_1^n V_i$ – the sum of all water displacing volumes of the submerged submarine. As we have noted, this equation is used for finding the pressure hull volume.

The equation of volumes can be also written as a three-term expression, similar to that of masses, with all component volumes subdivided into three groups:
- volumes which depend on V_0 (service and living quarters, systems and gears, variable ballast tanks, hull material, etc.);
- volumes which depend on $V_0^{2/3}$ (the power plant);
- independent volumes (weapons, sensors, etc.).

Then:

$$V_0 = \alpha V_0 + \beta V_0^{2/3} + V_{IND} \qquad (5.38)$$

where coefficient α is the sum of volume indices proportional to V_0, and β – the sum for those depending on $V_0^{2/3}$.

As we already know, main components of the CBV are volumes of the pressure hull V_{PH}, external pressure-proof tanks V_{EPT}, external fuel tanks V_{EFT}, containers and silos V_{SC} (for missile submarines). Using statistic coefficients, we can determine the sought CBV value of the submarine from:

$$V_0 = \frac{V_{PH} + V_{EPT} + V_{EFT} + V_{SC}}{\chi} \qquad (5.39)$$

The equation of volumes is in any case much more approximate than the equation of load.

First, it ignores the actual geometry of the installed equipment and numerous requirements to its mutual arrangement, and operates only with their volumes.

Secondly, finding coefficients for this equation by the analysis of prototype characteristics is more difficult than for the equation of load because it is hard to identify volumes defined not only by arrangement requirements but also by buoyancy, stability, trimming, etc. considerations.

Solutions obtainable from either of these equations of volumes show the minimum required CBV value at which the arrangement of all equipment is possible.

The CBV value found using this method as a rule does not agree with the normal displacement determined from the equation of load. Possible design cases are:

$$\text{a) } D_0 = \rho V_0$$
$$\text{b) } D_0 > \rho V_0 \quad (5.40)$$
$$\text{c) } D_0 < \rho V_0$$

Case (a), when the divergence of D_0 and V_0 is minor, is a rather seldom case. The difference between D_0 and V_0 values can be balanced by the solid ballast and the displacement margin at later design stages.

Case (b), when the normal displacement is dictated by weights, requires finding new principal solutions.

The designer should consider applying an alternative material, practical ways to reduce the weight of the power plant, going for some decrease in the diving depth (if approved by the customer), and adding buoyant materials.

If such measures fail to produce the desired result, it is necessary to increase V_0. This can be done by increasing PH compartment volumes but that is, generally speaking, undesirable, as any increase in the volume of the submarine as a body moving through the water inevitably increases all signatures and respectively decreases the stealthiness, not to mention reducing the speed and increasing the cost. It would be appropriate to mention that creating a submarine with the least possible displacement, advanced tactical and technical features and minimum cost, is the most valuable skill of the submarine designer. Fitting in all equipment and trimming the submarine at the expense of increasing the length or the diameter of the pressure hull does not require a university-grade specialist. Therefore, a real designer should find such CBV components that could be increased with minimum increment of the submarine weight. E.g., an increase in the PH volume increases its weight as:

$$\Delta P = g_{PH} \Delta V_{PH} \qquad (5.41)$$

Excessive weight compensation should satisfy the equation:

$$\rho \sum_{1}^{m} V_i + \rho \Delta V_{PH} = \sum_{1}^{n} P_i + g_{PH} \Delta V_{PH} \qquad (5.42)$$

Then the required additional volume will be:

$$\Delta V_{PH} = \frac{\sum_{1}^{n} P_i - \rho \sum_{1}^{m} V_i}{\rho - g_{PH}} \qquad (5.43)$$

From expression (5.41) it is evident that the larger the value of the pressure hull weight index g_{PH}, the more difficult it is to compensate for the excess of D_0 over V_0.

In the third case (c), when the submarine displacement is defined by volumes, it is necessary either to find a possibility to reduce them without deterioration of basic tactical and technical particulars or to go for a justified increase in the submarine weight.

A possible solution is to move some of the external pressure-proof tanks into the PH, and thus to decrease the CBV value. Another option, considering that the greater portion of the hull weight is made up by the plating, may be to increase plate thickness as it would be accompanied by only a minor increase in the volume. The increase of the PH weight, in turn, allows the yield strength of the hull material and, hence, the cost to be reduced.

It should be mentioned that satisfying the (5.40) equality by extensive addition of solid ballast can hardly be viewed as a rational solution of the task.

There may be other solutions that depend exclusively on the designer's trade-offs, a vital skill especially in submarine design.

Having established the normal displacement balanced in terms of both weights and volumes, we can make approximate estimations of other displacements of the submarine.

Thus the relationship between the full submerged and the normal displacements may be written as:

$$k_{FS} = \frac{V_{FS}}{V_0} = 1 + \sum_1^n V_{MBT} + \sum_1^n V_{PEP} = 1 + \varepsilon + \varepsilon_{PEP} \qquad (5.44)$$

where $\varepsilon = \dfrac{\sum_1^n V_{MB}}{V_0}$ – relative buoyancy reserve;

$\varepsilon_{PEP} = \dfrac{\sum_1^n V_{PEP}}{V_0}$ – relative volume of permeable structures of the submarine.

Analysis of (5.44) for a number of submarine designs shows that k_{FS} depends on the following factors:
- architectural type of a submarine;
- relative buoyancy reserve;
- amount of permeable volumes of the hull, i.e. the presence of large sonar array bays, extensive superstructures or large permeable volumes in the between-hull space.

For submarines of similar architecture and similar amounts of permeable hull volumes, k_{FS} increases together with the buoyancy reserve, though not in direct proportion to it (Fig. 5.23)

From Fig.5.23 we can see that for the whole family of traditional double-hull submarines k_{FS} varies within wide limits, usually from 1.4 to 1.8. Nevertheless, for individual types of submarines the spread of k_{FS} values is much less, and therefore these values may be used for first-approximation assessments of the relationship between V_0 and V_{FS}.

The bare hull volume V_{BC} is very important since it directly defines the main dimensions and the design wetted surface:

$$V_{BH} = V_{FS} - V_{SAIL}^{GR} - V_{AP}^{GR} \qquad (5.45)$$

where V_{SAIL}^{GR} and V_{AP}^{GR} – gross volumes of the sail and appendages.

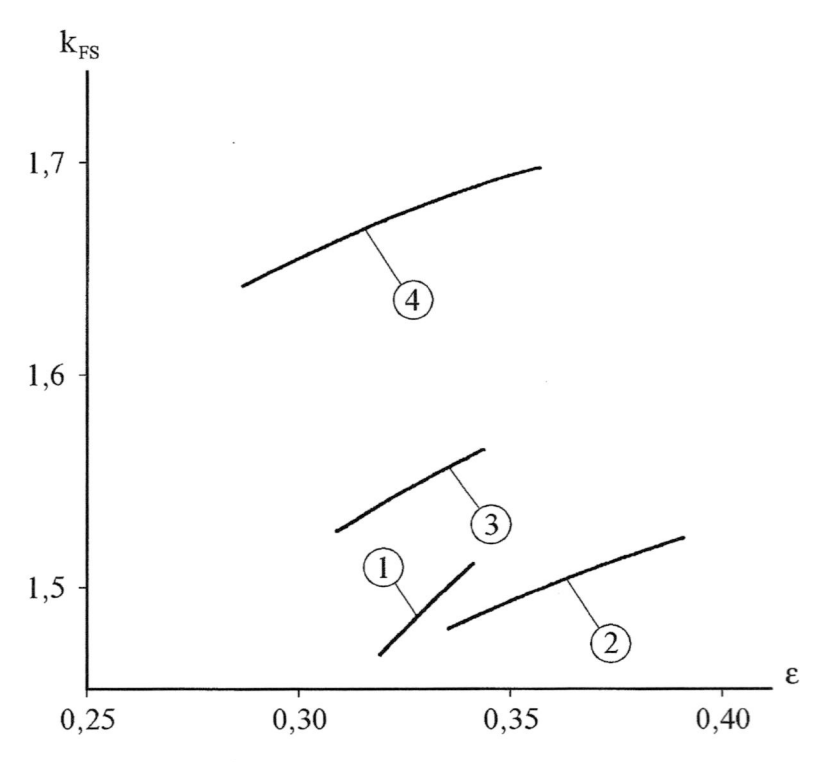

Fig. 5.23. Coefficient k_{FS} Versus Submarine
Relative Buoyancy Reserve
1 – SS, 2 – SSN, 3 – SSBN, 4 – SSGN

Coefficient $k_{BH} = \dfrac{V_{BH}}{V_{FS}} = 0.96\sim0.98$ shows the share of the bare hull in the full submerged volume of the submarine. The fact that this coefficient is close to zero allows us at the early design stages to assume $V_{BH} = V_{FS}$.

The coefficient that determines the ratio of the surface displacement to the normal displacement

$$k_{SFB} = \frac{V_{SFB}}{V_0} = 1 + \frac{\sum V_{PEP}^{SFB}}{V_0} = 1 + \varepsilon_{PEP}^{SFB} \qquad (5.46)$$

depends on the amount of permeable parts below the full-buoyancy waterline and varies within 1.2~1.25.

It should be noted that IVth and Vth generation submarines intended for service in the 21st century, will typically have single-hull or sidetank architectures [70], [86], [88], [94]. Therefore, the above relationships should be derived for these submarine designs separately.

The only exception is expected to be made for cargo submarines that will retain multiple hulls for the sake of environmental safety considerations [105].

6. SUBMARINE TRIMMING

After we have established the normal displacement and approximately estimated co-ordinates of the centre of gravity and the centre of buoyancy, it is necessary to check submarine stability and trimming for both the full-buoyancy and the submerged conditions. It is a very important issue and one should not proceed any further with design work until the submarine stability and trimming under different conditions has been checked because the outcome of these checks may require drastic changes in the whole general arrangement. Trimming should be performed in terms of both forces and moments since both conditions (5.1) and (5.2) have to be satisfied.

The greater part of available publications and research studies on submarine loads are devoted to the analysis of weights, a smaller number considers the analysis of volumes, and only a tiny portion deals with the analysis of levers and moments.

Therefore, we should first of all consider the issue of approximate estimations of the levers of weights and volumes.

6.1. Approximate Estimations of Weight Levers

Reliable submarine trimming depends not only on the accuracy of estimated forces but also on the accuracy of design values of moments of these forces, and, hence it is necessary to find their levers at the early design stages.

The first study devoted to the analysis of levers was made by professor B.M.Malinin (1950s). He considered the issue of relationships between the CB and CG co-ordinates positions of the submarine pressure hull. Nevertheless, even today there is still no sufficiently developed set of tools for approximate estimations of levers of weights and volumes applicable to submarines.

For the early stages of project development, when the structures are not yet designed, we can recommend that levers be scaled from prototypes. At those stages when the load is calculated by groups and sub-groups all weights can, for the sake of convenience, be segregated into two categories: distributed and concentrated weights.

Weights that are distributed, though not uniformly, along the entire length of the submarine may be put in the first category. Examples of such weights include weights of the hull, systems, cables, paints, coatings, etc. Longitudinal and vertical levers of these weights can be conventionally referred to the lever of the pressure hull centre of buoyancy of the prototype submarine.

Thus, e.g., CG coordinates of Group 500 X_{500} and Z_{500} may be using a close prototype represented as:

$$\frac{X_{500}}{X_{C_{PH}}} = \frac{X^0_{500}}{X^0_{C_{PH}}} = \overline{X}_{500} = \text{const} \qquad (6.1)$$

From which it follows that

$$X_{500} = \overline{X}_{500} X_{C_{PH}} \qquad (6.1a)$$

where $\overline{X}_{500}$ – relative lever of Group 500;

$X_{C_{PH}}$ – CB abscissa of the submarine pressure hull.

Similarly, ordinate X_{500} for the subject design is

$$Z_{500} = \overline{Z}_{500} Z_{C_{PH}} \qquad (6.1b)$$

The second category (concentrated weights) may include such weights that are in accordance with general arrangement conditions concentrated in certain points of the submarine.

They include weights of various weapons, weights of MPP components, etc. It is reasonable to refer CG offsets of such weights to individually assigned planes of reference representative for each particular case.

Thus, the longitudinal CG of the diesel plant X_{DP} can be derived from a close prototype in fractions of the diesel compartment length from either the fore or the aft bulkhead assuming $X_{DP} / \ell_{DP} = \text{const}$, and the vertical CG – in fractions of the PH diameter in way of the compartment assuming $Z_{DP} / d_{DP} = \text{const}$.

In a similar way we can find co-ordinates for electric propulsion motors.

For nuclear submarines SGP and STP CG coordinates can also be scaled from the prototype in fractions of the compartment length from either bulkhead of the relevant compartment. CG coordinates of missiles in silos, X_{MC} and Z_{MC}, may be counted from vertical and horizontal planes passed through the centre of buoyancy of the silo.

At early design stages CG coordinates for individual hull structures may be scaled from the prototype and measured from planes representative for each subject structure.

E.g., the sail CG can be approximately re-calculated assuming that relative ordinates $X_{SAIL} = \dfrac{X_{SAIL}}{\ell_{SAIL}}$ and $Z_{SAIL} = \dfrac{Z_{SAIL}}{h_{SAIL}}$ for the subject design and the prototype (with a similar shape and design of the sail) are the same. CG coordinates for the superstructure can be found in a similar way.

6.2. Approximate Estimations of CBV Component Levers

Statistic data analysis has shown that for submarines of similar (in terms of both power plants and weapons) types there is a steady relationship that may be used for the purpose discussed here. Let us assume that coordinates of the pressure hull CB $X_{C_{PH}}$ and $Z_{C_{PH}}$ are known from design studies on the equipment arrangement in the pressure hull. Then the CBV CG coordinates can be approximately determined knowing that within one type of submarines the ratios of longitudinal and vertical moments of major CBV components to the total CBV moment are approximately constant. E.g., for a torpedo diesel-electric submarine it is enough if such major components include the pressure hull, the external pressure and the external fuel tanks. Then, as follows from the above comments and demonstrated by statistics, the subject design and its close prototype have:

$$\frac{Z_{C(PH+EPT+FPT)}}{Z_{C_{CBV}}} = \text{const} \qquad (6.2)$$

$$\frac{X_{C(PH+EPT+FPT)}}{X_{C_{CBH}}} = \text{const} \qquad (6.2a)$$

This enables approximate Z_C and X_C of the submarine CBV to be found.

For missile nuclear submarines these major components can be defined as the pressure hull, the external pressure tanks and containers or external portions of missile silos.

In early-stage CG and CBV estimations the Y_0 ordinate is not calculated because the load and the CBV are close in terms of the symmetry. This offset is checked at a later stage and if it is found that $Y_C \neq 0$, Y_C should also be estimated. As a rule, Y_C is zeroed by ballast.

When design studies on equipment arrangement in the PH have been performed, and its dimensions and shape, as well as a schematic of general arrangement outside the PH, are already available, it becomes possible to update not only the CBV value but also the coordinates of the centre of buoyancy.

6.3. Submarine Trimming

Already at the earliest design stage serious attention should be paid to trimming. The submarine should have an even keel trim both on the surface and submerged and at the same time maintain the required level of stability. We shall skip issues associated with submarine surface trimming since in this case one can apply any surface ship design formulae [6]. Let us consider those issues that pertain to submerged submarine trimming.

To trim a submarine, it is necessary to observe the following conditions:

$$X_g = X_C \tag{6.3}$$

$$Z_C - Z_g = h_{\downarrow 0} \tag{6.3a}$$

At initial design stages abscissas X_C and X_g are usually different, i.e., the submarine is not trimmed. In this case we have to consider two alternatives:

1. The difference between X_g and X_C is minor and the task can be solved by shifting the solid ballast along the submarine.

2. If the difference between longitudinal centres of gravity and buoyancy is large, then it is necessary to re-arrange equipment inside the pressure hull or, sometimes, reconcile compartment subdivision.

In case the initial submerged metacentric height $h_{\downarrow 0}$ fails to meet requirements, its value can be corrected by changing either Z_g or Z_C. The Z_g coordinate is changed by vertical relocation of the heaviest equipment or solid ballast. Z_C can be changed by alterations in the configuration of PH end compartments, by height-wise relocations of pressure tanks and other CBV components.

6.4. Solid Ballast Weight and Longitudinal Profile Updating

At initial stages of the design the solid ballast (SB) weight is as a rule determined as a percentage of the normal displacement D_0 (see 3.6). Later, as the design becomes more elaborate, it becomes possible to update the amount of solid ballast and determine its longitudinal centre of gravity refined for submarine trimming. Since with the progress of design work the submarine load balance is continuously updated, the amount of solid ballast and its longitudinal profile also change. This process continues through all the design stages. It should be emphasised that some solid ballast should always be available, even if the design calculations show $D_0 = \rho V_0$.

This is necessary for later re-ballasting based on the outcome of stop trim dive trials of the constructed submarine in order to:
- achieve submerged trimming;
- ensure submarine stability;
- balance the upgrading displacement margin and unused design and construction margins.

The solid ballast may be located in different parts of the submarine and, depending on its location, its CG is considered using different methods:
- if the SB is located in the PH, in the load balance it is shown in the line titled «Solid Ballast, weight in air»;
- if the SB is located in the main ballast tanks, it is also listed in the load balance as «Solid Ballast, weight in air», but in the CBV Table they introduce an additional component «Solid ballast buoyancy»;
- if the SB is located in permeable hull structures, its line in the load balance becomes «Solid Ballast, weight in water».

Let us for the sake of an example consider the latter case: it is necessary to arrange the solid ballast in permeable structures along the submarine (the vertical lever is dictated by stability conditions).

The amount of solid ballast can be found from the stability equation:

$$\sum_1^n P_i + P_{BAL} = \rho \sum_1^m V_i + \rho \frac{P_{BAL}}{\rho_{BAL}} \qquad (6.4)$$

where $\sum_1^n P_i$ — submarine displacement without ballast;

P_{BAL} — solid ballast weight;

$\sum_1^m V_i$ — volume displacement of the submarine without the ballast;

$\rho \dfrac{P_{BAL}}{\rho_{BAL}}$ — solid ballast buoyancy.

After relevant transformations of (6.4) we can derive an expression for the solid ballast weight:

$$P_{BAL} = \frac{\rho_{BAL}}{\rho_{BAL} - \rho} \left(\rho \sum_1^m V_i - \sum_1^n P_i \right) \qquad (6.5)$$

Obviously, if the solid ballast is located inside the PH, multiplier

$$\frac{\rho_{BAL}}{\rho_{BAL} - \rho} = 1$$

The solid ballast longitudinal lever is determined by the equation of moments that in the given case may be written as:

$$\sum_1^n P_i x_{g_i} + P_{BAL} x_{BAL} = \rho \sum_1^m V_i x_{C_i} - \rho \frac{P_{BAL}}{\rho_{BAL}} x_{BAL} \qquad (6.6)$$

After transformations, the SB longitudinal lever formula becomes:

$$x_{BAL} = \frac{\rho_{BAL}}{\rho_{BAL} - \rho} \cdot \frac{\rho \sum_1^m V_i x_{C_i} - \sum_1^n P_i x_{g_i}}{P_{BAL}} \qquad (6.7)$$

When estimating the X_{BAL} lever, it is necessary to pay attention to the following factors.

First, X_{BAL} is not the location of the solid ballast itself but only the point of its geometrical centre of gravity.

Secondly, one should minimise the ballast lever X_{BAL} at early design stages as much as possible because later it may be required to move the ballast to the ends of the ship.

The final weight of the solid ballast and its arrangement are established during stop trim dive trials after the submarine is completed.

MGK-400EM
Upgraded submarine sonar system

Tasks
- Detection, classification, tracking and acquired target (submarine, surface ship, torpedo) assignment to torpedo and missile weapon in passive mode.
- Search of low-noise targets, detection of moored mines and navigation obstacles in active mode.
- Intercept, classification of the signals radiated by sonars, homing torpedoes and mines-torpedoes.
- LF/HF telephone, telegraph and coded communication with interacting submarines and ships, identification of correspondents (friend-foe) and measuring distance to them.
- Sonar background noise control, evaluation of operation range, self-monitoring of sonar operability and fault diagnostics.

Array
- The main receive array composed of 1008 hydrophones is several times larger compared to all known prototypes in respect of its active operating surface.
- High-sensitive array units of advanced design provide the stable characteristics and service reliability.

Processing
High-speed processors form 540 beam patterns and realize efficient algorithms of automated data processing optimized to hydrology and acoustic environment.

The MGK-400EM may be completed by flexible tow array and conditioning equipment according to customer's requirement.
Owing to unique technologies, know-how and engineering solutions used in the MGK-400EM its reliability and life specifications satisfy modern submarine sonars.

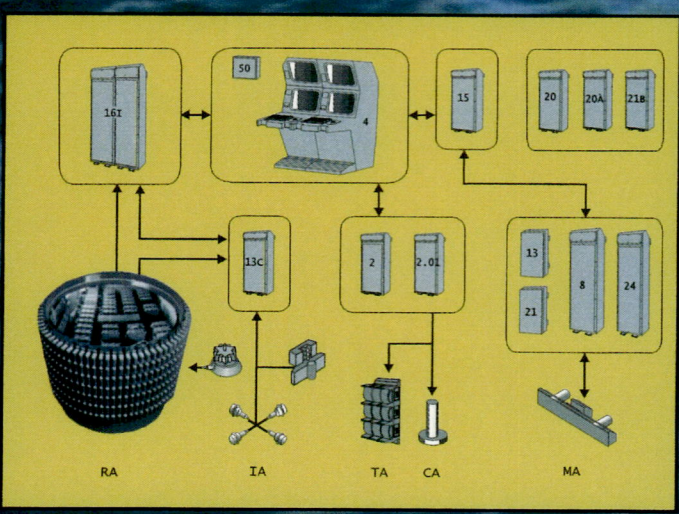

16I - digital signal processors
4, 50 - control consoles
15 - conditioning unit
20, 20A, 21B - power supply units
13C - communication subsystem
2, 2.01 - generator units
13, 21, 8, 24 - mine detection system
RA - main receive array
IA - intercept arrays
TA - transmitting array
CA - communication array
MA - mine detection array

MORPHYSPRIBOR
46, Chkalovsky Prospect
197376, St.Petersburg, Russia
Tel: (812) 320 8040/41

7. COMPENSATION OF SUBMARINE VARIABLE WEIGHTS

7.1 General Issues of Variable Weight Compensation

As has been pointed out in Chapter 5, any commercial or naval ship should be designed in such a way as to meet static equilibrium conditions with neither heel nor trim, i.e. the ship must always be trimmed (see 5.1):

$$D_0 = \rho V_0$$
$$\text{at } X_g = X_C, Y_g = Y_C.$$

Let us now deliberate on differences in the behaviour of a surface ship and a submarine when this static balance is disturbed.

For a surface ship, the balance of forces described by the fundamental buoyancy equation is maintained automatically. Losing or gaining any weight under normal operation conditions is accompanied by respective changes in the draught and, accordingly, in seakeeping qualities of a surface ship. This is due to the fact that only two parameters in the equation are variable: the load and the water density, while the third one, the volume, is free (at least within reserve buoyancy limits) and compensates for changes in the former two [21], [59].

This is also true for submarines as long as they stay on the surface. However, since submarines should always be able to dive, we need a different approach to the issue.

For submerged submarines the reserve buoyancy is zero and therefore there is no force that would, by itself, compensate for the submarine load change. Even the slightest load variation makes the submarine dive or surface.

From these postulates it follows that submerged hovering is achievable only when $D_0 = \rho V_0$, i.e. with zero residual buoyancy.

It should be also remembered that for a submerged submarine we should more carefully observe the balance of moments, first of all the trimming ones.

It is known from submarine statics that the submerged longitudinal metacentric height of the boat is equal to the transverse one and is not very high ($H_{0↓} \approx h_{0↓} \approx 0.3 \sim 0.5$ m) [30]. This means that the stability of a submerged submarine is poor and moments acquired in operation will result in large trim and heel angles.

Thus, since a submerged submarine is unable to compensate for the residual buoyancy and trimming moments herself, her static equilibrium has to be maintained by appropriate design provisions and crew actions.

Such design provisions include auxiliary ballast tanks for which the layout and the capacity depend on the involved type of consumable weights, as well as on the chosen methods of their compensation [47].

All weights constituting the submarine normal load balance can be subdivided into constant and variable. Constant weights do not change in the process of submarine operation. This group of weights includes hull structures, the power plant, systems, devices, etc.

Variable weights are those weights that are included into the normal displacement and then change either their values (taken aboard or consumed) or their locations in the submarine. Such weights include: weapons, fuel and oil, distilled water, provisions and fresh water, and some other. The static balance is also disturbed by water transfer from water- round-weapon tanks to weapon launchers, by changes in seawater density, by hull and external coating compression. This means that these factors should also be taken into account when considering issues pertaining to submarine trimming.

All variable weights can be subdivided into three groups depending on the compensation method:
- variable weights compensated from dedicated tanks. This group mainly includes various weapons;
- variable weights compensated in their own tanks, e.g. the fuel;
- variable weights compensated with the help of auxiliary ballast tanks. These tanks are also used to suppress all non-compensated forces and moments that may be generated due to the compensation of the first two variable weight groups, e.g. differences in forces and moments associated with fuel weight compensation.

It is impossible to predict exactly when, which weights and in what amounts they will be consumed because this depends each time on the current mission of the submarine. However, for some variable weights it is possible to plan the sequence of their consumption from different storages. Such weights include provisions, fresh water, fuel and some others.

Scheduled consumption of these weights allows the amount of ballast required for trimming to be reduced, and hence saves the volume of auxiliary ballast tanks.

A schematic arrangement of submarine main and auxiliary ballast tanks is shown in Fig.7.1.

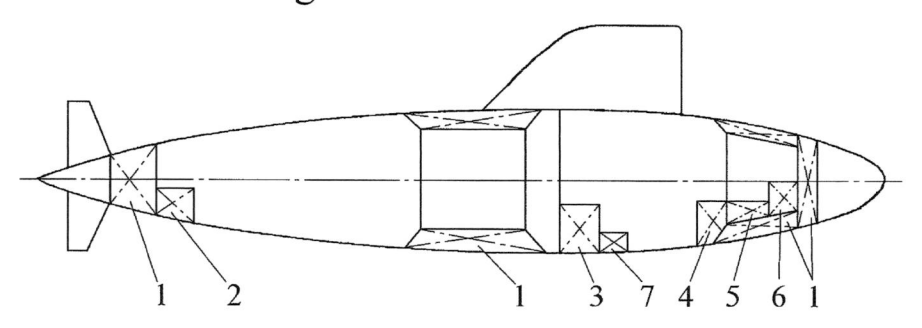

Fig. 7.1. A Schematised Arrangement of Submarine Main and Auxiliary Ballast Tanks

1 – main ballast tank; 2 – aft trim tank; 3 – regulating tank; 4 – torpedo tube flooding tank; 5 – water-round-torpedo tank; 6 – forward trim tank; 7 – quick diving tank.

When designing a submarine, it is advisable to place trim tanks as close to submarine ends as possible in order to increase the lever due to water transfer from one trim tank to another. This allows required trimming moments with lesser amounts of water to be produced, thus reducing the capacity of these tanks. As a rule, there are 2 to 4 trim tanks on a submarine. According to statistic data, the total capacity of trim tanks is usually about one per cent of the submarine normal displacement.

There are usually two regulating tanks on medium- and large-size diesel-electric submarines, and even more on nuclear ones. In terms of the submarine length, one of the regulating tanks is usually placed at the centre of buoyancy and intended mainly for compensating water density (salinity) variations. Sometimes they even call it the «salinity tank». With this position of the salinity tank, intaking or discharging water for its density variation compensation does not cause trimming moments and it is not necessary to apply trim tanks [62].

The second regulating tank is intended to compensate for the variable weights.

It would be most reasonable to put it close to the centre of gravity of all variable weights to be compensated with the help of this tank. However, in practical design the second tank is as a rule located close to the first one, towards the bow or the stern. The total capacity of regulating tanks according to statistics is about 3 to 4% of the submarine normal displacement.

On nuclear-powered boats the submarine centre of gravity is in way of the steam generating plant. That place is occupied by biological shield tanks, and therefore it is impossible to arrange regulating tanks described above.

On nuclear submarines regulating tanks have to be placed towards the bow from the centre of gravity, and in order to balance the moments this arrangement makes it necessary to put an additional regulating tank in the aft end, otherwise trim tank volumes would be too large.

Basically, the above-described arrangement is not by all means mandatory. It is possible to have only two trim-type tanks that cater for all submarine trimming needs. This approach was implemented on several submarines.

Regulating and trim tank capacities are determined based on the following principle: none of these tanks should ever be completely drained or flooded even in the most difficult cases of variable weight compensation, so that there would always be a little margin of both the water and the capacity since all calculations are to some extent provisional and unforeseen situations are possible. Thus, the auxiliary ballast capacity includes three components :

$$V_{ABT} = V_{ITW} + V_{VW} + \Delta V \qquad (7.1)$$

where V_{ITW} – initial trimming water;
V_{VW} – volume required to compensate for variable weights;
ΔV – a margin of free capacity.

The baseline for determining the initial amount of trimming water is the volume of water that according to variable weight compensation calculations must be discharged from the tank in order to balance the submarine. The value VITW obtained by this calculation is included in the submarine normal load balance.

Following on from the above comments, the first two components in (7.1) are found by a special calculation which is called the variable weight compensation analysis while the third one is assumed by the designer [84]. As a rule, ΔV is 0.25 to 0.50% of D_0.

The method of tank capacity calculations depends on the trimming model chosen for the designed submarine. The trimming model may be:

1. A segregated (self-contained) arrangement, i.e. when the entire volume of water necessary to compensate for the consumed weight is taken into the regulating tank while the trim is adjusted by pumping water from one trim tank to another (Fig.7.2).

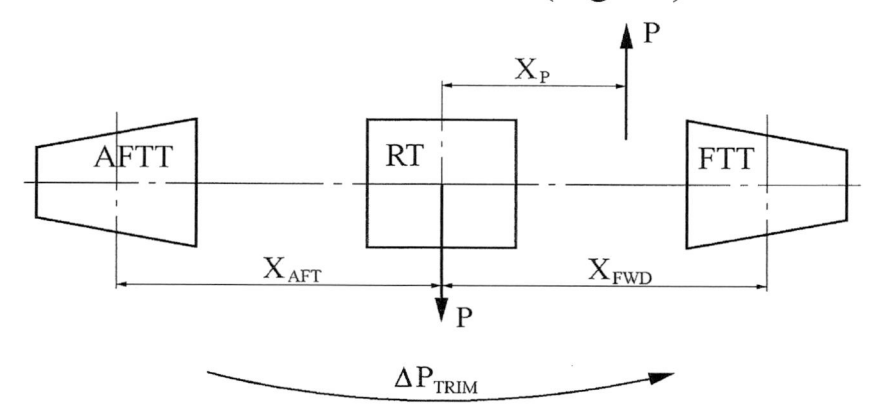

Fig.7.2. The Segregated (Self-Contained) Model of Submarine Trimming

This trimming model has the advantage of a clear division of functions between regulating and trim tanks which compensates for variable weights quickly and accurately. However, it requires a large capacity of regulating tanks. Trim tanks may in this case be smaller.

2. A shared arrangement, i.e. when compensating water is taken simultaneously into a regulating tank and into one of the trim tanks (Fig.7.3).

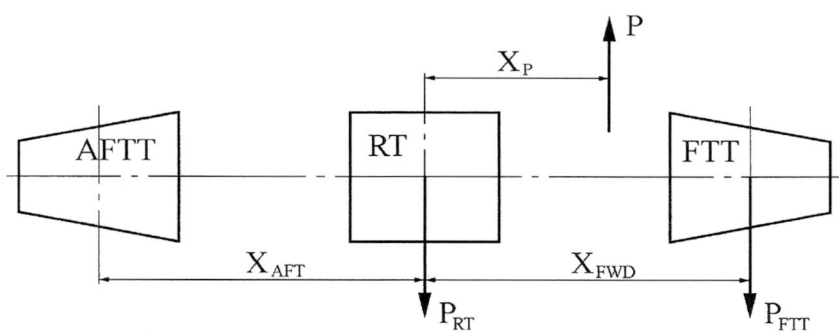

Fig.7.3. The Shared Model of Submarine Trimming

In this case the regulating tank needs somewhat less capacity than in the previous model but the trimming process becomes more complicated because it is necessary to estimate the amount of water to be taken into each of the involved tanks. With the modern level of automation this task is totally feasible.

The main rules of submarine variable weight compensation are as follows [51], [84].

1. Weights consumed in large amounts at a time (weapons) are compensated by flooding tubes, silos, containers or special compensating tanks placed by the designer so that their centres of volume as far as possible coincide (by length) with the centre of gravity of these weights. If the process generates any trimming moments, they are to be eliminated by water transfer between trim tanks. Such weights should be compensated immediately as they are lost.

2. When snorkelling for a long time, SSDEs lose weight due to fuel consumption. The fuel weight is to be compensated depending on the consumption rate.

3. All other weights are consumed gradually and they are compensated with the help of trim and regulating tanks as required, e.g. prior to hovering or when it is discovered that the submarine is out of trim.

It should be remembered that mistakes in variable weight compensation lead to accidents. E.g., a mistake in variable weight compensation made on the surface may cause only a slight change in the trim, but when the submarine dives, it can result in acquiring intolerable trim angles, dropping beyond the maximum diving depth and losing the submarine [30].

7.2. Weapon Weight Compensation

Torpedoes. Modern submarines may carry various types of torpedoes. Torpedoes of different weight, size and negative buoyancy put together in different combinations are used to compile alternative combat load packages. Therefore, in weight compensation calculations it is assumed that with the heaviest combat load package the submarine is trimmed by the solid ballast. This weight and the associated moment, taking into account the water in the water-round-torpedo tank, are included into the submarine normal load balance. When the submarine takes any other combination of torpedoes, it is necessary to calculate the difference in weights and moments and to trim the submarine accordingly, i.e. to take as much water into the WRTT tank and other auxiliary ballast tanks that the sums of weights and moments of the added water and the new set of torpedoes are equal to the final sums of weapons specified in the normal status tables of variable weights.

Let us review how the torpedo complex operates during firing. (Fig.7.4).

Prior to firing, water from the water-round-torpedo tank is pumped to the torpedo tubes.

Thus, in the torpedo tube we have the combined weight of the torpedo plus the water from the WRTT tank ($P_{TOR} + P_{WPTT}$). The resulting trimming moment is eliminated by water transfer between trim tanks. The amount of water to be transferred from one trim tank to another for every particular torpedo model is found by calculations.

When the torpedo leaves the tube, the latter is flooded and its load changes by the value of the torpedo negative buoyancy, i.e. $\Delta P_{TOR} = P_{TOR} - \rho V_{TOR}$.

To observe the equality of forces, we have to satisfy:

$$P_{TOR} + P_{WPTT} = \rho V_{TOR} + P_{WPTT} + \Delta P_{TOR}. \qquad (7.2)$$

The torpedo negative buoyancy ΔP_{TOR} is at the instant of launching transferred into the poppet valve drain tank (PVDT) through the poppet valve, adjusted in such a way that, along with the air, it admits some specific amount of water.

When reloading torpedo tubes on a submerged submarine, it is necessary to perform two operations that disturb submarine trimming:

– after firing, unless it was the last launch, the torpedo tube water is transferred to the WRTT tank and, when the latter is filled, to the torpedo tube flooding tank. This results in a trimming moment to the aft and it is necessary to apply the trimming system to balance it.

– when moving a reload torpedo from the rack into the torpedo tube, it is necessary to transfer water from the fore trim tank to the aft one. When torpedo tubes are reloaded on the surface, moments should be balanced after completing all operations.

It should also be pointed out that the amount of water to be transferred between the trim tanks depends not only on the torpedo model but also on how close the torpedo tube flooding tank, as well as torpedo tubes and the WRTT tank, are located to the centre of gravity of reload torpedoes.

The weight of the last launched torpedoes is compensated directly by flooding the torpedo tubes (the water remains there) and the water to compensate for the torpedo negative buoyancy is taken into the WRTT tank via the PVDT tank. On some submarines torpedo tubes may be drained into the PVDT tank after the last launch.

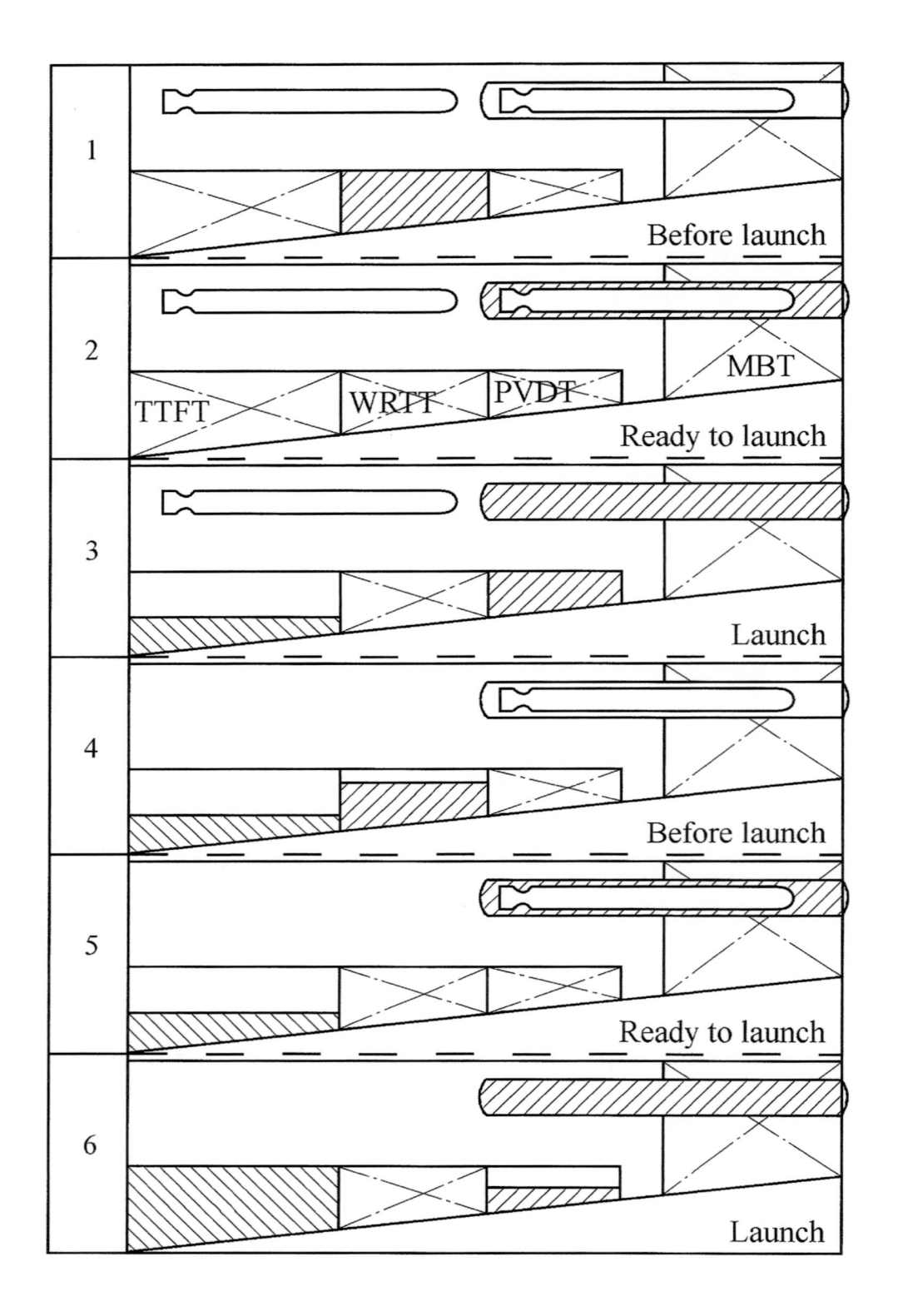

Fig.7.4. Torpedo Weight Compensation
TTFT – torpedo tube flooding tank; WRTT – water- round-torpedo tank;
PVDT – poppet valve drain tank, MBT - main ballast tank

Mines

Submarines may take mines instead of torpedoes and lay them through torpedo tubes. All operations related to their weight compensation are similar to those for torpedo weight compensation.

On dedicated (mine-laying) submarines, mines are stowed in tubes or trunks either in dry or in wet condition.

The dry method requires a water-round-mine tank and, if there are reload mines, a mine tube flooding tank as well.

In this case the balance of forces can be written as follows:

$$\sum_1^n P_{MINE} + P_{WRTT} = P_{WRTT} + \rho \sum_1^n V_{MINE} + \sum_1^n \Delta P_{MINE} \qquad (7.3)$$

where $\sum_1^n P_{MINE}$ and $\sum_1^n V_{MINE}$ – total weight and volume of mines;

$\sum_1^n \Delta P_{MINE}$ – total negative buoyancy of mines.

The water that compensates for the total negative buoyancy of mines is transferred to the water-round-mine tank or to the mine tube flooding tank. In case the submarine carries reload mines, mine tubes or trunks are drained into the same tanks.

With the wet storage it is necessary to compensate only for the negative buoyancy of mines. In this case there is no need for a water-round-mine tank and a mine tube flooding tank. The lost weight is compensated by regulating and trim tanks.

Missiles

The number of tanks required to compensate for the weight of launched missiles depends on the launch position (surface or submerged) and the launching method (dry or wet).

After a surface launch, silo (container) caps are closed and under normal conditions the silos remain dry. In this case the weight of launched missiles is compensated by taking water into dedicated missile weight compensating tanks or, if there are no such tanks, into the regulating tank.

For dry underwater launching the silos or containers are not flooded, but they are flooded after the missile is launched and then the submarine load balance changes by

$$\Delta P = \rho V_{SILO} - P_{MIS.} \qquad (7.4)$$

Since the weight of the water within the silo volume is usually more than that of the missile, it is necessary to have special flooded tanks to be blown after launching the missile and flooding the silo.

Water in these tanks must be included into the submarine normal load balance.

For wet launching, the silo is flooded before the launch. In this case it is necessary to have water-round-missile tanks of a capacity sufficient to fill annular volumes of all silos. The difference between the missile weight and the water weight within the missile volume is compensated into water- round-missile tanks.

7.3. Fuel and Oil Weight Compensation

Diesel fuel weight is compensated by taking seawater into fuel tanks and it is considered that 98% of the full fuel capacity must be compensated. The resulting increase in the submarine weight can be expressed as:

$$\Delta P = P_W - P_F = (\rho - \rho_F)V_F = \frac{\rho - \rho_F}{\rho_F}P_F \qquad (7.5)$$

where P_F and V_F – weight and volume of compensated fuel;
$\rho = 1.02 - 1.03$ t/m³ water density;
$\rho_F = 0.85 - 0.88$ t/m³ fuel density.

When calculating the weight compensation capacity, it is necessary to assume the maximum water density at the intended naval theatre and the minimum fuel density because otherwise the auxiliary ballast tank volume may turn out to be insufficient.

E.g., with the water density of $\rho = 1.03$ the increase in submarine weight is:

$$\Delta P = \frac{\rho - \rho_F}{\rho_F}P_F = \frac{1.03 - 0.85}{0.85}P_F = 0.212 P_F \qquad (7.6)$$

This means that when the diesel fuel is consumed, it is necessary to discharge large amounts of water from auxiliary ballast tanks. The diesel oil is consumed simultaneously with the fuel, and therefore when compensating negative buoyancy due to fuel consumption it is also necessary to take into account the consumed diesel oil weight. The amount of diesel oil carried by a submarine is as a rule $P_{OIL} = 0.05$ to $0.07 P_F$ and almost all of it is consumed due to burning in engine cylinders. The oil weight is not compensated by taking water into oil tanks. Therefore, the total amount of water discharged during fuel and oil consumption is:

$$\Delta P = \frac{\rho - \rho_F}{\rho_F}P_F - k_{OIL}P_F \approx (0.142 - 0.162)P_F \qquad (7.7)$$

Let us note some specific aspects associated with taking excess-capacity fuel into fuel-ballast tanks. This guarantees positive buoyancy that may be estimated as:

$$\Delta P = \rho \sum_1^n V_{FBT} - \rho_F \sum_1^n V_{FBT} = V_{FBT}(\rho - \rho_F) \qquad (7.8)$$

Thus, for maintaining the submarine equilibrium and ability to dive it is necessary to take some additional amount of water ΔP into the regulating tank.

On nuclear submarines the amount of fuel and oil for diesel-generators is not very large, and principles of its weight compensation are similar to those for diesel-electric submarines. The main oil stock on nuclear submarines is used for turbine lubrication. In the variable weight compensation analysis it is usually assumed that the consumed and compensated weight of the main turbine oil is 75% of the lub oil tank capacity. The remaining part foams.

7.4. Compensation for Submarine Buoyancy Variations Due to Hydrological Effects

Submarine buoyancy variations at a constant load are due to changes of the water density ρ. The seawater density in turn depends on temperature, pressure and the amount of salts in it, i.e. the salinity. Any submarine is designed to a water salinity value specified in SDS and its increases or decreases around the design value are compensated by taking or discharging water into/from auxiliary ballast tanks.

At the same time pressure and temperature variations change the submarine submerged volume due to the compression of pressure structures and external coatings. Taking into account all these factors, submarine buoyancy variations can be expressed as:

$$\Delta P = \rho V \left[\frac{\rho_1 - \rho_0}{\rho_0} + \rho(\alpha_P + \alpha_\rho)(H_1 - H_0) + \alpha_t(t_1^0 - t_0) \right] \qquad (7.9)$$

where α_P – relative reduction of the watertight hull volume due to pressure increase by one unit;

α_ρ – relative change of the water density due to pressure increase by one unit;

α_t – submarine volume expansion coefficient, i.e. the relative change of the submarine volume due to temperature t change by 1°C.

Numerical values of the above parameters as published by V.G.Vlasov [22] are:

$$\alpha_p = -\frac{13}{H_P} \cdot 10^{-4} \text{ m}^3/\text{t}$$

$$\alpha_\rho = 4.8 \cdot 10^{-6} \text{ m}^3/\text{t} \qquad (7.10)$$

$$\alpha_t = 37 \cdot 10^{-6} \frac{1}{1°C}$$

It should be mentioned that for residual buoyancy ΔP estimations, the displacement V should be taken as a function of submarine sailing conditions associated with the subject change in these factors. The trimming moment, which may appear along with a buoyancy variation, must be calculated based on the condition that the residual buoyancy ΔP is applied at the centre of submarine buoyancy X_C with displacement V:

$$M = \Delta P X_C \qquad (7.11)$$

7.5. Variable Weight Compensation Analysis

Variable weight compensation analysis is performed in order to establish the required auxiliary ballast capacity. It is either a check calculation, when, e.g., tank capacities and their longitudinal levers have been assigned based on general arrangement considerations and data from prototypes, or it serves as a basis for finding auxiliary ballast tank capacities when the longitudinal levers have already been specified due to some design considerations.

Compensation calculations are usually performed for various submarine load cases. The basic initial case is the normal displacement at the water density specified in SDS.

For diesel-electric submarines it is also obligatory to consider the case of the displacement with excess-capacity fuel at the specified water density. They also consider load cases related to weapon changes (e.g. torpedoes for mines), etc.

Depending on submarine designation and type, the following cases may be considered during the analysis of variable weight consumption and compensation:

– weapon utilisation (torpedoes, missiles, mines);
– fuel and oil consumption at various water densities;

- provisions and fresh water consumption;
- transfer of annular space flooding water from tanks to launchers;
- compression of the hull and coatings, etc.;

The analysis of variable weight compensation starts from the description of the relevant principles and assumptions.

E.g., it is indicated which tanks are to be used to compensate for this or other variable weight:
- reload torpedo weight is compensated into the torpedo tube flooding tank;
- weight of last salvo torpedoes is compensated into torpedo tubes and the PVDT tank;
- fuel and oil weights are compensated into fuel tanks, etc. [47].

The following examples may serve as an illustration of assumptions made for such calculations:
- consumed and compensated fuel weight is assumed to be 98% of the total fuel weight;
- diesel oil is consumed and compensated within 5 to 7% of the useable fuel, etc.

In this section it is also necessary to indicate the chosen model of variable weight compensation (segregated or shared).

It is necessary to provide main data on the submarine relevant for these calculations:
- volume displacement according to the CBV table V_0, m^3;
- centre of buoyancy offset from the midship section X_C, m;
- oil density $\rho_{OIL} = 0.9$ t/m^3;
- fuel density $\rho_F = 0.85{\sim}0.88$ t/m^3.
- water density $\rho = 1.015{\sim}1.030$ t/m^3.

Weight and volume data on weapons and launchers are given in tables, as well as their longitudinal levers with respect to the midship section.

Data on auxiliary ballast tanks should include their volumes (if already chosen) and longitudinal levers which are assumed, as has already been mentioned, taking into account design and general arrangement considerations (Table 7.1). Apart from the regulating tanks, all other auxiliary ballast tanks are placed on either side.

Then, in the form of a table they show data on variable weights on board the submarine (Table 7.2).

Table 7.1.

Tank Volumes and Positions

Description	Net volume, m³	Tank CG offset from midship section, m
Regulating tank No.1		
Regulating tank No.2		
Regulating tank No.3		
Fore trim tank No.1		
Fore trim tank No.2		
Aft trim tank No.3		
Aft trim tank No.4		
Torpedo tube flooding tank No. 1		
Torpedo tube flooding tank No.2		
Water-round-weapon tank No.1		
Water-round-weapon tank No. 2		
PVD tank No. 1		
PVD tank No.2		

Table 7.2.

Variable Weights on Board the Submarine

No	Description of weights	Weight t	Lever from midship m	Moment tm
1	**Weapons** Torpedoes in torpedo tubes Torpedoes on racks Water round torpedoes in tubes Residual water in WRTT tanks.	$P_{TOR}n_{TT}$ $P_{TOR}m_{RELT}$ $P_{WRT}n_{TT}$ ΔP_{WRT}	x_{TT} x_{RELT} x_{TT} x_{WRTT}	$P_{TOR}n_{TA}x_{TA}$ $P_{TOR}m_{RELT}$ x_{RELT} $P_{WRT}n_{TT}x_{TT}$ $\Delta P_{WRT}x_{WRTT}$
	Total	ΣP_{TS}		ΣM_{TS}
2	**Fuel** in fuel tank No.1 in fuel tank No.n	P_{TOR1} P_{TORn}	x_{TOR1} x_{TORn}	$P_{TOR1} x_{TOR1}$ $P_{TORn} x_{TORn}$
	Total	ΣP_{TOR}		ΣM_{TOR}

Table 7.2. (continued)

No	Description of weights	Weight t	Lever from midship m	Moment tm
3	Oil in oil store tank in clean diesel oil tank	P_{OIL} P_{DOIL}	x_{OIL} x_{DOIL}	$P_{OIL} x_{OIL}$ $P_{DOIL} x_{DOIL}$
	Total	ΣP_{OIL}		ΣM_{OIL}
4	Provisions and water b) fresh water in tank No.1 in tank No.2 in tank No.n	P_{PROV} P_{W1} P_{W2} P_{Wn}	x_{PROV} x_{W1} x_{W2} x_{Wn}	$P_{PROV} x_{PROV}$ $P_{W1} x_{W1}$ $P_{W2} x_{W2}$ $P_{Wn} x_{Wn}$
	Total for provisions and water	$\Sigma P_{P,W}$	$x_{P,W}$	$\Sigma M_{P,W}$
5	Feed water in tank No.1 in tank No.2 distillate in tank	P_{FW1} P_{FWn} P_{DIST}	x_{FW1} x_{FWn} x_{DIST}	$P_{FW1} x_{FW1}$ $P_{FWn} x_{FWn}$ $P_{DIST} x_{DIST}$
	Total	ΣP_{FWi}		ΣM_{FWi}
—				
n	Regeneration cartridges	P_i	x_i	$P_i x_i$
	Total:	ΣP_{VW}	x_{VW}	ΣM_{VW}

7.6. Estimation of the Required Volume of Auxiliary Ballast Tanks

Let us consider, as an example, trimming calculations for a submarine with a compensation model according to which the water is taken not only into regulating tanks but also into the fore or aft trim tank. The calculation is made in the form of a table (Table 7.3).

It is obvious that in order to keep the submarine balanced while compensating for variable weights, it is necessary to choose an amount of the auxiliary ballast and to distribute it among tanks in such a way as to achieve:

$$\sum_{1}^{n} P_{VW} = \sum_{1}^{n} P_{ABL}$$
$$\sum_{1}^{n} P_{VW} \cdot x_{VW} = \sum_{1}^{n} P_{ABL} \cdot x_{ABL}$$

7.12

Table 7.3.

Load Changes Due to the Consumption of Variable Weights

No	Description	Weight t	Lever from midship m	Moment tm
1	**Weapons** All ammunition spent Water to torpedo tubes Water to torpedo tube flooding tank Water to water-round-torpedo tank	$-P_{TOR}n_{TOR}$ $(P_{TOR}-\Delta P_{TOR})n_{TT}$ $P_{TOR}m_{RELT}$ $n_{TT}\Delta P_{TOR}$	x_{TT} x_{TT} x_{RELT} x_{WRT}	$-P_{TOR}n_{TOR}x_{TT}$ $(P_{TOR}-\Delta P_{TOR})n_{TT}x_{TT}$ $P_{TOR}m_{RELT}x_{RELT}$ $n_{TT}\Delta P_{TOR}x_{WRT}$
	Total	0		$-\Delta M_B$
2	Transfer of annular space water from torpedo tubes to WRT tanks.	$-P_{WRT}$ $+P_{WRT}$	x_{TT} x_{WRT}	$-P_{WRT}x_{TT}$ $P_{WRT}x_{WRT}$
	Total	0		$-\Delta M_{WRT}$
3	Fuel and oil Consumed fuel Difference between weight of compensation water and fuel weight at $\rho_T=0.85$ t/m³ and $\rho=1.03$ t/m³; $\Delta P=0.212P_T$ Consumed oil	ΔP $-k_{DOIL}P_F$	x_{FT} x_{FT}	$\Delta P\, x_{FT}$ $-k_{DOIL}P_F x_{FT}$
	Total	$\Delta P_1 = \Delta P - k_{DOIL}P_F$	$x_{TЦ}$	ΔM_T
4	Provisions and water	$-P_{PW}$	x_{PW}	$-\Delta M_{PW}$
5	Feed water	$-P_{FW}$	x_{FW}	$-\Delta M_{FW}$
—				
n	Regeneration cartridges	$-P_{RC}$	x_{RC}	$-\Delta M_{RC}$
	Total	ΣP_{ABL}	x_{ABL}	ΣM_{ABL}

where $\sum_{1}^{n}P_{ABL}$ and $\sum_{1}^{n}M_{ABL}$ – weight and trimming moment due to taken auxiliary ballast.

To achieve this balance and to find the auxiliary ballast volume, let us compensate for variable weights in Table 7.4.

Table 7.4.

Compensation of Variable Weights

No.	Description of loads and tanks	Weight t	Lever from midship m	Moment tm
1	Weapons Water to fore trim tank Water from aft trim tank	0 $+P_{FTT}$ $-P_{AFTT}$	 x_{FTT} x_{AFTT}	$-\Delta M_{TS}$ M_{FTT} $-M_{AFTT}$
	Total	0		$\Delta M_{TS=0}$
2	Transfer of annular space water Water to fore trim tank Water from aft trim tank	0 $+P_{FTT}$ $-P_{AFTT}$	 x_{FTT} x_{AFTT}	$-\Delta M_{WRT}$ M_{FTT} $-M_{AFTT}$
	Total	0		0
3	Fuel and oil Water from regulating tank No.2 Water from aft trim tank Water to fore trim tank	ΔP $-P_{RT2}$ $-P_{AFTT}$ $+P_{FTT}$	 x_{FTT} x_{AFTT} x_{FTT}	ΔM_F $-M_{RT2}$ $-M_{AFTT}$ M_{FTT}
	Total	0		0
4	Provisions and water Water to regulating tank No.2 Water to fore trim tank Water from aft trim tank	$-P_{PW}$ P_{RT2} P_{FTT} $-P_{AFTT}$	 x_{RT2} x_{AFTT} x_{AFTT}	$-M_{PW}$ M_{RT2} M_{FTT} $-M_{AFTT}$
	Total	0		0
5	Feed water Water to regulating tank No.2	$-P_{FW}$ P_{RT2}	 x_{RT2}	$-M_{FW}$ M_{FW}
—				
n	Regeneration cartridges Water to regulating tank No.2	$-P_{RC}$ P_{RT2}	 x_{RT2}	$-M_{RC}$ M_{RT2}
	Total	0		0

Table 7.4. makes it clear that when any individual variable weight is consumed and compensated for, the equilibrium of forces and moments is maintained. This means that in all cases the submarine will be trimmed.

Based on the variable weight compensation table, we now can draw up the final table which will allow us to determine capacities of auxiliary ballast tanks and the amount of initial trim water in them (Table 7.5).

Table 7.5

Tank Status Summary for Compensated Consumed Variable Weights

No.	Tank	Compensation cases						
		1	2	3	4	5	—	n
		Weapons	Annular space	Fuel and oil	Provisions and water	Feed water		Regeneration cartridges
1	Regulating tank No.1							
2	Regulating tank No.2	—	—	$-P_{RT2}$	P_{RT2}	P_{RT2}		P_{RT2}
3	Fore trim tank	P_{FTT}	P_{FTT}	P_{FTT}	P_{FTT}	—		—
4	Aft trim tank	$-P_{AFTT}$	$-P_{AFTT}$	$-P_{AFTT}$	$-P_{AFTT}$			—

This table covers all possible cases of both the consumption and the compensation of variable weights. Analysis of this table allows us to determine required tank capacities and the amount of initial trim water for every tank. When analysing the table it is necessary to pay special attention to the possibility of overlapping compensation cases, i.e. it is important to specify incompatible cases too.

Regulating tank No.1 capacity depends on variations of hydrological factors.

It should be noted that some of above-listed cases of variable weight compensation happen simultaneously (e.g., fuel and oil, provisions and fresh water, etc.). Therefore, it is impossible to consider them separately. This latter aspect calls for important corrections to estimated capacities of auxiliary ballast tanks and initial trim water amounts.

The initial trim water amount (ITW) is found as the sum of all cases when the water is removed from the subject tank plus a certain water margin assumed by the designer based on statistic data:

$$P_{ITW} = \sum_{i=1}^{i=n} (-P_i) + \Delta P \qquad (7.13)$$

Free volumes of each tank required for variable weight compensation are calculated as the sum of all compensation cases, when the water is taken into the subject tank plus a certain volume margin:

$$V_i = \sum_1^n \left(\frac{P_{vw}}{\rho}\right) + \Delta V \qquad (7.14)$$

The variable weight compensation analysis is completed by making general conclusions that must prove that the chosen number of auxiliary ballast tanks and their capacities will ensure submarine trimming in all practically possible cases of variable weight consumption.

Based on this calculation, they generate a Manual on variable weight consumption and compensation that becomes a part of the submarine document package. It is intended as a guide for the crew with regard to consumption and compensation of variable weights during submarine operation.

In particular, the Manual describes two methods for submarine trimming during operation [84]:

1) method of comparison against the previous trimming;
2) method of comparison against the normal load balance.

The first method is used routinely because it is very simple. It can be described as follows.

When calculating total changes in loads and their moments ΔP and ΔM, only those values of variable weights are entered into the trimming table that were consumed or taken aboard since the previous trimming. Summation of ΔP and ΔM values, may result in:

$$\Delta P \gtrless 0$$

$$\Delta M \gtrless 0 \qquad (7.15)$$

The amount of water in auxiliary ballast tanks is determined based on the solution of equations of forces and moments.

The second method is a verification procedure, and trimming with this method is performed periodically.

Into the trimming table are entered all variable weights present on the submarine at the time of calculation with indications of their longitudinal and transverse moments. They also calculate and enter the submarine residual buoyancy due to changes in water density and ballast, hull and coatings.

The total of actual weights as calculated in the trimming table is compared against the sum of all variable weights under the normal load balance. The difference between these values defines the total amount of water in regulating and trimming tanks.

In order to compensate for the trimming moment, the estimated amount of water must be distributed between the tanks in such a way that the total moment of this water weight would be equal to the difference between the constant moment of all variable weights under normal load and the final actual moment of weights on the submarine. Obviously, water distribution among auxiliary ballast tanks should be done taking into account the trimming model (segregated or shared). One should remember that with the segregated arrangement, the amount of water in trim tanks is always constant.

8. SUBMARINE HULLFORM DESIGN

8.1. Selection of the Architectural and Structural Type

The determination of submarine main dimensions is preceded by one of the most crucial decisions - the choice of hull architecture and principles of weapon and equipment general arrangement. These design choices almost completely govern all later opportunities to vary individual elements to obtain the highest combat efficiency without exceeding the assigned displacement. The architectural type is selected based not only on the requirements set for the submarine, and individual service conditions, but also on the designer's experience and design school.

In practical underwater shipbuilding, «architecture» is understood as the total combination of general design solutions which determine specific features of the overall configuration, hullform and structures, sail, appendages, arrangement of rooms, tanks, major weapons and equipment of the submarine 28.

The major submarine architecture elements are usually considered to be:
– architectural and structural type of submarine;
– shape of the outer hull and appendages;
– number and location of shaft lines.

The term «architecture» may also include other specific features of the submarine which affect her external appearance [82].

Depending on whether the submarine has an outer (light) hull, the architectural and structural type may be:
– *single-hull*, when there is no outer hull anywhere along the submarine entire length. Ballast tanks on submarines of this type

are located in the ends (Fig.8.1a) and, if necessary for surface floodability, inside the pressure hull;
- *double-hull*, when the outer hull completely covers the pressure hull along its entire length. In this case the main ballast tanks are placed between the hulls (Fig.8.1b);
- *side-tanks submarine* (better known internationally as «combined» type), when single-hull and double-hull structures are arranged alternatively along the pressure hull. (Fig.8.1c);
- *multihull*, when the outer hull completely covers two or more pressure hulls. This is typical for underwater cargo vessels (Fig.8.1d).

a)

b)

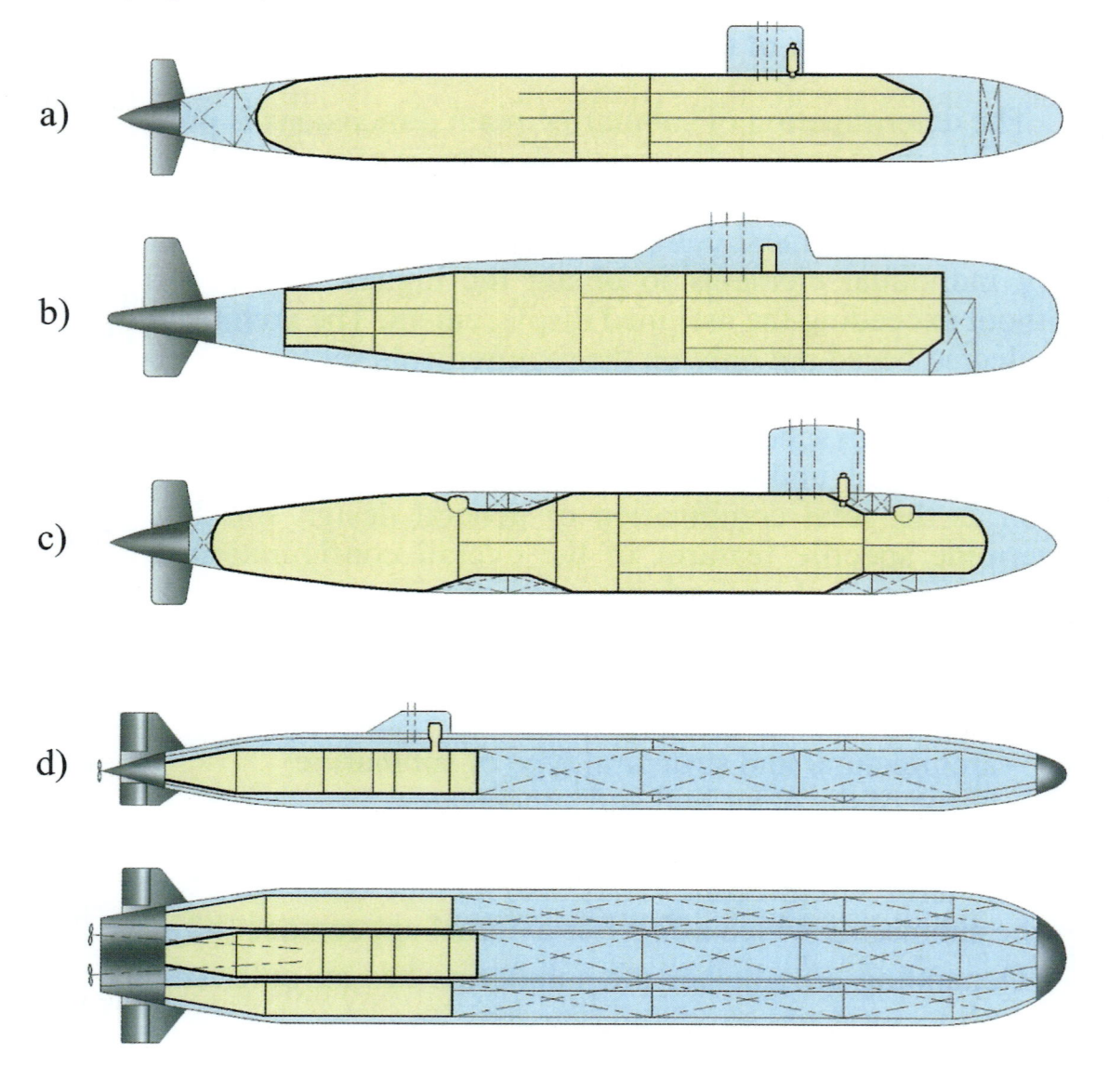

c)

d)

Fig.8.1. Architectural and Structural Types of Modern Submarines

An architectural and structural type of a submarine is to be determined at the initial design stage because it directly influences the displacement and other principal particulars of the boat.

Let us consider some aspects which should be taken into account when selecting an architectural type. First of all, it is necessary to make a decision about the surface floodability in order to be able to assign a first-approximation value of reserve buoyancy.

Submarines of any architectural and structural type have end main ballast tanks. The share of reserve buoyancy provided in end MBTs varies with the displacement. Smaller submarines ($D_0 \approx 400$ to 500 t) may have of up to 50% of reserve buoyancy in the end tanks, larger submarines - much less. It is obvious that to design a single-hull submarine with sufficient surface floodability is rather a difficult task.

There are also problems with distributing the reserve buoyancy on double-hull submarines of smaller displacements. The minimum width of the between-hull space depends on design and technological aspects and this results in high reserve buoyancy and, consequently, in limitations on some technical and tactical characteristics of the submarine.

International practice reveals two approaches to submarine surface floodability.

When developing nuclear submarines of the first generation «Nautilus», «Skate» (Fig.8.2), their American designers have chosen the side-tanks architecture.

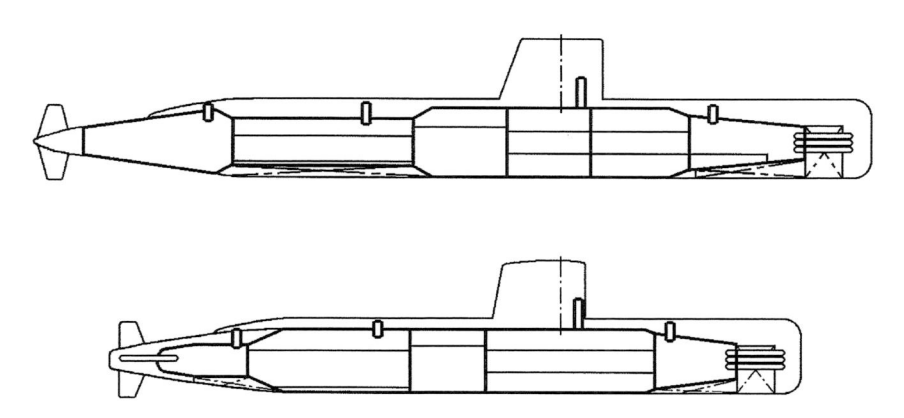

Fig.8.2. Schematic Architecture of First-Generation US Submarines

That decision was motivated by the desire to reduce the full submerged displacement for the sake of higher speeds with the available power of the plant. Besides, it was no longer necessary to provide the submarine with a sizeable fuel capacity.

The choice of the side-tanks architecture has forced us to reconsider the approach to surface floodability and, as a result, to reject the single compartment flooding standard. While other factors have sort of evolved naturally, in this aspect there was a certain jump in submarine architecture philosophy leading to a new quality level.

As a result, the reserve buoyancy was reduced from 25~35% of the normal displacement, which was typical for diesel submarines, to 14~16%.

Unlike in the USA, Soviet nuclear submarines of the first generation, Type «Leninsky Komsomol» (Fig.8.3), have retained the double-hull architecture.

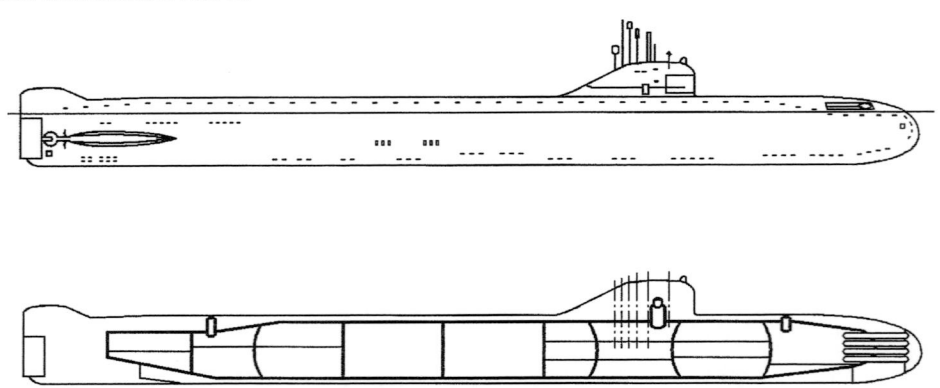

Fig. 8.3 General Appearance and Hull Configuration of the First Nuclear Submarine «Leninsky Komsomol»

This was determined by the fact that there were no doubts about the need to meet the single compartment flooding standard. Actually, Navy specialists even insisted that a damaged submarine should be able to surface from the sea floor with any one compartment of the pressure hull completely flooded. This has resulted in the need not only to increase the reserve buoyancy, but also to increase the strength of compartment bulkheads and the amount of HP air 5.

Another stimulus for continued double-hull design of Soviet submarines was the possibility to achieve well-streamlined outer hull shapes. This type allows greater freedom in designing both the pressure and outer hulls.

When selecting the architectural type, it is also necessary to consider the stability. Single-hull submarines normally have good submerged stability.

The reduction of single-hull submarine stability on the surface is due to the small area of the effective waterline and, as a result, small values of the moment of inertia. It is especially difficult to ensure surface stability for submarines of small displacements. E.g., on the WWII XIIth series of «Malyutka» submarines, in order to get satisfactory surface stability, they had to put such an amount of solid ballast that it was $P_{BAL} \approx 12\%D_0$ [35].

On double-hull submarines the width of the effective waterline is greater thanks to main ballast tanks in the between-hull space, and therefore surface stability characteristics are better. The submerged stability of double-hull submarines is achieved by proper equipment and ballast arrangement.

The choice of submarine architecture and structural design may also be influenced by design considerations. For example, the double-hull option allows external frames to be used on the pressure hull, and thus means some advantages for locating equipment inside the pressure hull.

The double-hull also reduces the effects of some antisubmarine weapons, and thus enhances the survivability.

Design specifications for smaller submarines may include a requirement to transport them by railway. It is not easy to decide which architectural type has better advantages in this particular case because such transportation may involve dismantling outer structures or cutting the submarine into blocks.

The majority of submarines constructed in the world during the last two decades have either single- or double-hull architecture. Each school followed the way of its own nation. In the USA they continued to develop the single-hull option while in Russia they stuck with the double-hull one.

«Los Angeles»- type submarines (Fig.8.1a) have become the first truly single-hull nuclear submarines. There is neither an outer hull nor a superstructure along the entire pressure hull. The pressure hull has conical shell ends forward and aft to which end portions are secured. This design also reveals a change in the approach to pressure bulkheads: their number was reduced.

The weight which used to be allocated for pressure bulkheads can now be utilised more rationally for other tasks, particularly, to improve habitability and to provide better arrangement and servicing conditions for machinery and systems 28.

Russian submarines have till recently always been of the double-hull architecture.

However, for the IVth generation submarines they have chosen the side-tanks (combined) architecture. This has allowed surface floodability with lower reserve buoyancy (18 to 20% D_0) values.

8.2. External Hullform and Appendages

Submarine hydrodynamic qualities determine the capabilities to exercise 3D manoeuvres at a required speed. These qualities (performance and manoeuvrability) in turn determine the submarine's ability to reach the area of combat operations within the specified time, to fight and to avoid enemy antisubmarine forces and weapons. Therefore, hydrodynamic aspects, which usually are in good accord with acoustic ones, are dominant priorities in submarine hullform design and location of appendages [63].

The process of submarine evolution has generated certain design principles that by now have became fundamentals for everything associated with hydrodynamic qualities:
- indisputable priority of submerged performance;
- maximum possible submerged speed at the specified power plant output;
- high manoeuvrability within the entire speed range;
- optimum conditions for the propulsion;
- minimum flow noise contribution to the submarine acoustic signature and platform noise hampering the onboard sonar performance.

Hydrodynamic qualities depend on the submarine external configuration and on elements of the so-called external architecture: the hullform and the location of appendages. These features are described by a set of parameters that may be subdivided into two groups. The first group describes the hullform as a whole and the other characterises the shape of hull details and architectural elements (sail, stern control surfaces) and their location on the hull.

The main parameters (the first group) include:
- hullform as a whole;
- hull relative length defined either as the ratio of the length to the full submerged displacement $\ell = L/V_{FS}^{1/3}$ or as the length-to-beam ratio $\ell = L/B$, as a particular case characteristic of the submerged condition;
- midsection aspect ratio: hull height to breadth H/B;
- relative lengths of fore and aft ends and of the parallel middlebody L_{FWD}/L; L_{AFT}/L; L_{SYL}/L;
- form factors (block coefficient δ, prismatic coefficient φ and midship coefficient β).

Local parameters of individual hull detail shapes (the second group) include:
- angle of run α;
- fore and aft end coefficients δ_{FWD} and δ_{AFT};
- shape, dimensions and longitudinal position of the sail;
- type, dimensions and longitudinal positions of stern control surfaces and forward planes.

The whole evolution of submarine architecture vividly illustrates trends of external appearance development and may serve as a didactic example of designers' efforts to achieve dialectical unity between the engineering contents and the form into which the contents are shaped.

This requires a timely-developed, scientifically-justified concept of future hullform design. T.K.Gilmer, a well-known US expert in the field of design also believes that, in spite of the multiplicity and complexity of various systems and subsystems and of their importance for the ship, there is no characteristic more important than the geometrical shape of the hull [23].

Analysing the development of approaches to hullform and appendage design, we can identify several stages in which the composition of factors affecting the design and related criteria have been considerably changed.

At the first stage, which may be regarded as the period before the wide use of nuclear power plants, the hullform was determined primarily by surface performance, manoeuvrability and seakeeping requirements. This was a direct result of limited power plant outputs that were available for submerged propulsion and dictated submarine tactics [85].

Another important factor was the poor development of antisubmarine defences that allowed submarines to actively operate on the surface.

Theoretical grounds for submarine hullform parameters were at that time derived from numerous studies on fast surface ship performance and manoeuvrability.

The aim of this approach to hullform design may be expressed as obtaining the maximum surface speed.

$$\vartheta \Rightarrow \max \vartheta_{SURF}(X_i)$$

$$X_i \in X_{LIM} \qquad (8.1)$$

where ϑ_{SURF} – full surface speed;
X_i – outer hullform parameters.

The result of such an approach was that submarines had hull-forms typical of fast ships (Fig.8.4)

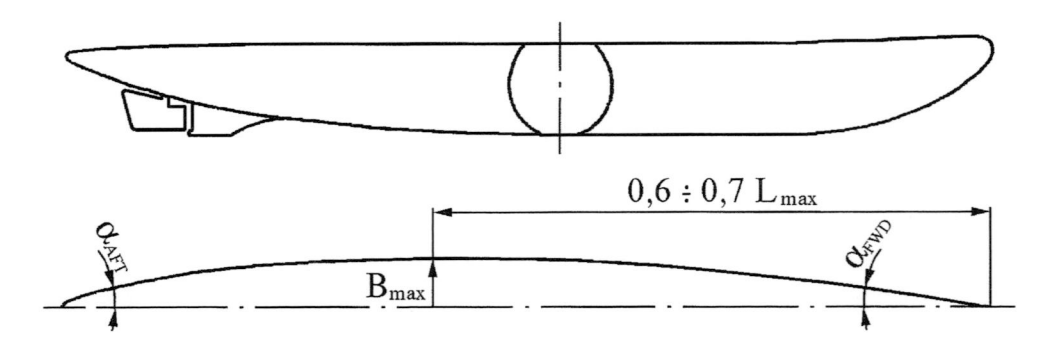

Fig.8.4. Stem-Shaped Submarine Hullform

The outer hull used to feature:
- stem-shaped ends;
- high length-to-beam ratios, L/B up to 13 to 14, and relative lengths $L/V_{FS}^{1/3} > 5.5\sim6.0$;
- sharp fore and aft lines (entrance angle α_{FWD} = 4 to 5°, run angle α_{AFT} = 7 to 8°);
- the largest-area cross-section frame offset aftwards from midship;
- flared and elevated sides in the bow;
- large flat decks, numerous parts poorly streamlined for submerged conditions.

The breakthrough in converting a submarine from a diving boat into a true underwater ship has been possible because of nuclear power plants. Their introduction has offered numerous possibilities to improve performance and manoeuvring capabilities which could only be fully implemented on submarines with hullforms adapted to the maximum possible extent for sailing under the water. The maximum underwater speed has come to be considered as the main dynamic characteristic. Later this design approach was extended to diesel submarines of the IIIrd and IVth generations because today a submarine sitting on the surface is doomed to be destroyed. Fig.8.5. illustrates this trend in maximum surface and underwater speeds for diesel submarines of various generations [48].

Thus, in accordance with the concept of high performance and manoeuvring qualities, designers tried first of all to increase the maximum underwater speed, at the same time satisfying certain ship handling criteria [63]:

$$\vartheta \Rightarrow \max \vartheta_{SUBM}[X_i(X_j)]$$

$$X_i \in X_{LIM} \qquad (8.2)$$

$$X_j \in X_{jLIM}$$

where ϑ_{SUBM} – full submerged speed;
X_i – parameters of hullform and control surfaces;
X_j – manoeuvrability criteria.

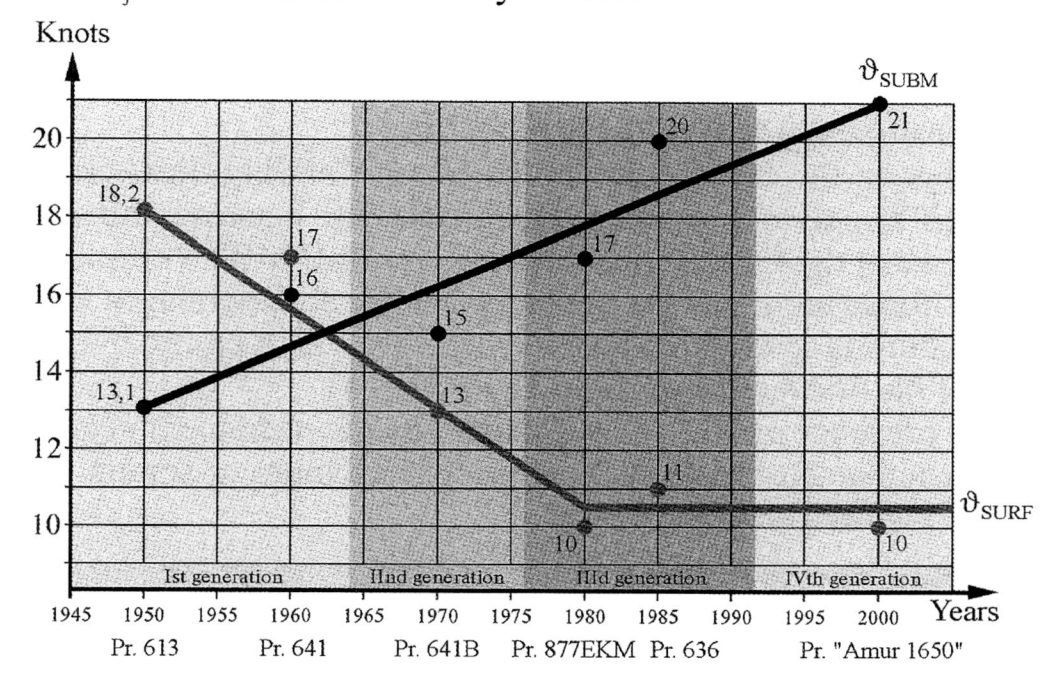

Fig.8.5. Maximum Surface and Submerged Speed Variation Patterns for Diesel Submarines of Different Generations

The optimum shape for submerged propulsion has long been known to specialists of hydromechanics (Fig.8.6.).

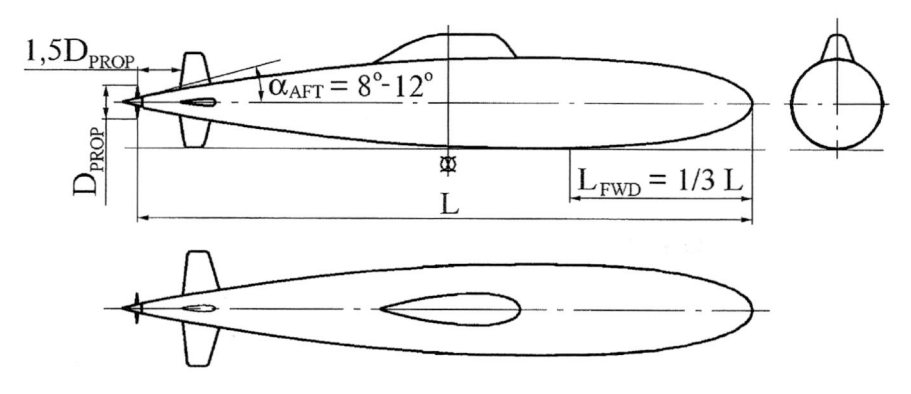

Fig.8.6. Axisymmetric Submarine Hullform

This is a body of revolution with its fore end shape close to a tri-axial ellipsoid. The aft end is very sharp $\alpha_{AFT} \approx 8$ to $12°$. The fullest section is shifted forward from midship and located approximately at 0.3 to 0.4 L from the bow perpendicular. L/B ratios of such hulls are quite low and amount to ≈ 7 to 9 at $L/V_{FS}^{1/3} \approx 4.8$ to 5.3.

However, it is much easier to draw an ideal shape that will provide the least submerged resistance than to design a real submarine hull. There are several reasons for this.

First, there are problems with arranging equipment in compartments of a hull optimised in terms of the L/B ratio.

Secondly, the narrow aft end makes it difficult to fit rudder and plane actuators together with the shaft line.

Besides, in the case of aft compartment flooding there are problems with negative buoyancy compensation because the narrow space between the hulls in the aft means a small MBT volume.

The submarine outer hull usually has local appendages like the sail, the conning tower and control surfaces, which contradict general laws of hullform optimisation (Fig.8.7) [45]

Their presence on the hull is dictated by the need to solve a whole set of particular problems such as the arrangement of sonars, masts, weapons, rudders and planes, etc.

Fig.8.7. Sail and Conning Tower Arrangement on an Axisymmetric Submarine Hull

Such deviations from the ideal naturally increase submarine resistance and reduce stealth qualities, but they are necessary for normal operation.

Let us formulate the task of selecting submarine hullform main parameters and dimension ratios as follows:
1. The optimum hullform for underwater propulsion is known.
2. Some hullform design choices inevitably have to disagree with performance-wise recommendations.
3. It is necessary to investigate effects of variations in main parameters upon performance and manoeuvring qualities so that the designer would realise the price to be paid for deviations from optimum values.

8.3. Effects of L/B and H/B Ratios on Submarine Performance and Manoeuvring Qualities

In the theory of ship design the relative length ℓ is understood as the relationship describing the waterline run

$$\ell = \frac{L}{\sqrt[3]{V}} = \sqrt{\frac{L^3}{\delta LBT}} = \ell_B^{2/3} \cdot (b_T/\delta)^{1/3} \qquad (8.3)$$

It is evident from this formula that the relative length is affected by such characteristics as hull fullness and fineness as ratios $\ell_B = L/B$, $b_T = B/T$ and the block coefficient δ.

V.L. Pozdyunin has noted [67] that none of these characteristics separately could give a complete idea about hullform fineness while the relative length ℓ is a characteristic of fineness, and therefore should be used to estimate submarine performance.

At the same time, many publications applicable to submarines often refer not to the relative length ℓ, but to the L/B ratio which they call «submarine elongation». To what extent is such a substitution justified for performance investigations? As a matter of fact, when investigating the relative length influence on submarine submerged seakeeping abilities, the draught in (8.3) is substituted by the hull height H, which for a body of revolution is equal to the beam (H = B). In this case formula (8.3) is transformed into

$$\ell = \frac{L}{\sqrt[3]{V_{FS}}} = \delta^{-1/3} \cdot \left(\frac{L}{B}\right)^{2/3} \qquad (8.4)$$

that enables us to analyse L/B variation effects upon submarine sea-keeping abilities.

It is known that the performance is understood as the submarine's ability to run at a specified speed with minimum required power.

In general, submarine speed, and the power plant output required to reach it, are related as:

$$N = \xi \frac{\rho \vartheta^2}{2} \frac{\Omega_0}{\eta_H \eta_{PROP}} \quad (8.5)$$

where N – shaft power;
ξ – resistance coefficient;
Ω_0 – bare hull wetted surface area;
η_H – hull efficiency factor;
η_{PROP} – propeller efficiency.

Thus, performance qualities of the submarine depend on the resistance (ξ), on the propulsor efficiency and interaction with the hull (η_H; η_{PROP}), and on the area of wetted surface (Ω_0), i.e. on submarine dimensions.

The most essential components for the submarine «bare hull» (without appendages and the sail) are submerged friction resistance and form resistances. For modern submarines with well-streamlined axisymmetric hulls, the friction resistance coefficient represents 50 to 60% of the total drag. From naval architecture we know that the friction resistance is found from Schlichting's formula for an equivalent plate and then is corrected with the help of coefficients for hull surface roughness and curvature:

$$R_{FRIC} = \frac{0,455}{(\log Re)^{2,58}} \frac{\rho \vartheta^2}{2} \Omega_0 \quad (8.6)$$

where $Re - \frac{\vartheta L}{\nu}$ – Reynolds number;
ν – kinematic viscosity coefficient, m^2/s

The second component - the form resistance is due to water viscosity and reflects the influence of the hullform and main dimension ratios on the pressure distribution along the ship.

Taking into account the common viscosity-related origins of friction and form resistances, for streamlined bodies the form resistance is usually expressed through the friction resistance:

$$R_F = k_F R_{FRIC} \quad (8.7)$$

where k_F – non-dimensional coefficient relating the form resistance to the friction resistance. This factor may be found from V.F.Droblenkov's nomogram (Fig.8.8) [29].

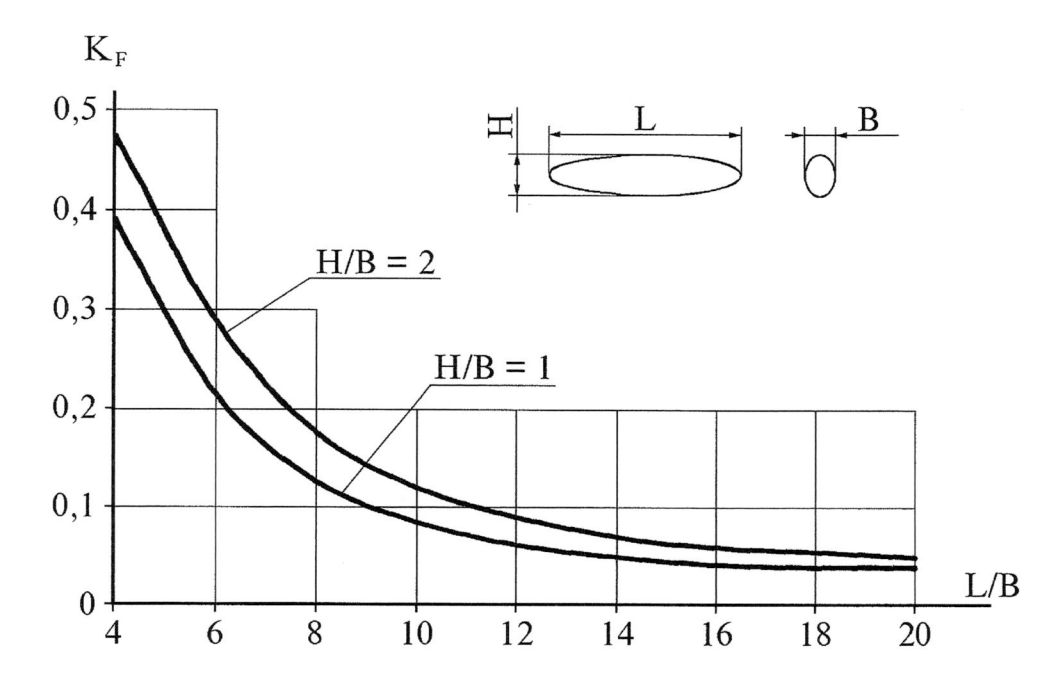

Fig.8.8. V.F.Droblenkov's Curves of Design Values of the kF Coefficient for Streamlined Bodies Depending on their Relative Elongation L/B and Aspect Ratio H/B.

From the above Figure we may see that the range of $k_Ф$ variations is 0.1 to 0.2 for the subject range of submarine elongations. This means that the form resistance value does not exceed 10 to 20% of the friction resistance. It is also obvious that an increase in hull elongation up to L/B = 10 to 14 results in form resistance reduction.

At the same time, the elongation increase results in the growth of the wetted surface and, consequently, the friction resistance.

Submarine elongation growth results in increasing the Reynolds number and decreasing the friction resistance but its influence is considerably less than the effect of the wetted surface area. These oppositely-directed effects of submarine hull elongation on viscous resistance components suggest that there should be such a value of elongation when the resistance value is minimum (Fig.8.9) [28].

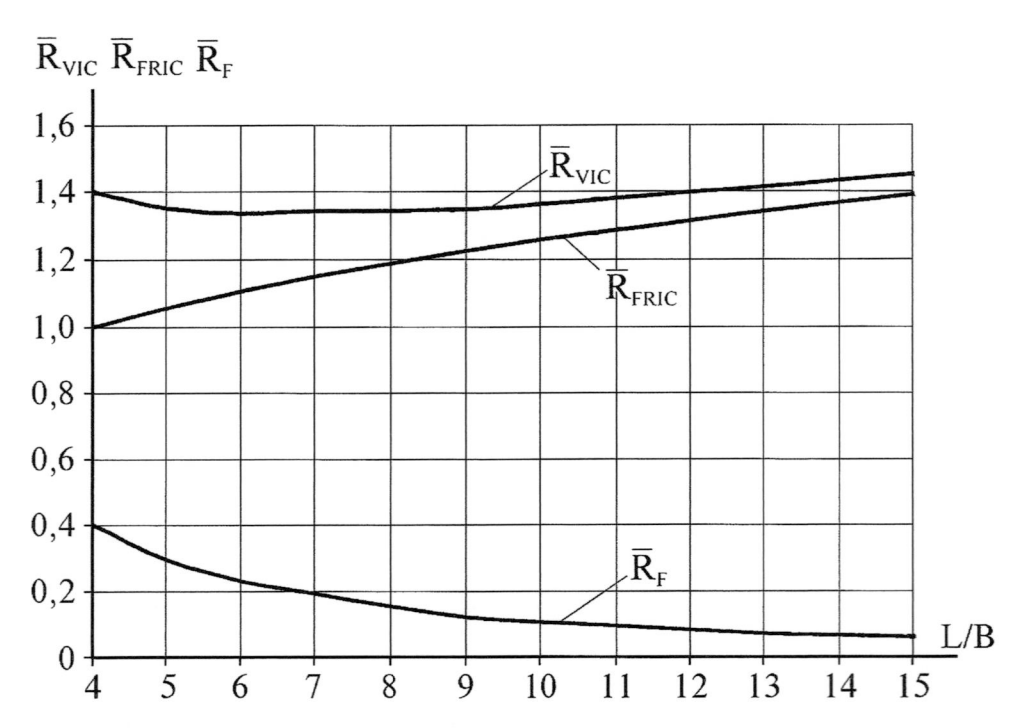

Fig.8.9. Submarine Hull Viscous Resistance Versus Relative Elongation L/B

$\overline{R}_{FRIC}$ – relative friction resistance;
$\overline{R}_F$ – relative form resistance;
$\overline{R}_{VIS}$ – relative viscous resistance;
$\overline{R}_{VIS} = \overline{R}_{FRIC} + \overline{R}_F$;

Notes: 1. Relative values of all resistance components were obtained as their ratios to the friction resistance at hull elongation L/B = 4;
2. These curves were plotted assuming a constant full submerged displacement.

The hull oval cross-section aspect ratio (H/B ≳ 1) also increases the wetted surface. Quantitatively, the relative wetted surface area $\overline{\Omega} = \Omega/V^{2/3}$ as a function of H/B at different L/B ratios of streamlined bodies may be seen in Fig.8.10 [29].

The Figure shows that $\overline{\Omega}$ and, with equal V, Ω increase both with the elongation and with deviations from the circular cross-sections.

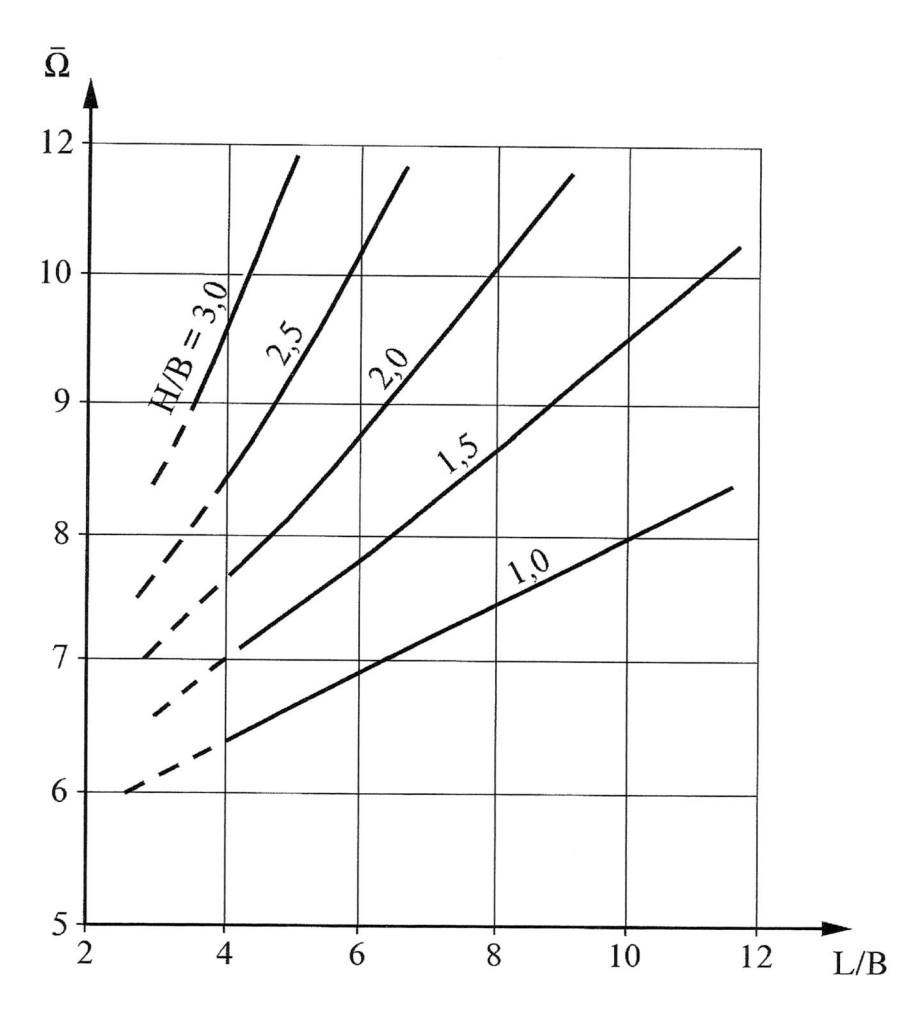

Fig.8.10. Relative Wetted Surface Areas Versus Main Dimension Ratios L/B and H/B for Streamlined Bodies

Comparative calculations show that when hull cross-sections deviate from the circular form, full viscous resistance values increase and optimum values of L/B are noticeably shifted towards lower end (Fig.8.11).

Summarising the above comments, we may say that from the point of view of the performance there exists a certain optimum submarine hull elongation and any deviation from it to either side results in greater resistance and, consequently, higher required power. However, the smooth $R_{vis} = f(L/B)$ curve in the minimum domain (see Fig.8.9) allows hull elongations to vary within L/B = 7 to 12 without considerable increases in the power because in this case changes in the resistance do not exceed 10% [51].

199

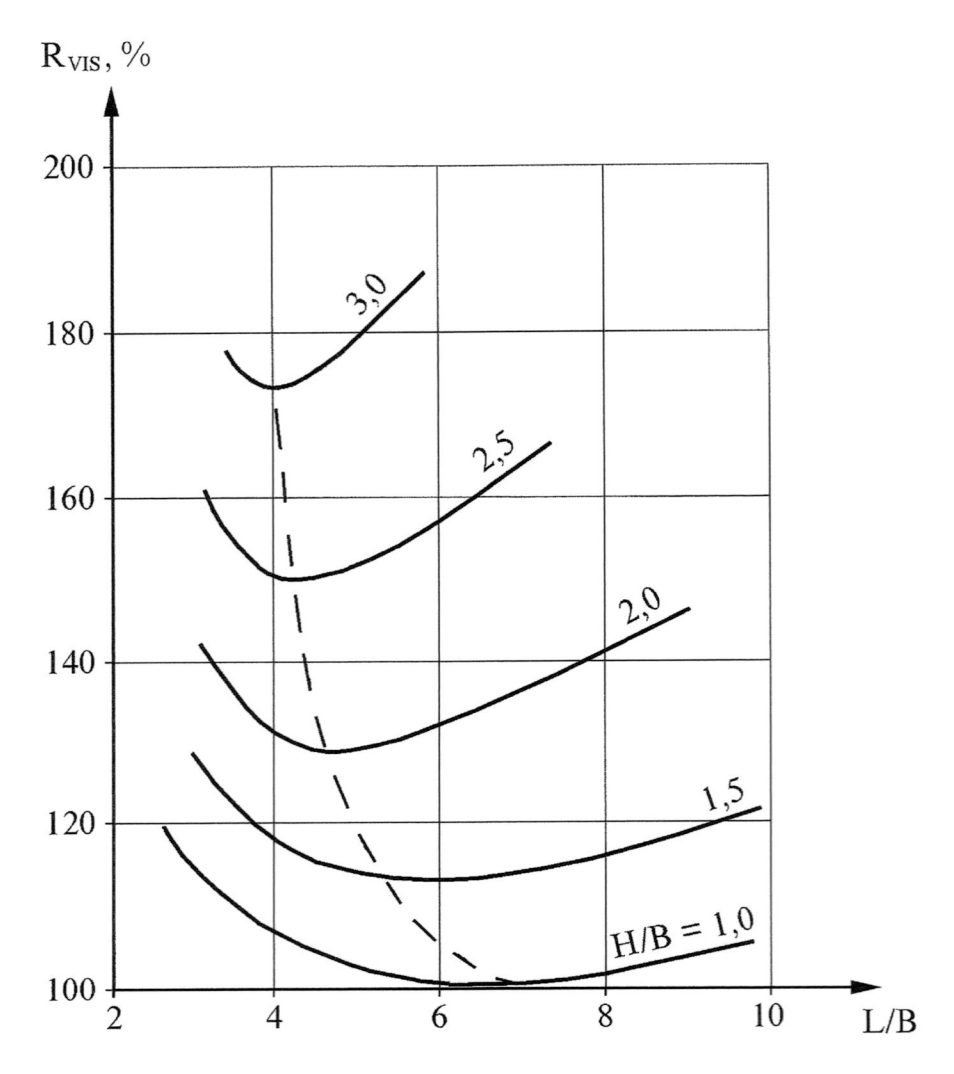

Fig.8.11. Viscous Resistance Percentage Increments due to Main Dimension Ratio Variations around Optimum Values (H/B ≈ 1.0 and L/B ≈ 7.0) for Streamlined Bodies.

The outer hull elongation also affects submarine manoeuvrability parameters. Thus, with the increase in the L/B ratios submarine static and dynamic stability at steady motion also increase. And this means that with an increase in the elongation, all other conditions, including the dynamic stability coefficient $k = \dfrac{\overline{b}^{\omega}}{\overline{b}} = $ const (where $\overline{b}^{\omega} - \overline{b}$ reduced lever of damping forces; $\overline{b}$ – lever of resultant hydrodynamic forces), unchanged, the area of control surfaces should go down.

Fig. 8.12 shows variations in the relative area of submarine control surfaces with the L/B value at k = const for plane surfaces.

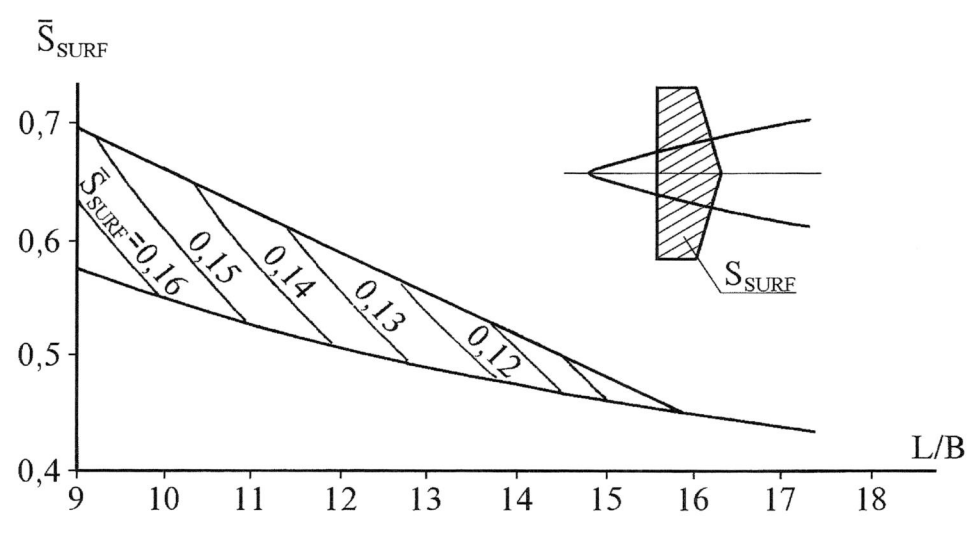

Fig.8.12. Submarine Elongation Effects upon
the Relative Area of Hydroplanes

As regards the influence of the H/B ratio on submarine manoeuvrability, we should note the following. For a body of revolution with H/B=1, bare hull hydrodynamic characteristics in both the vertical and the horizontal planes are the same. However, it is usually impossible to make a completely axisymmetric hull. Due to the need of arranging certain equipment, a sort of superstructure always appears which is especially large on missile submarines due to their specific arrangement requirements. Nobody has so far managed to do without a sail for masts and a pressure access trunk on any kind of submarine. To meet operating requirements we have to provide a flat superstructure deck. The breadth of such a deck should be as small as possible. This is due to the fact that transitions from the flat deck to hull sides inevitably produce systems of vortices that increase the resistance and change hydrodynamic characteristics. E.g., when surfacing with large angles of attack (emergency surfacing when the speed is lost), i.e. actually when the flow is transverse, there appear hydrodynamic forces which may result in high heel angles, especially if the sail is big. Cross-section shape variations $H/B \gtrless 1.0$ considerably affect hydrodynamic and course stability characteristics, which at specified criteria of static and dynamic stability affect the required area of control surfaces. Calculations indicate that when the hull is flattened in the vertical direction, $H/B <1$, the vertical plane course stability increases and, consequently, the required area of fixed horizontal control surfaces is reduced (Fig.8.13). However, it should be kept in mind that then the area of tiltable control surfaces (hydroplanes) has to be increased to save the original turnability [72].

$$\bar{S}_{SURF} = \frac{S_{SURF}}{V^{2/3}}$$

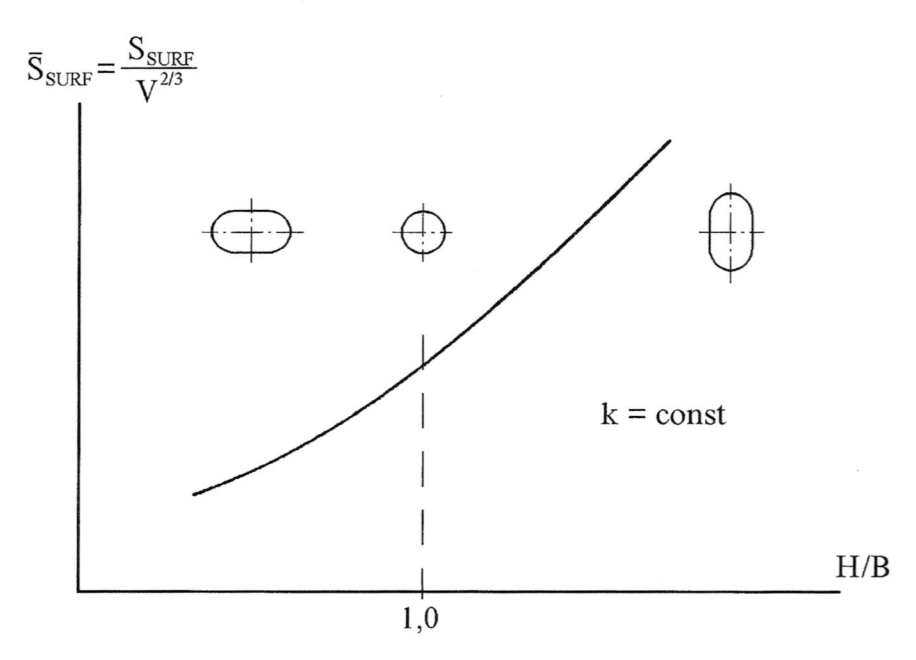

Fig.8.13. Relative Area of Horizontal Control Surfaces Versus H/B.

Naturally, similar comments could be made about the influence of H/B on the area of vertical control surfaces.

8.4. Effects of Relative Lengths of the Parallel Middlebody, Fore and Aft Ends upon Submarine Performance

Besides the effects of main ratios, the form resistance is sensitive to hull fullness and the shapes of its ends.

This is due to the fact that in its physical nature the form resistance represents the resultant of pressure forces distributed in the flow around the hull. The pressure force profile is determined by the hullform, but even for streamlined bodies of revolution with attached flow there exists considerable pressure gradients in the fore and aft hull portions. In the bow portion, where the shape is characterised by an abrupt increase in cross-section area, the inflow velocity notably increases leading to flow pressure reduction and turbulisation of the boundary layer. At the stern this process is reversed: the velocity reduces with a smooth gradient and the pressure respectively increases. Due to water viscosity and velocity losses in the boundary layer, pressures in the aft portion turn out to be less than in the forward one. Pressure gradients are also generated in way of appendages (the sail, aft control surfaces, domes of various devices, etc.) that cause vortexes. This difference in hull fore and aft end pressures results in the form resistance.

Fig.8.14. shows pressure profiles for bodies of revolution of three shapes: parabolic, elliptic and elliptic with a cylindrical middlebody. As may be seen from these curves, the least pressure gradients and, consequently, the least form resistance belong to the elliptical option without the cylindrical middlebody[5].

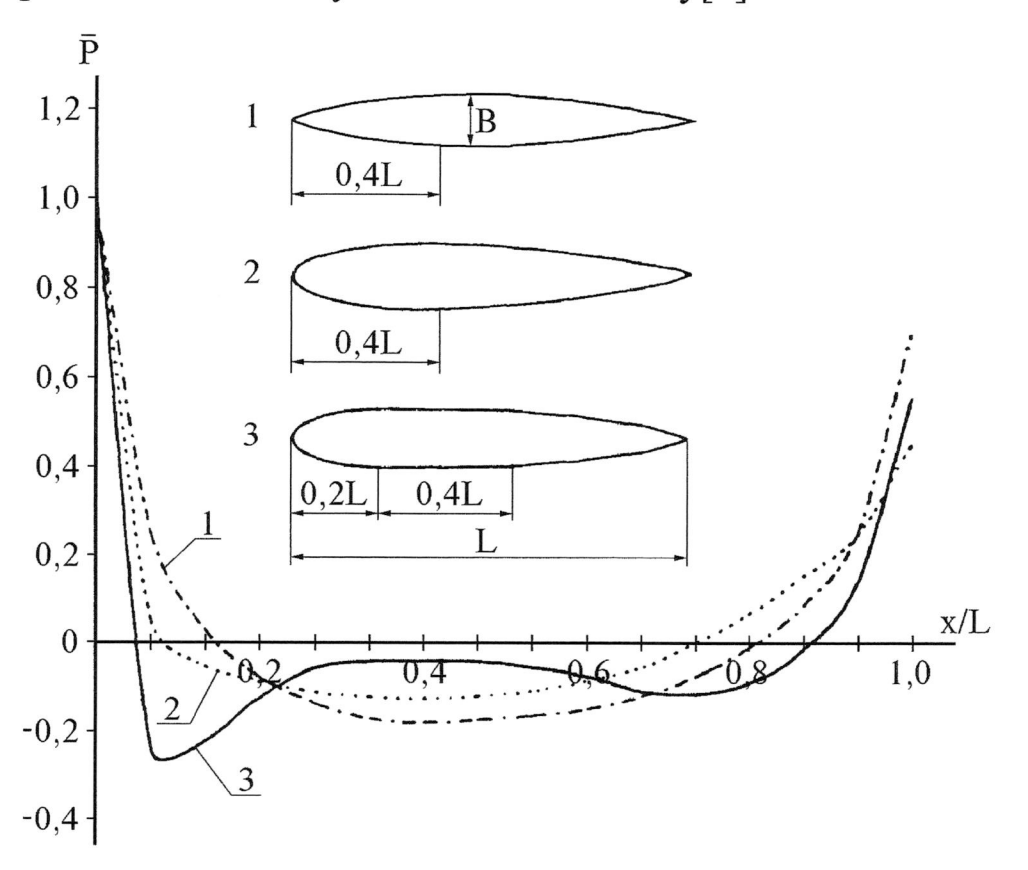

Fig.8.14. Flow Pressure Profiles for Bodies of Revolution, Relative Length L/B = 7.

1 –parabolic shape ; 2 –elliptic shape ; 3 –elliptic shape with cylindrical middlebody.

Notes: $\overline{P}$ – relative flow pressure that is the ratio of the pressure at an arbitrary cross-section to the dynamic head;

x/L – relative longitudinal co-ordinate.

On the other hand it should be noted that if the outer hull volume has been specified, the cylindrical middlebody reduces the submarine length, and thus the wetted surface area and the friction resistance.

Introducing a cylindrical middlebody and reducing the length of variable-section fore and aft portions results in increasing pressure differentials at submarine ends (Fig.8.15.), and hence in increasing the form resistance. [82].

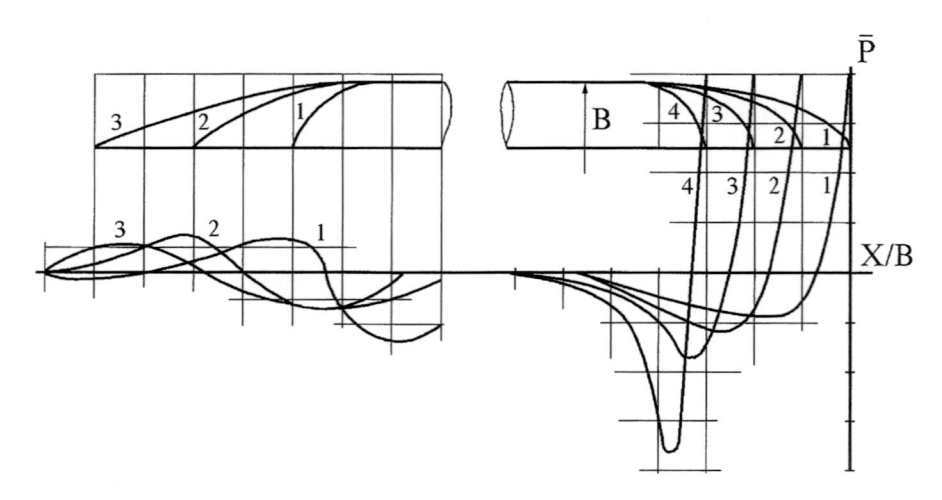

Fig.8.15. Flow Pressure Variations for a Body of Revolution due to Fore and Aft End Length Variations

However, practical experience shows that the reduction of the wetted surface area exceeds the form resistance increase due to the increase in the prismatic coefficient φ when L_{CYL}/L increases and eventually the resistance R is reduced.

The availability of the cylindrical middlebody and the possibility to shift it along the hull also allows us to influence submarine full-buoyancy trimming in the process of design.

Besides the resistance, submarine propulsive performance qualities also depend on propulsor-hull interaction. For a submarine with propellers (and submarines are mostly equipped with such propulsors) the propulsion coefficient in expression (8.8) is:

$$\eta = \eta_{PROP} \cdot \eta_H = \eta_{PROP} \cdot \frac{1-t}{1-w} \qquad (8.8)$$

where η_{PROP} — propeller efficiency;
η_H — hull-efficiency factor;
t — thrust deduction coefficient;
w — wake coefficient.

The propeller-hull interaction is determined by two processes: propeller operation in the wake due to the water mass travelling together with the hull, and stern flow acceleration due to propeller thrust deduction [51].

The wake in the propeller disk area is determined by the submarine aft portion boundary layer which in turn depends on the submarine hullform and absolute dimensions. The wake makes the propeller inflow velocity less than the absolute speed of the submarine. This has a positive impact on propeller operation conditions and leads to some increase in its efficiency as compared to the efficiency of the same propeller in «open water».

This propeller efficiency increment depends on the ratio of its diameter to the wake width. In practical design calculations this ratio is substituted by the propeller diameter to hull beam ratio. With the increase of this ratio within the realistic range from 0.3 to 0.6~0.7, the wake effect increases and results in a higher propeller efficiency. However, practical chances to increase the propeller diameter depend on power plant parameters (shaft revolutions) and there is only a limited range for optimisation.

The thrust deduction, contrary to the wake, has a negative effect because it leads to some increase in the resistance. The reason for this phenomenon is the effect of stern flow acceleration due to the operating propeller. This results in the growth of friction resistance at the stern and the reduction of the flow pressure on the hull aft portion makes the form resistance higher. The domain of the thrust deduction effect is not large and is limited to a distance equal approximately to two propeller diameters.

For submarines, the hull-efficiency factor depends mostly on the aft end shape and on the hull elongation. For single-shaft submarines, the main parameter which determines the aft end fullness and length is the angle of run, i.e. the angle formed by the hull line and the longitudinal axis. As a rule, the angle of run is limited by the range of α_{AFT} 8 to 12° (to one side) due to the desire to achieve separation-free flow.

The thrust deduction coefficient increases together with the angle of run (Fig.8.16) [5].

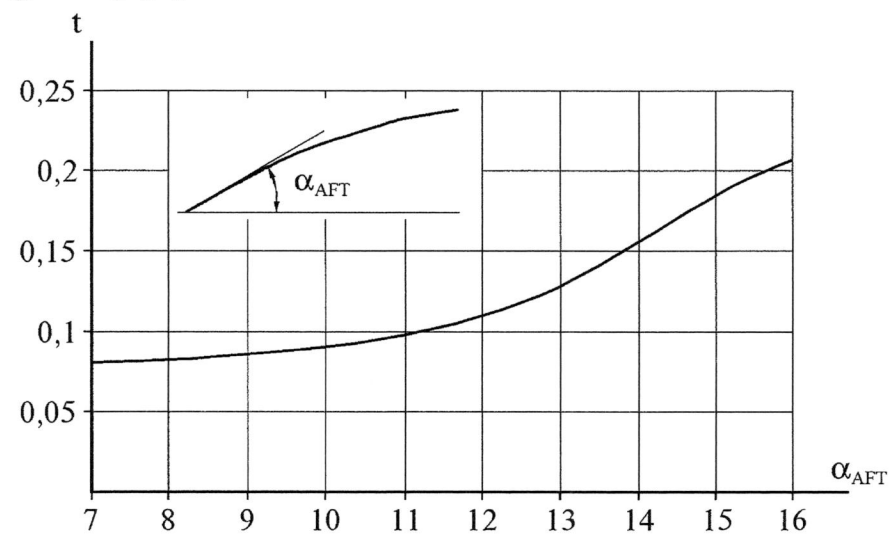

Fig.8.16. Thrust Deduction Coefficient t versus Angle of Run aAFT

Note: This function was plotted for a propeller with a relative diameter $\bar{d} = d/B = 0.4$
where d – propeller diameter;
       B – submarine hull beam.

As may be seen from this Figure, the thrust deduction coefficient is minimum within angles of run $\alpha_{AFT} = 7$ to $9°$.

The wake is mostly governed by the hull elongation (Fig.8.17).

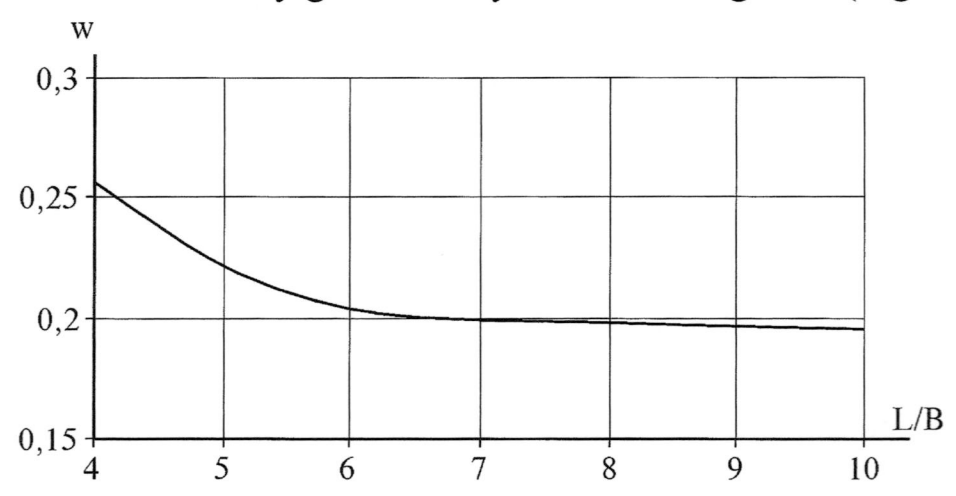

Fig.8.17. Wake Coefficient w versus Hull Relative Elongation L/B.

Note: This function was plotted for a propeller with a relative diameter $\bar{d} = d/B = 0.4$

With increases in the submarine hull relative length, the wake coefficient becomes less and reaches its minimum at L/B within 8 to 10. Therefore, it is advisable to aim at L/B ratios from 6 to 8 that are also optimal in terms of friction and form resistances.

Provided all these conditions for thrust deduction and wake coefficients are observed, the hull-efficiency factor that describes their joint effect will, for single-shaft submarines, be within 1.14 to 1.16 (Fig.8.18) and will ensure propulsion coefficient values of 0.75 to 0.80 [5].

It should be noted that with the increase in the aft end angle of run and fullness its stiffness also grows and that has a favourable effect on vibration characteristics. At the same time, however, it complicates the task of arranging stern control surfaces. The greatest problems arise with cruciform fins because the lower vertical rudder may protrude below the base plane, i.e. go beyond hull dimensions and increase the submarine navigational draught. Besides, the submarine overall beam over horizontal control surfaces also increases, causing some problems during operation. Taking this into account, aft end shape parameters are selected based on trade-offs among propulsion qualities, controllability, design and operational aspects.

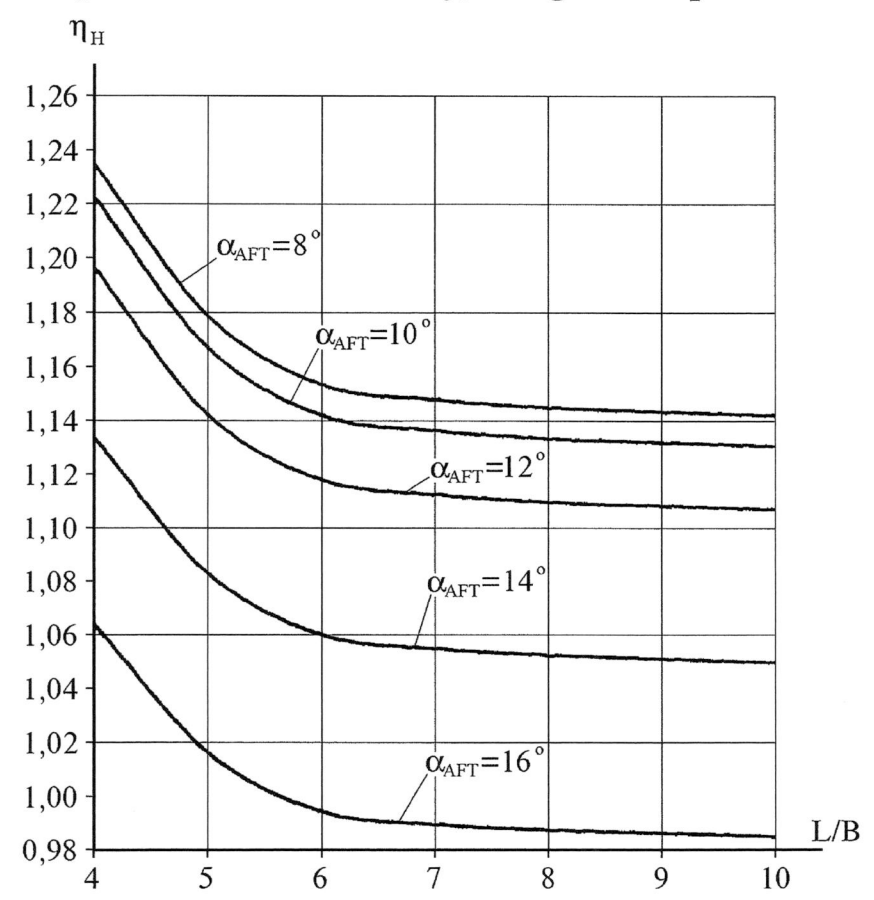

Fig.8.18. Hull-Efficiency Coefficient H versus Relative Length L/B at Different Angles of Run α_{AFT}.

In conclusion it should be noted that these deliberations on the optimum ratio of hull main dimensions were made considering only major bare hull resistance components (friction and form). However, for the sake of required manoeuvrability and course-keeping qualities, we have to install rudders, planes and stabilisers. For normal operation the submarine also needs a sail for the masts. Additional resistance components due to these details, as well as due to hull penetrations and roughness, result in shifting the optimum elongation towards larger L/B values and, as design practice shows, this increase averages 10 to 15%.

On the other hand, when a submarine is designed for specified weapon and power plant packages, from the point of view of equipment arrangement, non-optimal elongation may turn out to require a smaller displacement than the optimum one.

In view of the last comment it may be appropriate to indicate some near-optimum interval of submarine hull elongation values: L/B =7 to 12.

8.5. Outer Hullform Factors

For the purposes of submerged operation the outer hull shape is characterised by three fullness coefficients:
- block coefficient δ;
- prismatic coefficient φ;
- midship coefficient β.

The prismatic coefficient $\varphi = V/\omega L$ is a crucial design characteristic, since it affects the form resistance. Theoretically, the form resistance for a submarine hull with L/B = 7 to 8 without a cylindrical middlebody should be minimal at $\varphi_{OPT} = 0.60$. A deviation from the indicated value to either side means an increase in the form resistance. Since with axisymmetric hullforms, as we have mentioned above, the fullest cross-section is shifted forward and the fore end is fuller than the aft one, prismatic coefficients of the ends are different. Thus, in the subject case of $\varphi = 0.60$ it is $\varphi_{FWD} = 0.68$ to 0.70 for the fore end and $\varphi_{AFT} = 0.53$ to 0.55 for the aft end. However, in practical design the prismatic coefficient is often higher than φ_{OPT}. This is due first of all to the cylindrical middlebody that increases φ to values of 0.7 to 0.8, although this means a higher form resistance. At the same time, the form resistance contribution to the total resistance is, as was mentioned above, not more than 10%. This allows us to somewhat increase it to solve other tasks.

The same can be said about the block coefficient $\delta = \varphi\beta$, which for modern submarines is 0.60 to 0.65 and is numerically equal to the product of the prismatic coefficient φ and the midship coefficient β. The latter as a rule has a value close to 0.785 because almost all submarines have outer hulls with close to the circular cross-sections.

In practical submarine design, for the sake of various calculations and investigations, the actual outer hull is often substituted by an equivalent triaxial ellipsoid. Such a substitution considerably simplifies many ship dynamic calculations [72]. The equivalent ellipsoid is understood as an ellipsoid with the same main dimensions as the real submarine hull. Therefore, let us compare triaxial ellipsoid form factors with those of real submarines (Table 8.1).

It should be noted that ellipsoid block coefficients given in the Table remain unchanged at any $L/V^{1/3}$ and H/B ratios, and therefore such substitutions are quite permissible at the initial design stages.

Table 8.1

Submarine Outer Hullform Parameters

Period of construction	δ	β	φ	$L/V^{1/3}$	H/B
Before 1945	0.5	-	-	6~7	0.9~1.2
1946~60	0.52~0.65	0.8~0.82	0.65~0.7	5.5~7.0	0.9~1.2
1960~80	0.52~0.65	0.8~0.85	0.67~0.7	4.9~5.5	1.0~1.4
1980~90	0.55~0.75	0.8~0.85	0.65~0.8	4.5~5.0	0.7~1.4
Triaxial ellipsoid	0.525	0.785	0.670	-	-

8.6. Sail Shape Selection

Appendages (the sail, stern control surfaces, forward or fin hydroplanes, domes of various devices, etc.) make a considerable contribution to submarine resistance. For modern submarines this share reaches 15 to 30% of the total resistance.

The major components for appendages, as for the hull, are friction and form resistances. However, taking into account the difference in Reynolds numbers, under design procedures the resistance of appendages is calculated separately from that of the hull, though the friction resistance approach is the same as for the hull. Accordingly, the earlier stated conclusion on the necessity to minimise the wetted surface area is still true.

To determine the form resistance of appendages, they extensively use model test data. Therefore, based on accumulated experience it is possible to make a comparative analysis of qualities of various shapes of appendages used in submarine design practice.

The largest (in size) appendage on the submarine hull is the sail. Practical design has evolved two main types of sails: foil and sedan (Fig.8.19).

a)

b)

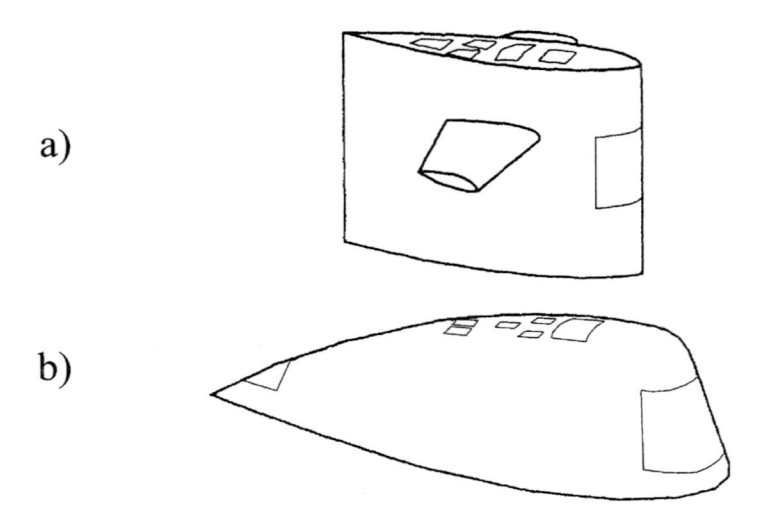

Fig.8.19. Typical Sail Shapes

a) foil-type sail b) sedan-type sail

Foil-type sails typically have relatively small lengths (ℓ_{SAIL}/h_{SAIL} = 1.0 to 1.5), comparatively large absolute heights (over 5 to 7m), vertical walls and sections along waterlines shaped as a symmetrical foil with ℓ_{SAIL}/b_{SAIL} = 4 to 5. The sedan shape is characterised by an elongated profile of the centreline (ratio ℓ_{SAIL}/h_{SAILY} > 4~5) with the front edge inclined backwards and a smooth run of the aft edge from the sail top to the superstructure deck (incidence angles of about 30°). In the cross-section, walls of sedan sails are usually inclined to the centreline. In real life sail shapes often combine elements of both the foil and the sedan types.

When designing a submarine, the choice of this or other sail shape is determined by its internal outfitting (number of masts, dimensions of their aerials, the escape capsule, etc.) and by dimensions of the sail required to accommodate this equipment. It may be noticed that mast sails on Russian submarines have rather sizeable dimensions, especially in terms of the length which on attack submarines reaches 20 to 25% of the hull length.

US Navy submarines have compact sails, small in length and big in height. At the expense of providing no access to the masts and their aerials or any passages inside the sail and due to some other factors, the sail width on US submarines was reduced to 2 m. Thus, US Navy submarines have foil-type sails while Russian ones display both sedan sails and some shapes close to foils.

Due to some original design solutions, on Russian non-nuclear attack submarines of the IVth generation it has become possible to attain sail width $\ell_{SAIL} < 2m$ with the length being $\ell_{SAIL} \approx 0.11$ to 0.13 L_{OH}. This design is known among Russian designers as the «knife-type» sail.

Fig.8.20 shows resistance data for sails of various shapes. It may be seen that the least resistance, all other things being equal, can be expected from foil shapes with soft (rounded) edges and the sedan [5].

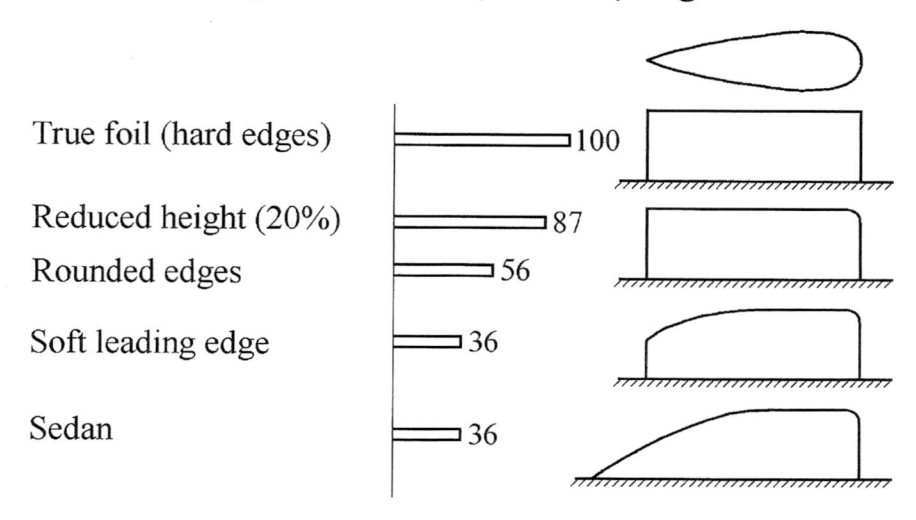

Shape	Resistance
True foil (hard edges)	100
Reduced height (20%)	87
Rounded edges	56
Soft leading edge	36
Sedan	36

Fig.8.20. Resistance of Sails of Various Shapes

Sail resistance is quite sensitive to the interaction with the hull because the attachment stimulates vortexes: backwater at the front portion and side along sail walls. To reduce this resistance component, the area of attachment is very carefully mated to remove any abrupt transitions and right angles (Fig.8.21) [99].

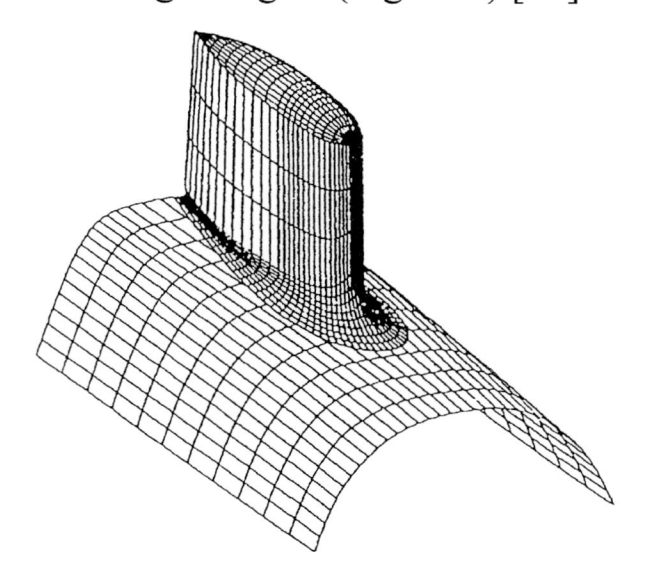

Fig.8.21. A Smooth-Mated Sail to Hull Connection

Fig.8.22 shows test results for foil-type sails of the same height and breadth but with different lengths [5]. It may be seen that the total sail resistance can be reduced if we decrease the interaction-induced component by optimising the relative length of the sail.

The resistance component value due to hull interaction can also be reduced if the sail is located at a distance of 0.2 to 0.3 of the hull length from the forward perpendicular.

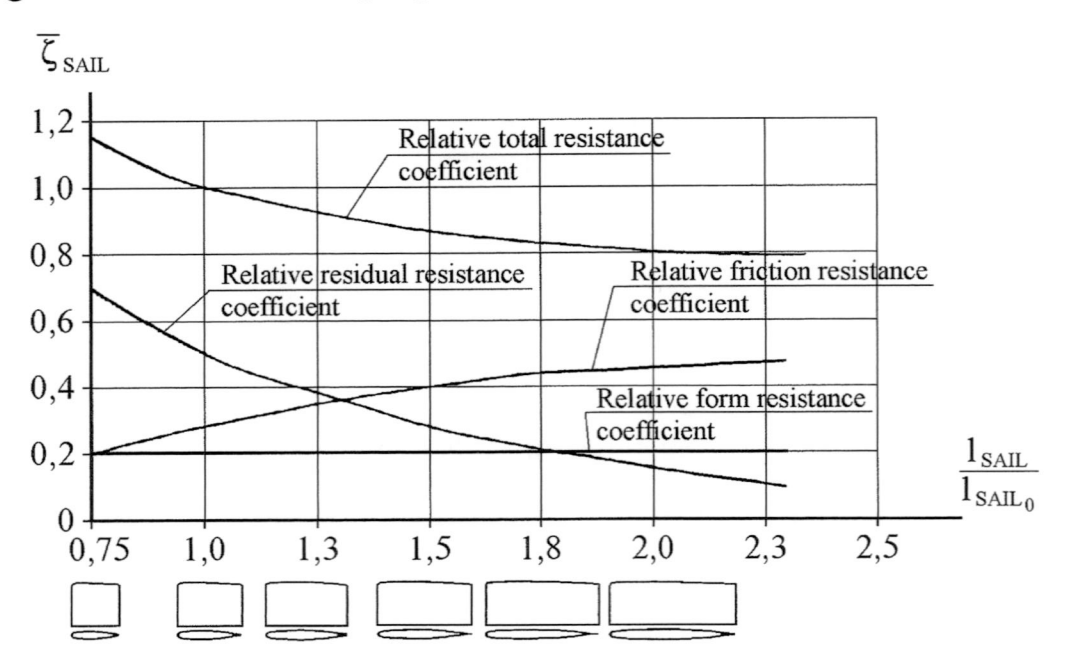

Fig. 8.22. Sail Resistance versus Length

Note:

1. $\overline{\varsigma}_{mey} = \dfrac{\varsigma_{SAIL}}{\varsigma_{SAIL_0}}$ – relative resistance coefficient with respect to the basic sail

2. These curves were plotted assuming constant sail height and width

3. Relative elongation of the basic option $\dfrac{l_{SAIL}}{b_{SAIL}} = 8$.

However, shifting the sail to the submarine bow increases the lever (of the submarine longitudinal centre of gravity) and the lateral force on the sail. This increases the angle of kick and, consequently, reduces the turning circle radius. On the other hand, moving the sail towards the bow somewhat increases the submarine resistance and increases boundary layer turbulence, thus worsening propeller acoustic characteristics.

Eventually, the sail position with respect to the submarine length is dictated by the arrangement of equipment and frames in the pressure hull and, first of all by the location of the main control room and masts. Introduction of masts that do not penetrate the pressure hull will allow this limiting factor to be excluded. Location of the sail will then be defined by the location of the access hatch trunk with regard to the main control room and by hydrodynamic considerations.

When selecting the sail type it is also necessary to take into account the lateral hydrodynamic force arising due to oblique flow past the sail. Other things (first of all, dimensions) being equal, the foil-type sail generates a higher hydrodynamic lift when the submarine changes course or drifts.

With the sedan sail, the derivative of the lateral force coefficient, with regard to the angle of kick, is less than with the foil-type sail of the same area on the profile plan of the lines drawing. With the foil-type sail this force has a longer lever to the submarine vertical centre of gravity. This causes higher heeling moments when turning.

To conclude this section, it should be noted that irrespective of the sail shape its contribution to the total submarine resistance is minor. Therefore, the sail type choice is defined more by preferences of the design school, equipment composition and arrangement, than by any resistance minimisation considerations.

8.7. Shaft Number Effects upon Submarine Hullform and Hydrodynamic Qualities

The number of shafts of the submarine power plant is one of the factors that determine the aft body shape and equipment arrangement in the stern (Fig.8.23).

The number of shafts affects such submarine features as survivability, displacement, propulsive qualities and noise level.

The first post-war generations of diesel-electric submarines usually had power plants with two or three shafts. The main motivation behind the choice of the number of shafts was the set of parameters of engine series available to the designer. Taking into account the speed specified in SDS, it could easily occur that due to these parameters, particularly the output, a twin-shaft plant was insufficient and there was no alternative to the three-shaft option. The first nuclear submarines were also twin-shaft ones. This was explained by the traditional approach inherited from diesel-electric submarines and by the lack of confidence in the reliability of new power plants. And indeed, in terms of survivability, multishaft plants are better.

The Great Patriotic War experience confirmed this: submarines did return to their bases with a damaged shaft. Service records show that failures and damages in shaft lines occur in peace time too. Thus, it might be considered that to increase the power plant survivability and, hence that of the submarine as a whole, the submarine should be designed to have several shafts. But at the same time it should be kept in mind that all these arguments are correct only the submarine is attacked with a conventional weapon.

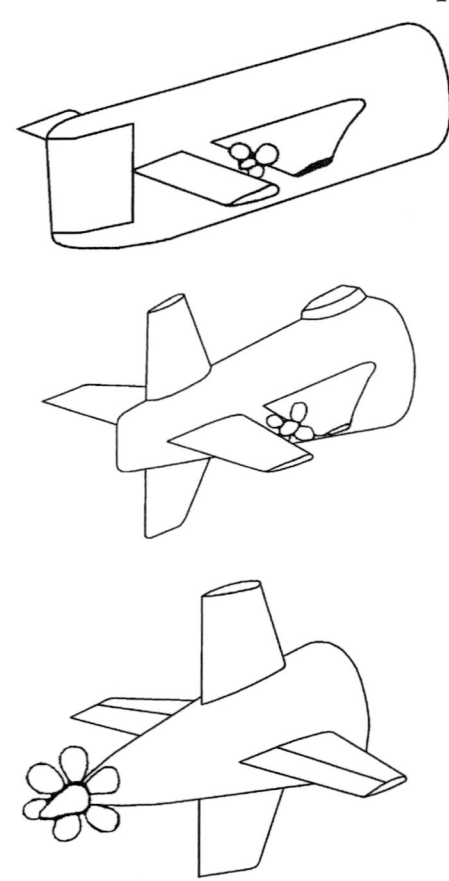

Fig.8.23. Evolution of Submarine Aft End Architecture [51].

Starting with the second-generation submarines, US designers preferred single-shaft power plants. Besides hydrodynamic factors, this was due to favourable basing and operating conditions of US Navy submarines. It would not be correct to consider that hydrodynamic advantages of the single-shaft architecture type have not been known and properly estimated by our national specialists. However, at that time they could not reach a decisive argument against a reliable back-up for the case when the propulsion plant suffers a shaft failure.

Design experience proves that, all other things being equal, the least power plant weight, and hence the submarine least normal displacement D_0, is reached with a single-shaft power plant. With the twin-shaft arrangement the weight of the plant increases due to duplication of the machinery. So, considering the power plant weight share in the normal displacement (~ 20%), and assuming the Normand's number is $\eta_{NOR} = 2.5$, changing from a single-shaft power plant to a twin-shaft one of the same output means 10 to 15% increase in the normal displacement. This is of course an important factor and it must be taken into account.

A large number of investigations and experiments, as well as evaluation of data on submarines constructed all over the world, definitely show that submarines with single-shaft plants undoubtedly have undoubtful advantages in terms of propulsive qualities.

Fig.8.24 [5] demonstrates how displacements and speeds vary with shaft power for twin-shaft and single-shaft submarines. These functions were calculated for an assumed «basic» submarine, and therefore are, to some extent, arbitrary.

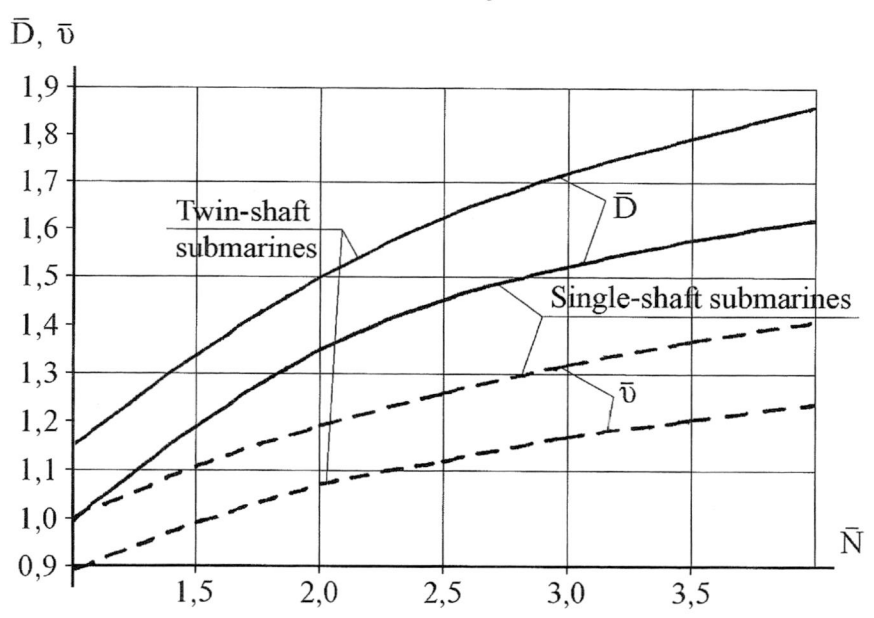

Fig.8.24. Displacement and Speed versus Main Propulsion Plant Power for Twin-Shaft and Single-Shaft Submarines

$\overline{N}$ – relative output of the main power plant; $\overline{D}$ – relative displacement. $\overline{D} = 1$ is assumed for a single-shaft submarine displacement with MPP of $\overline{N} = 1$; $\overline{\vartheta}$ – submarine relative full speed. $\overline{\vartheta} = 1$ is assumed for a single-shaft submarine speed with MPP of $\overline{N} = 1$.

It may be seen from the plot that single-shaft submarines have 10 to 12% higher speed than twin-shaft ones. This speed increase, besides the lesser displacement, is due to two hydrodynamic reasons.

1. The single-shaft design enables a larger propeller to be installed than on a twin-shaft submarine (in the latter case the propeller diameter is limited by its small distance from hull side). The possibility to increase the propeller diameter, along with the increase in the number of blades, allows both the specific load on propeller blades and the propeller revolutions to be reduced. This, in turn, increases the propeller efficiency which is higher on single-shaft submarines than on twin-shaft ones by 15 to 20%.

2. A single-shaft arrangement naturally fits the axisymmetric aft end shaped as a body of revolution. This produces favourable conditions for propeller operation. Wake and thrust deduction coefficients associated with the axisymmetric aft end ensure that a single-shaft submarine has a hull efficiency coefficient of considerably higher than 1. It may reach 1.05 to 1.20 increasing the total propulsion coefficient to $0.75 \sim 0.80$. For submarines with twin-shaft power plants the hull efficiency coefficient is usually close to 1, and the total propulsion coefficient does not exceed $0.60 \sim 0.65$ (i.e. 1.25 times lower than for single-shaft submarines).

A way to improve the hull efficiency coefficient, and hence the total propulsion coefficient, for twin-shaft submarines is to apply an aft end with axisymmetric propeller shaft cones [82].

The above is true only for the case when the outputs of single- and twin-shaft plants are equal. Actually, the twin-shaft architecture allows the main plant power to double by arranging two propulsion plants instead of one, increasing the twin-shaft submarine speed by 15% as compared to the single-shaft one.

A no less important consideration than the performance, and today even more important, affecting the MPP shaft number choice is the requirement of the acoustic signature level. Within certain frequency ranges it is determined by the propeller radiated noise. This depends on the propeller speed and on the degree of nonuniformity of the flow in way of the propeller. The nonuniformity is caused by the sail, by stabilisers and other control surfaces, by the elliptical cross-section of the outer hull, and thus depends on the general arrangement of appendages and the propeller and, consequently, on the number of shafts of the power plant.

Naturally, an axisymmetric-stern single-shaft submarine offers better chances to form a more uniform propeller inflow than a twin-shaft one.

As regards twin-shaft submarines, it is still possible to achieve better conditions for the propeller by extending the aft end into extra-long propeller shaft cones (Fig.8.25.). However, it appears that all resources of the twin-shaft architecture are already exhausted and this obsolete shape contradicts new requirements put forward for modern submarines.

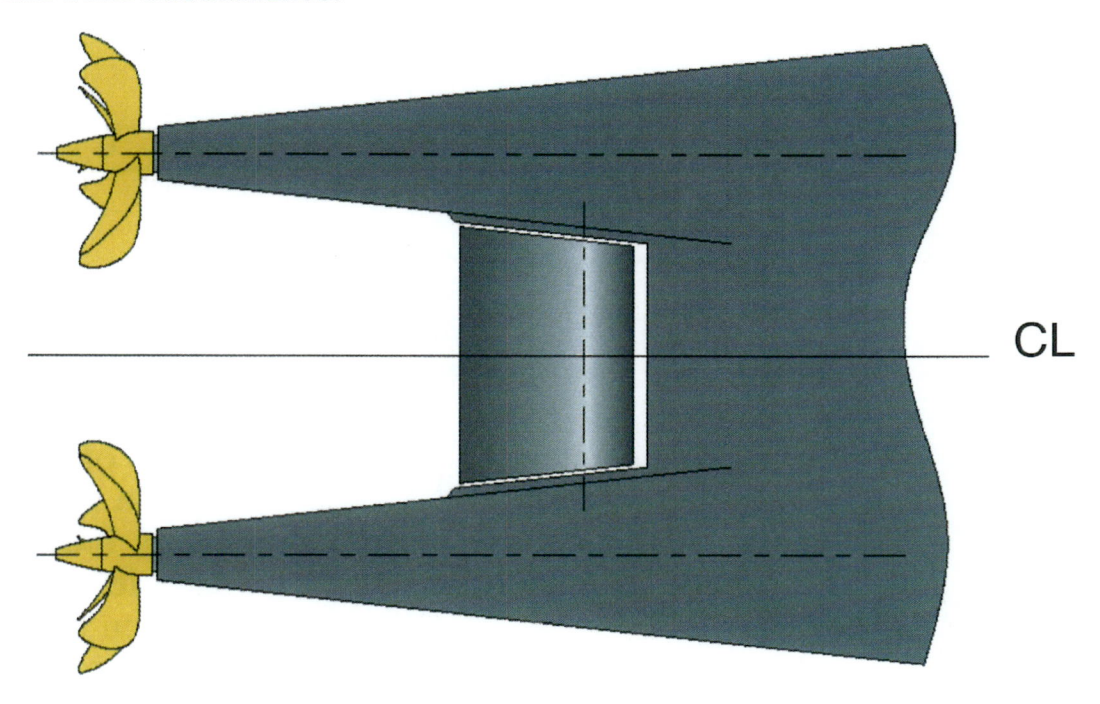

Fig.8.25. A Twin-Shaft Submarine Stern with Extra-Long Propeller Shaft Cones (informally known as «pants»).

Fig.8.26. shows diagrams of velocity fields within the propeller disk plane for single-shaft and twin-shaft submarines. The same Figure provides aft end schematics of subject designs [82].

As we may see from Fig.8.26, the flow is more uniform in the propeller disk of the single-shaft submarine. It should also be noted that the propeller speed and load reduction enable the single-shaft submarine critical speed to increase, i.e. the speed when cavitation becomes audible, by 25 to 30%. This tactical feature is especially important for attack boats; when hunting other submarines they should be able to accelerate with as low as possible penalties in terms of noise and interference to their own sonars.

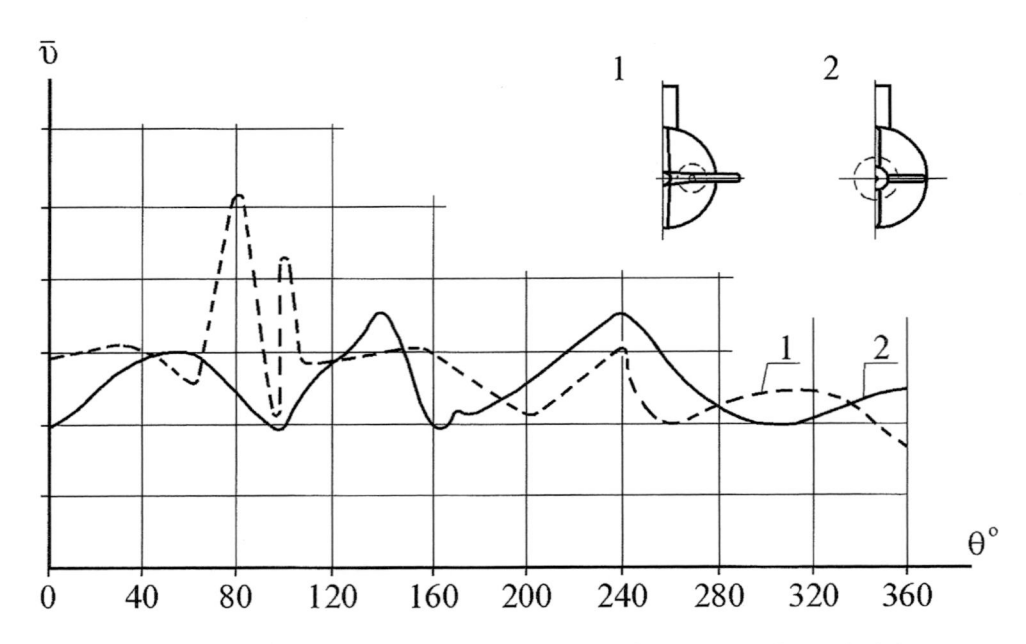

Fig.8.26. Diagrams of Propeller Disk Velocity Fields
1 – twin-shaft power plant 2 – single-shaft power plant

Thus, except for survivability and maximum speed at limited outputs of available engines, all criteria indicate that single-shaft submarines can offer more advantages. Regarding the issue of survivability, it can be resolved by installing reserve propulsion facilities. Russian designers usually choose reserve propulsors hidden inside the hull. These may be tunnel propellers, retractable pods, waterjets, etc. They may additionally serve as thrusters when navigating in narrow passages, mooring, etc. Submarine reserve propulsors should provide a certain submerged speed (about 5 knots, as a rule) and enable the submarine to return to the base for repairs. A possible configuration of such a reserve propulsion system is shown in Fig.8.27 [48].

Fig.8.27 Project 877 EKM "KILO" Submarine Reserve Propulsion System

8.8. Effects of Layout Considerations upon the Choice of Outer Hullform Parameters

In submarine outer hull design and hullform parameter selection, one should, besides all other earlier described considerations, remember that they are influenced by many design factors, primarily by general arrangement and layout of key equipment and weapons. In the majority of practical designs main dimension ratios were dominated in the first place by weapons, sonars and power plants.

Sometimes a designer, although aware that a certain decision will adversely affect one or other submarine characteristic or a group of characteristics, is forced to make a decision for the sake of fitting the required weapon package.

As an example of such a forced decision we can mention SSBN elevated superstructure in way of missile silos (Fig.8.28). This particular design has adversely affected full submerged displacement, stealth, manoeuvrability, submerged speed, surface and especially surfacing stability of the submarine, but allowed the main task to be met: to install a new, more powerful, though more bulky, missile package using an economically justified method.

Fig. 8.28. A Nuclear Missile Submarine

A similar effect on submarine outer hullform design may be imposed by main dimension restrictions due to shipyard limitations, by certain full-buoyancy draught limits and some others. Such constraints force designers to deviate from recommended optimum hullforms.

Similar deviations may occur in the design of local hull portions too. E.g., the fore end design may be driven by sonar array demands that require changes in the bow shape leading to a change in flow conditions (see Fig.8.15).

The pressure hull, and sometimes there are several pressure hulls, as well as pressure structures in the between-hull space directly affect the outer hullform and structure.

Usually the process of submarine design is arranged in such a way that at first the pressure hull layout is established and only then work starts on the outer hull. With such an approach there is no chance to consider any significant number of outer hullform options [9]. The external dimensions are already largely limited by dimensions and specific features of the pressure hull. If the thus evolved outer hull fails to satisfy the requirements to submarine properties and qualities, it may become necessary to re-design the pressure hull. Thus, following the convergence method the two hulls are gradually fitted together.

It is then, when it is time to choose the final design, that we should see the designer's most valuable asset - the skill of finding trades-off by combining contradictory requirements into the single whole which is called an optimised submarine. It may be claimed that if the outer hullform and the entire submarine appearance are beautiful, the designer has achieved an optimum combination of all qualities in accord with natural purposefulness.

THE STATE RESEARCH CENTER OF THE RUSSIAN FEDERATION
CSRI ELEKTROPRIBOR

THE INSTITUTE PERFORMS A FULL OPERATIONAL CYCLE FROM FUNDAMENTAL INVESTIGATION TO PRODUCTION

NEW GENERATION OF NAVIGATION EQUIPMENT IS AVAILABLE:

- **INERTIAL NAVIGATION SYSTEM**
- **INERTIAL SYSTEM FOR STABILIZATION AND NAVIGATION**
- **MINIATURE INTEGRATED INERTIAL/GPS NAVIGATION SYSTEM**
- **MARINE GRAVIMETER**
- **ELECTRONIC CHART DISPLAY INFORMATION SYSTEM**

NEW SCOPE OF ACTIVITIES:

- **PERISCOPES AND OPTRONICS MASTS**

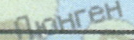

ЭЛЕКТРО ПРИБОР

30, Malaya Posadskaya str.,
Saint-Petersburg, 197046, Russia
Tel. (812) 232 59 15, 238 78 01, fax: (812) 232 33 76

9. DETERMINATION OF SUBMARINE PRINCIPAL DIMENSIONS

9.1. General

The next step after we have established the displacement and the architecture of the submarine is the selection of main dimensions: length, beam and height of the hull, as well as the draught (L, B, H, T).

The choice of main dimensions should be co-ordinated with all features and qualities of the submarine being designed. Main dimensions should correspond to the architectural type and meet a number of general design requirements. If the combination of main dimensions achievable with the subject pressure hull does not allow some items of the Submarine Design Specifications to be satisfied, it is necessary either to alter SDS or to reconcile earlier estimated pressure hull dimensions and submarine displacement.

The specific nature of submarines imposes certain constraints on the selection of principal dimensions. It means that when the architectural type is selected and pressure hull dimensions are specified, the submarine length is already largely determined and there is no way for any variations. The lower limit of the beam is also restricted. With the single-hull architecture the pressure hull completely dictates the beam and to a considerable extent governs the height and the draught. Main dimension variations are in this case related to variations of pressure hull dimensions and equipment arrangement in it.

Thus, submarine main dimensions are closely related to the shape and dimensions of the PH. Owing to this fact, purely analytical methods used so extensively to find main dimensions in surface ship design [6], [15], [60] are not fully applicable to submarine design because with them one may easily fail to match the pressure hull and the outer hull together.

Due to these reasons, submarine main dimensions are established by a sketch of the lines drawing for the chosen dimensions of the pressure hull. Then, the thus obtained dimensions are evaluated from the point of view of equipment and weapon arrangement in fore and aft ends and in the superstructure, as well as in terms of basic seakeeping abilities, and, if required, corrected. Aiming to find such a combination of main dimensions that would be best suited to meet SDS requirements, main dimensions are varied within certain tolerable limits. If it turns out to be necessary, they also consider alternative (shorter or longer) pressure hulls.

This method of finding main dimensions requires a rather sizeable amount of time for drawing a number of lines options and requires certain design experience. When sketching the lines drawing, they apply approximate formulae and statistic data on constructed submarines. This allows the correct decision to be found somewhat faster.

Today, when lines drawings can be generated and, especially, varied by computers, this task takes much less time and it is possible to consider a large number of options.

This chapter explains methods for tentative estimations of submarine main dimensions and their subsequent updating in the process of work on the lines drawing. These tentative estimations of main dimensions, as well as their variations, are made for a constant displacement established by preceding approximations.

9.2. Submarine Hull Length Estimations

In submarine design they distinguish several kinds of hull lengths (Fig.9.1).

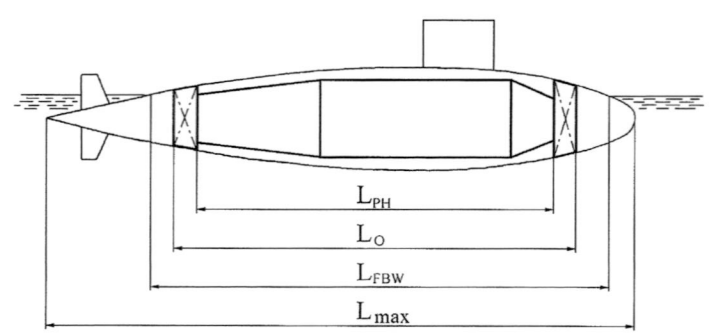

Fig.9.1. Classification of Submarine Hull Lengths

L_{PH} – pressure hull length;
L_0 – length of the watertight volume in surface trim;
L_{FBW} – full-buoyancy waterline length;
L_{MAX} – extreme length of the bare hull (without the propeller);
L_{OV} – overall length of the submarine.

The length of the watertight volume L_0 is the basic length for generating the lines drawing. Permeable end structures are located fore and aft of the watertight volume. As the pressure hull length is already known, the watertight length can be found from:

$$L_0 = L_{PH} + \ell_{FWDT} + \ell_{AFTT} \qquad (9.1)$$

where ℓ_{FWDT} and ℓ_{AFTT} – lengths of bow and stern end main ballast tanks.

The fore tank length ℓ_{FWDT} is defined mainly by the length of external portions of torpedo tubes located inside the end MBT and by the sonar bay. The tank length is taken from preliminary studies on the general arrangement, or from a prototype with the same torpedo tube calibre.

It should be remembered that on modern submarines of the IIId and IVth generations sonar bays are made water-tight and also provide surface floodability. For that purpose the sonar bay is equipped with a kingston valve and with a vent valve.

The length ℓ_{AFTT} can be found either based on the required volume of the aft MBT and equipment arrangement in the end body or from the prototype.

When establishing the fore end length, one should carefully consider torpedo tube muzzle shutters. Wartime experience has demonstrated low survivability of long shutters [9]. Ice, heavy weather and enemy weapons often damaged or jammed muzzle shutters, making the submarine unable to use weapons. Sometimes they even removed the shutters but this resulted in loosing 1.5 to 2.0 knots of the submerged speed due to increased resistance. In the light of these comments submarine designers should keep in mind that a longer fore end means larger shutters and that may lead to absolutely unacceptable structural solutions. In order to ensure unobstructed torpedo exit from the tube, it is necessary to arrange a firing cone with a certain angle α(Fig.9.2a) and that increases the shutter size even further. On WWII stem-type submarines they had to cut the length of muzzle shutters by introducing a local S-shape waterline at the level of the torpedo tubes (Fig.9.2b). Obviously, that increased the form resistance.

The shutter length issue is not always so critical for bodies of revolution, but the need to arrange a proper firing cone in a long end body may also result in inadvisable design solutions (Fig.9.2c).

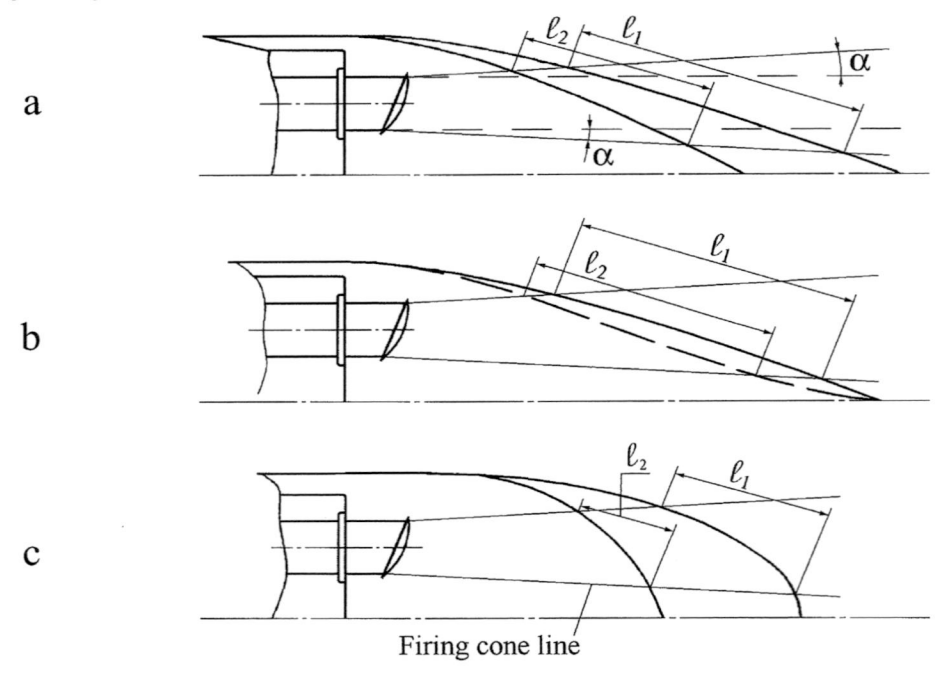

Fig.9.2. Submarine Torpedo Tube Muzzle Shutters

To determine the length of the fore end at the full-buoyancy waterline L_{FBW} it is necessary to draw the centre plane projection of the pressure and the outer hulls and to draw the end body equipment arrangement (or use a prototype).

Obviously, for stem-type submarines $L_{FBW} \approx L_{MAX}$ and for axisymmetric designs L_{MAX} is completely dictated by the drawn centre plane (see Fig.9.1).

9.3. Submarine Hull Beam Estimations

The hull breadth for single-hull submarines is completely determined by the PH diameter while for double-hull submarines the beam defines the between-hull volume and, thus, the reserve buoyancy. It should be mentioned that as soon as the pressure hull dimensions and the reserve buoyancy are established, the outer hull breadth is, for all practical purposes, determined and may change only insignificantly. In the first approximation it can be found from

$$V_{BH} = \delta_{LBH} \qquad (9.2)$$

The bare hull volume V_{BH} is made up by the constant buoyant volume (less volumes of the sail, stabilisers and control surfaces), reserve buoyancy and volumes of permeable structures:

$$V_{BH} = V'_0 + V_{MBT} + V_{PEP} \qquad (9.3)$$

where $V'_0 = kV_0 = V_0 - V_{SAIL} - V_{ST} - V_R$ – constant buoyant volume of the bare hull, m³

$k < 1.0$ – coefficient accounting for volumes of the sail, stabilisers and other appendages;

V_0 – constant buoyant volume, m³

Then the overall beam can be estimated as:

$$B_{max} = \frac{V_{BH}}{\delta LH} \qquad (9.4)$$

It should be kept in mind that the hull beam may be subject to construction limitations: minimum possible dimensions of the between-hull space, width of the construction place, width of the construction shed gate, width of the dock, etc.

As the beam is closely related to the submarine initial surface stability, it is necessary to compare the beam found from (9.4) with that obtained from the equation of stability (9.26), (9.27).

9.4. Submarine Hull Height Estimations

The ⌀ hull height with a chosen pressure hull diameter is defined by the superstructure and the keel (Fig.9.3).

$$H = d_{PH} + \Delta H_1 + \Delta H_2 \qquad (9.5)$$

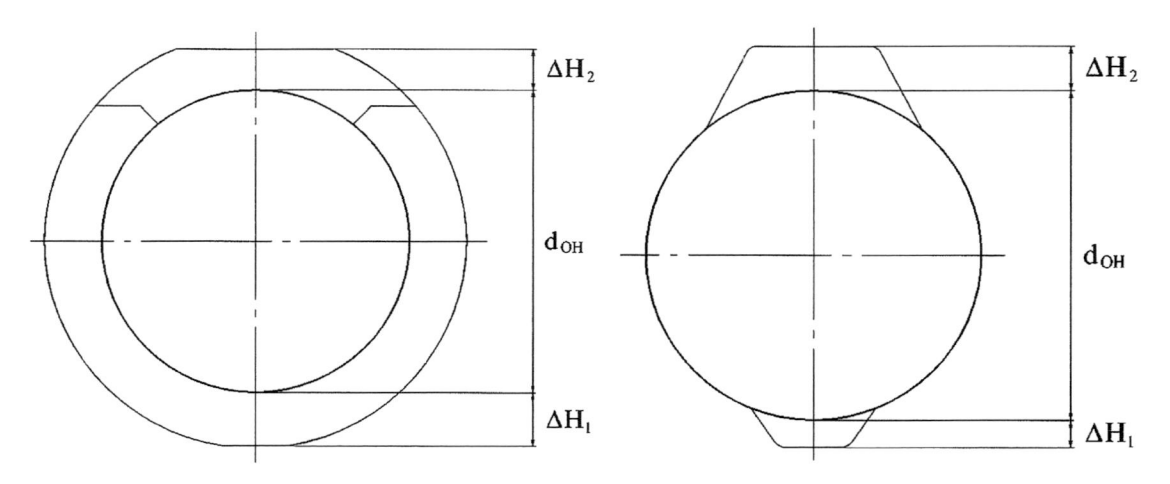

Fig.9.3. Submarine Hull Cross-Sections

The keel depth ΔH_1 is usually selected based on general design considerations and for double-hull submarines amounts to $\Delta H_1 \cong 0.9{\sim}1.2$ m. In case it is necessary to increase the reserve buoyancy but impossible to increase the beam, an additional margin of buoyancy can be obtained by increasing ΔH_1. It is necessary to check the result against any draught restrictions. Sometimes even single-hull submarines have a so-called ballast keel (Fig.9.3b) accommodating solid ballast needed for submarine stability. The keel height in this case is $\Delta H_1 \cong 0.5{\sim}0.6$ m. The resulting increase in the hull vertical dimension does not, as has been already said, cause any significant growth of the wetted surface area but decreases the surface stability.

The superstructure height ΔH_2 is found based on internal arrangement of HPA bottles, pipelines, vent valves, etc. On torpedo submarines it is usually $\Delta H_2 \cong 1.0{\sim}1.5$ m, on missile submarines the superstructure height is dictated by weapon package requirements (see Fig.8.28).

If the superstructure arrangement has not yet been elaborated, ΔH_2 may be derived from the prototype.

9.5. Reserve Buoyancy. Approximate Draught Estimations Based on the Specified Reserve Buoyancy

In any commercial or combat ship design main dimensions and form factors have to be such so as to ensure the required tonnage (cargo capacity), speed, stability, as well as reserve buoyancy and a certain level of floodability.

A.N.Krylov has pointed out that «the measure of the reserve buoyancy is the volume between the effective waterline of the ship and the upper watertight deck assuming that the boards are watertight» [6].

Hence, the total reserve buoyancy of a submarine in any surface trim is the volume of all watertight parts of the hull above the waterline from the one hand:

$$W = V_1 + W_2 \qquad (9.6)$$

and the volume of empty main ballast tanks on the other one (Fig.9.4):

$$W = W_1 + W_2 \qquad (9.7)$$

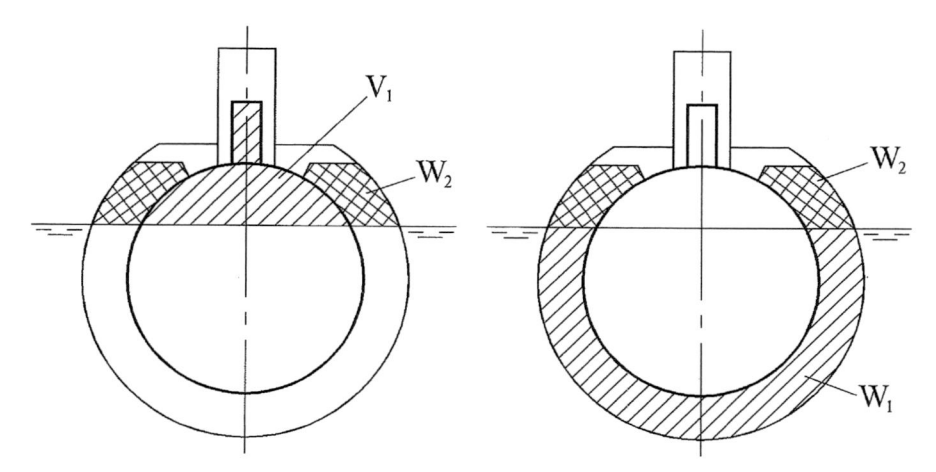

Fig.9.4. Distribution of the Reserve Buoyancy on Submarines.

From Fig.9.4 it may be seen that the volume of main ballast tanks located above the waterline W_2 is included in both equations.

Hence,
$$W_1 + W_2 = V_1 + W_2 \\ W_1 = V_1 \tag{9.8}$$

Thus, the main ballast tank volume above the waterline does not influence the submarine surface draught and is called the passive reserve buoyancy, in contrast to the volume of MBTs below the waterline called the active reserve buoyancy.

It is obvious that the draught remains unchanged irrespective of whether the MBT upper stringer is at the top level of the PH or in any other place above the waterline. The passive reserve buoyancy is useful because it increases the total reserve buoyancy, and therefore improves submarine seakeeping abilities in surface trim and surface floodability including at damaged condition heel and trim angles [62]. Nevertheless, it should be kept in mind that the increase in the reserve buoyancy, though favourable from the point of view of floodability and seakeeping abilities in the surface trim, has a number of disadvantages:

- the total underwater displacement increases entailing the growth of the wetted surface area and, hence, the submarine submerged resistance;
- the normal displacement of the submarine increases due to the larger light hull and its increased weight. The weight of some systems, e.g., the HPA system, increases as well;
- due to the increased MBT volume, the diving time increases too. Design measures aimed to save the original diving time may result in an increase in the weight of the diving system.

Due to above-listed reasons it is not desirable for the reserve buoyancy to exceed the required minimum. In submarine design it should be remembered that the displacement value has always been important for stealthiness and, in the future, this factor will become even more crucial as a result of developments in non-acoustic methods of submarine detection.

Knowing PH dimensions and compartment subdivision established based on the estimated displacement value, we can tentatively set a value of reserve buoyancy necessary to meet specified floodability standards. Actually, there are no general rules for reserve buoyancy required for floodability and satisfactory surface trim seakeeping. And there can hardly be any such rules. For every submarine it is necessary to determine the worst damage case and then, based on floodability requirements, establish the reserve buoyancy.

Under the present design philosophy the reserve buoyancy should be the minimum required for surface floodability and satisfactory seakeeping. When these targets are achieved, the reserve buoyancy usually turns out to be 25 to 35% of normal displacement for double-hull designs and 18 to 20% for single-hull and side-tank submarines.

Submarine designs presently produced in other countries as a rule ignore surface floodability standards and that allows the required reserve buoyancy to decrease to 12 to 15% of D_0.

Ballast tanks of single-hull submarines are usually located in fore and aft ends while on double-hull submarines they are placed in the between-hull space and, if the arrangement permits, in the fore and aft ends (Fig.9.5).

The subdivision of the between-hull space into MBTs, as well as the arrangement of tanks in the fore and aft ends, should be made in such a way that tank volumes are approximately equal. A difference in tank volumes beyond 30% is usually regarded as unacceptable. This approach enables more uniform flooding or blowing of MBTs. When deciding where to place MBT transverse bulkheads, it is necessary to consider surface floodability requirements. In particular, the tank length should be approximately half of the average compartment. The breadth of the between-hull space on side-tank and double-hull submarines should be at least 0.7 m due to construction requirements associated with light hull welding and other work to be performed in the between-hull space.

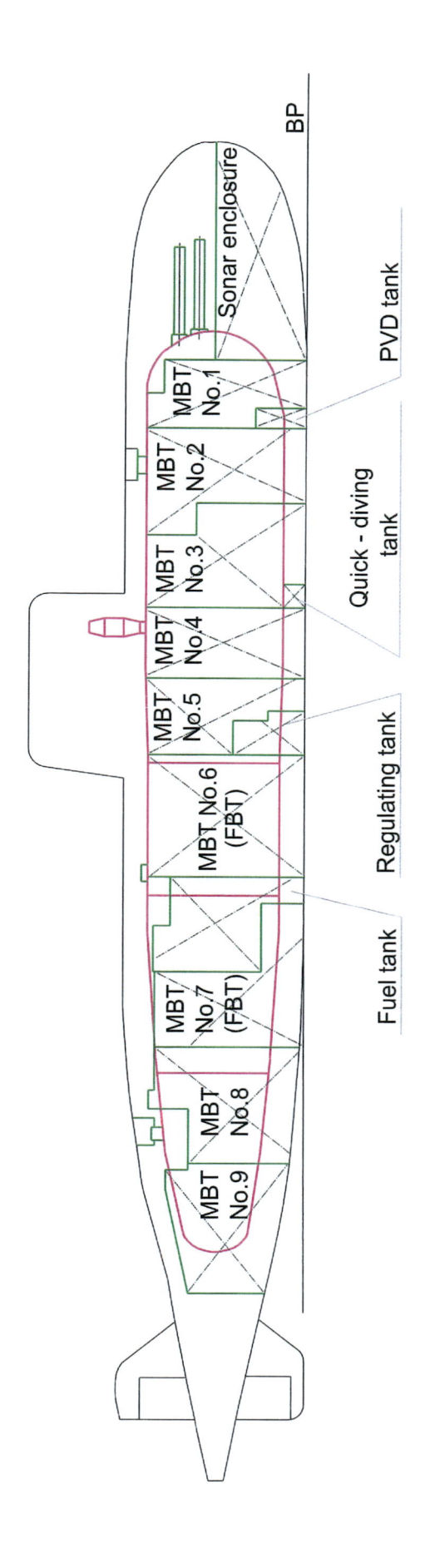

Fig.9.5 Main Ballast Tank Layout on SS

When allocating volumes for main ballast tanks along the submarine, one should try to bring reserve buoyancy (watertight volume above the water) and total MBT volume centres to one vertical line. This condition allows any designed surface trim of the submarine to be avoided. This should be done taking into account the layout of pressure tanks, bays and other volumes in the between-hull space.

Of all the MBTs, the designer should pay special attention to the middle group. The number of tanks in the middle group and their longitudinal location should ensure that the submarine will surface with the least possible trim by the stern and be safe to stay in the low-buoyancy trim with minimum allowable reserve buoyancy and freeboard.

According to their watertightness arrangements, MBTs are divided into:

– kingston-valve tanks: the main ballast tanks made watertight with the help of vent valves in the upper portion and kingston valves in the lower portion;

– floodport (free-flooding) tanks: those made watertight only by vent valves in the upper portion while in the lower portion they have grated flooding holes.

In submarine design it is preferable to use MBTs with kingston valves because (in spite of some increase in the submarine weight due to kingston valves and their control systems) they ensure higher reliability in accidents and when cruising in rough seas.

Thus, before we assign or estimate the required reserve buoyancy, it is necessary to decide what should be the extent of surface floodability. The reserve buoyancy value should be determined immediately after determining PH compartment sizes and machinery and equipment arrangement, because it largely defines both the architecture and the main dimensions of the submarine.

There are several ways to achieve approximate reserve buoyancy estimations:

1. Based on statistics from already constructed submarines – the most simple but at the same time the least reliable method, as the reserve buoyancy depends not only on the displacement but also on many parameters the most important of which have been discussed above.

2. Based on a prototype with a very similar equipment layout and compartment subdivision. This method is more reliable since it accounts for special features of the project.

3. Based on surface floodability curves suggested by naval architects I.S.Vasiliev and D.L.Garmashev. This surface floodability diagram (Fig.9.6) is built in rectangular co-ordinates and represents a totality of two families of curves for functions: $M_{TRIM} = f(V)$ at T = const and $M_{TRIM} = f_2(V)$ at ψ = const. The vertical axis is submarine displacement V (m^3); the horizontal axis shows trimming moments M_{TRIM} (m^4), by the bow to the right and by the stern to the left.

The fact that the diagram displays (·) O where lines of equal trims converge when the submarine is completely covered by water indicates that this submarine has no free-flooding ballast tanks. Such a floodability diagram is included into the set of technical documents of every submarine and enables damaged trim and reserve buoyancy to be found without complicated calculations by the lines. However, it should be kept in mind that this dimensional diagram is applicable only to the particular submarine it has been plotted for.

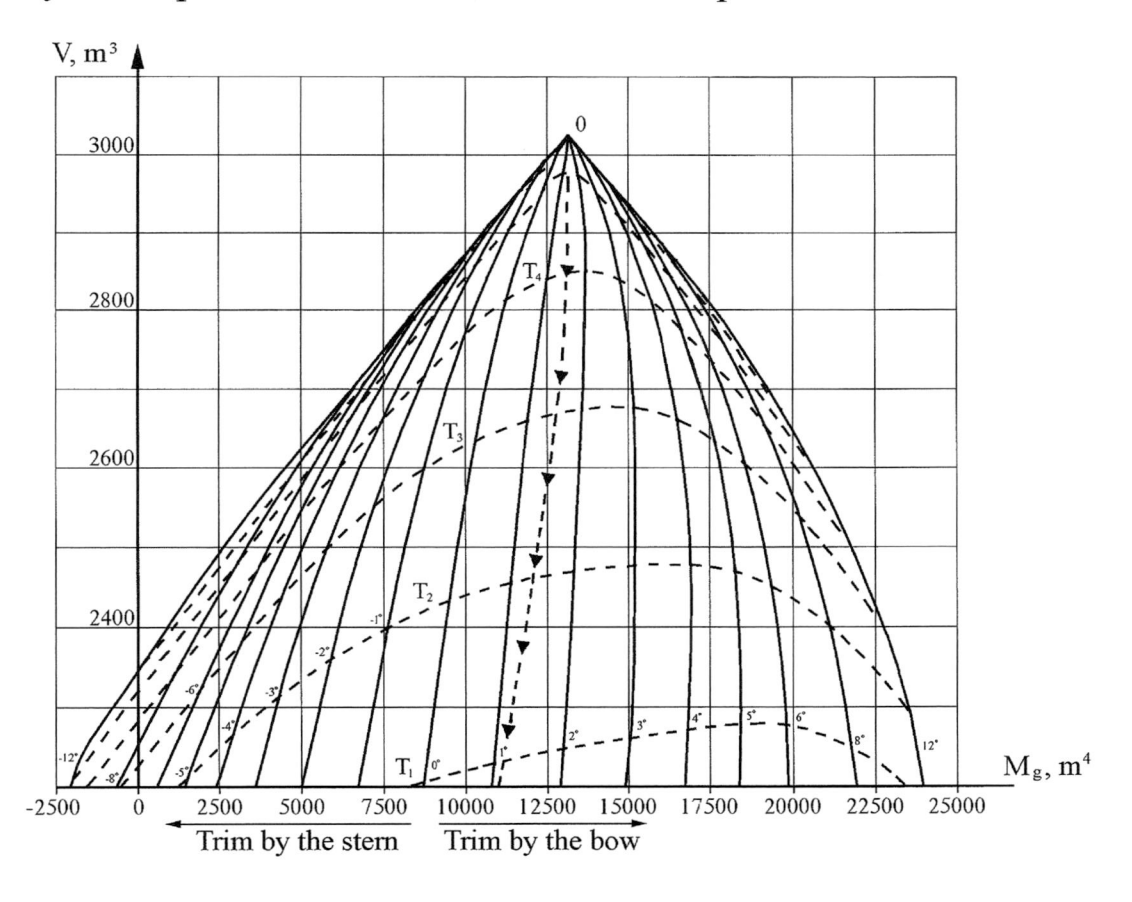

Fig. 9.6 Surface Floodability Diagram of a Submarine

Symbols: ——— – curves of equal trim angles;
– – – – curves of equal midship draughts; — ▶ – – – curves of the mean moment M_{TRIM}

Therefore, at early design stages it is more convenient to estimate the required reserve buoyancy from a non-dimensional surface floodability diagram as suggested by A.V.Bazilevich [9]. Such a diagram is plotted based on dimensional curves. Let us shift the origin of the diagram grid to (·) O which corresponds to the condition when the submarine is completely covered by water. In this case the above-water volume, i.e. the reserve buoyancy W' remaining after an accident, is the ordinate and its moment M is the abscissa (Fig.9.7). In our case, to determine the required reserve buoyancy it is enough to plot only the curves of limiting trim angles.

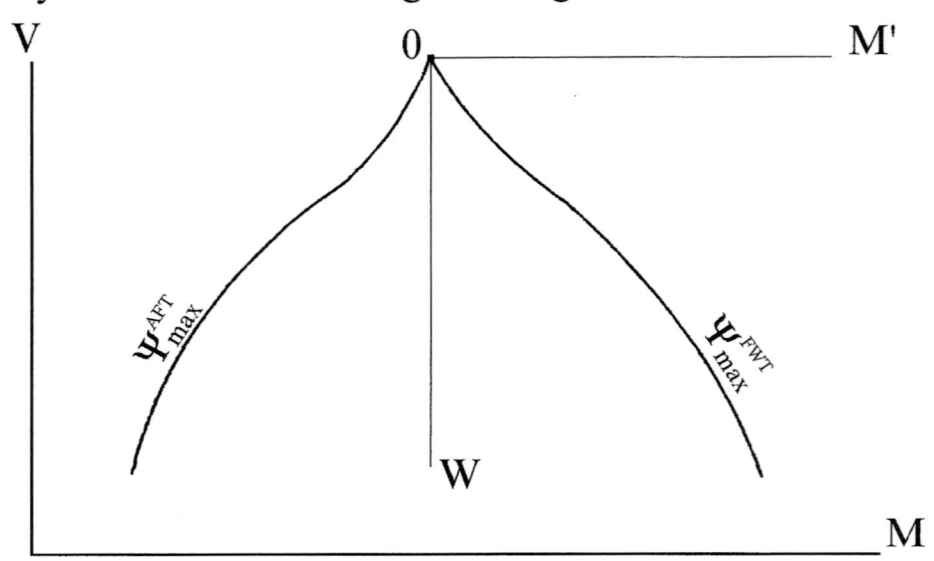

Fig.9.7. To Reserve of Buoyancy Estimations

Non-dimensional values of the reserve buoyancy and its moment may be found from:

$$\overline{W}' = \frac{W'}{V_0} \qquad (9.9)$$

$$\overline{M}' = \frac{M'}{V_0 L_0} \qquad (9.10)$$

where $\overline{W}'$ – relative reserve buoyancy remaining after the accident;
W' – reserve buoyancy remaining after the accident, m^3;
V_0 – constant buoyant volume of the submarine for every plotted floodability diagram, m^3;
$\overline{M}'$ – relative moment of the above-water volume;
M' – moment of the reserve buoyancy remaining after the accident in terms of the length from the midship section, m^4;
L_0 – length of the watertight part of the submarine, m.

Having thus converted dimensional diagrams of similar submarines, e.g., of airship architecture, into non-dimensional, we can obtain a family of curves that, as follows from Fig.9.8, may with an accuracy sufficient for initial design stages be replaced by a single curve. This allows us to apply the obtained non-dimensional diagram to first approximations for any submarine of the subject type.

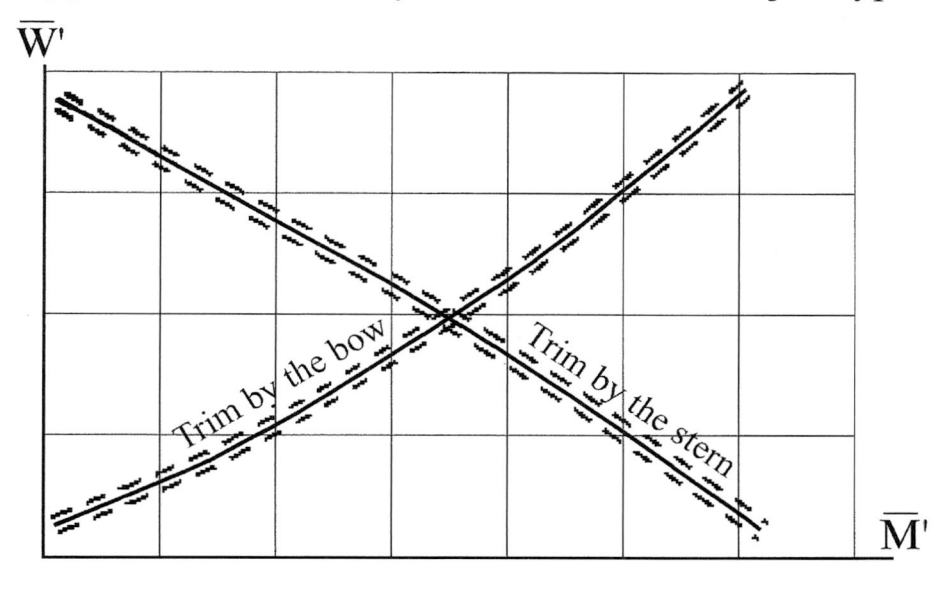

Fig.9 8. The Non-Dimensional Floodability Diagram Suggested by A.V.Bazilevich

The procedure for this method of submarine reserve buoyancy estimations is as follows: Firstly, we choose a prototype submarine similar in shape and arrangement of compartments and tanks. The lines of maximum tolerable equal trim angles from the prototype floodability diagram are converted into non-dimensional curves. Then, using the general arrangement plan of the new submarine, we find the worst damage case and make approximate estimations of the net flooded compartment and tank volume V_i and its moment with respect to the midship section $V_i x_i$ for this chosen damage case. Next we calculate the non-dimensional reserve buoyancy remaining after the accident $\overline{M}'$ and apply it to the plot (Fig.9.8).

$$\overline{M}' = \overline{M}'_0 - \frac{V_i x_i}{L_0 V_0} \qquad (9.11)$$

where $\overline{M}'_0$ – reserve buoyancy moment of the prototype submarine for which this diagram has been originally plotted.

From the diagram we take the non-dimensional $\overline{W}'$ of the reserve buoyancy remaining after the accident (assuming that the trim angle does not exceed the tolerable limit). Then the required reserve buoyancy as a percentage of V_0 will be:

$$\Delta \overline{V}^{FWD} \cdot 100\% = \left(V + \frac{V_i}{V_0}\right) \cdot 100\% \qquad (9.12)$$

After that we, in a similar way, find the required reserve buoyancy for the case of damaged trim by the stern $\Delta \overline{V}^{AFT}$ and take the greater of the two obtained values.

In case the designed submarine has some free-flooding tanks, the required reserve buoyancy increases. However, it is rather difficult to estimate this increase and for such a case the non-dimensional floodability diagram is usually inapplicable. This is explained by the fact that non-dimensional curves are quite sensitive to both the volume and the longitudinal position of free-flooding tanks.

4. When the required reserve buoyancy is estimated with design formulae, we should first of all remember that the aim is to ensure surface floodability based on other conditions than the above-stated. After the accident the damaged submarine is supposed to remain afloat with a positive longitudinal stability regulated in terms of the maximum lever of the longitudinal righting moment.

Let us for the sake of an example consider a method developed by Yu.K.Prytkov. Under his procedure the sought reserve buoyancy value depends on the following parameters:

$a = \frac{x_{DC}}{L_0} + \frac{V_i x_i}{L_0 V_0}$ – relative lever of the centre of gravity of the damaged submarine;

$b = 1 + \frac{V_i}{V_0}$ – relative submerged volume of the damaged submarine;

$k = \frac{\ell_{max} \cdot 10^3}{L_0}$ – relative lever of the longitudinal righting moment of the damaged submarine;

$\sigma = \frac{V_{FFT}}{V_{MBT}}$ – ratio of free-flooding tanks to the total number of MBTs MBTs on the submarine;

where V_i – total flooded volume (net) of compartments and tanks;
x_i – abscissa points of centres of buoyancy of these volumes;
ℓ_{max} – target maximum lever of the longitudinal righting moment of the damaged submarine;
V_{MBT} – volume of main ballast tanks;

V_{FFT} – volume of free-flooding tanks;
V_0 – normal displacement of the submarine;
L_0 – watertight volume length in surface trim.

Then, there are approximate formulae for the required reserve buoyancy $\Delta \overline{V}$ for damage cases of trim by the bow $\Delta \overline{V}^{FWD}$ and trim by the stern $\Delta \overline{V}^{AFT}$ for a submarine with only kingston-valve tanks.

$$\Delta \overline{V}_{KT}^{FWD} = f_1(a; b; k)$$
$$\Delta \overline{V}_{KT}^{AFT} = f_2(a; b; k) \qquad (9.13)$$

If for some reason the new submarine is designed with a part of the main ballast in free-flooding tanks, the required reserve buoyancy, as has been already mentioned, increases:

$$\Delta \overline{V}_{FFT}^{AFT(FWD)} = k_{FFT} \Delta \overline{V}_{KT}^{AFT(FWD)} \qquad (9.14)$$

where k_{FFT} – influence coefficient of free-flooding tanks on the reserve buoyancy that can be determined as:

$$k_{FFT} = 1 + 0.3 \left(1.0 - \frac{x_{AFT}}{L_0}\right) \sigma \qquad (9.15)$$

where x_{AFT} – abscissa of the centre of buoyancy of kingston-valve-main ballast tanks.

From this formula it is evident that k_{AFT} may vary from 1.0 in case all tanks are equipped with kingston valves to 1.3 when all tanks are free-flooding. It means that for the submarine with free-flooding tanks, all other conditions being equal, the reserve buoyancy required for surface floodability is 30% higher.

This increase in the reserve buoyancy can also be explained in physical terms. If a surfaced submarine was holed, due to the increased water head caused by the increased draught the water would flood not only the damaged compartment and MBT but the free-flooding tanks as well. This effect calls for a higher reserve buoyancy [89].

Today almost all Russian submarines are built to ensure surface floodability but there is a tendency towards the partial surface floodability standard, i.e. when a compartment is flooded only partially. This is possible if larger compartments are subdivided by watertight decks designed to pressures of 1.5 to 2 atm. In case there is no requirement for submarine surface floodability, reserve buoyancy estimations become much simpler as the sought value is mostly dictated by surface trim seakeeping abilities.

The specified (or target) value of the reserve buoyancy may be used as a condition in approximate determinations of the submarine draught.

Let V_1 be the reserve buoyancy of the submarine under design.

From the prototype we derive the non-dimensional coefficient:

$$\lambda_0 = \frac{V_{1_0}}{F_0 L_{PH_0}} \qquad (9.16)$$

where V_{1_0} – reserve buoyancy of the prototype;
   F_0 – area of the prototype pressure hull segment above the full-buoyancy waterline (Fig.9.9);
   L_{PH_0} – length of the prototype pressure hull.

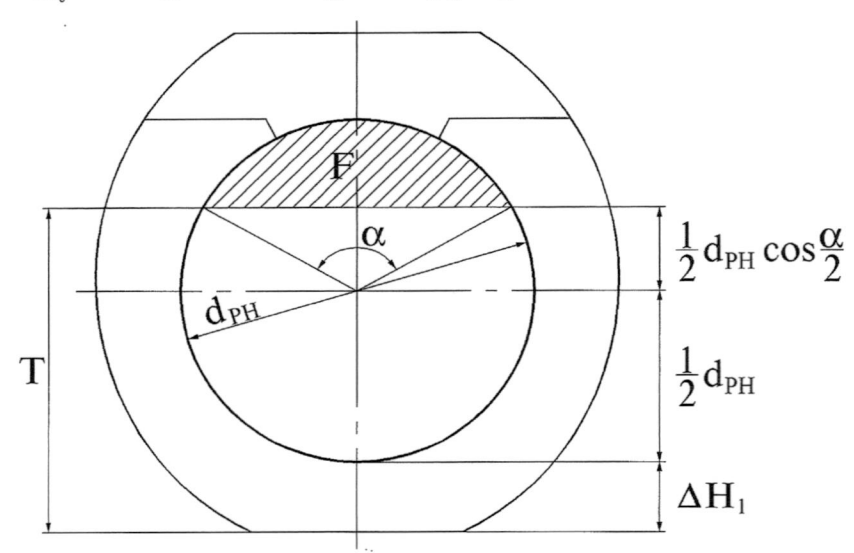

Fig.9.9. To Submarine Draught Estimations Based on the Specified Reserve Buoyancy

Assuming that for the new submarine this coefficient will be close to the found one, we estimate the prototype segment area:

$$F = \frac{V_1}{\lambda_0 L_{PH}} \qquad (9.17)$$

where V_1, L_{PH}, F refer to the new design.

According to Fig.9.9, the submarine draught is defined by:

$$T = \frac{d_{PH}}{2} \cos\frac{\alpha}{2} + \frac{d_{PH}}{2} + \Delta H_1 = d_{PH}\cos^2\frac{\alpha}{4} + \Delta H_1 \qquad (9.18)$$

where d_{PH} – pressure hull midship diameter;
   α – central angle corresponding to segment F.

With a given diameter of the circle, angle α for the segment F area can be found from any mathematical reference book.

9.6. Application of the Stability Equation in Estimations of Main Dimensions

In has been already stated that the main service condition for modern submarines is submerged operation. Nevertheless, this does not mean that surface trim and surface propulsion are not any more vital. It is in the surface trim that the submarine leaves the base and returns back. In emergencies submarines also come up onto the surface. Therefore, as early as the initial design stages it is necessary to make an approximate assessment of surface stability parameters and evaluate how the stability may influence the choice of main dimensions.

In order to apply the initial transverse metacentric height formula

$$h_0 = z_c - z_g + \rho \qquad (9.19)$$

to estimations of submarine main dimensions, let us represent the offsets of the centre of buoyancy z_c and the centre of gravity z_g, as well as the transverse metacentric radius ρ, as functions of main dimensions and form factors.

Approximate formulae for z_c and ρ are:

$$z_c = k_c T; \quad \rho = k_\rho \frac{B_0^2}{T} \qquad (9.20)$$

where B_0 – full-buoyancy waterline beam.

Coefficients k_c and k_ρ may be found with corrected (in terms of numerical factors) formulae suggested by professor B.M.Malinin [55]:

$$k_c = k_1 \left(1 - 0.415 \frac{\delta}{\alpha}\right) \qquad (9.21)$$

$$k_\rho = k_2 \frac{(2\alpha + 1)^3}{\delta} \qquad (9.22)$$

Numerical values of k_1 and k_2 obtained from statistic analysis of modern submarine data are:

$k_1 = 0.860; \quad k_2 = 0.0030$ – for double-hull submarines;

$k_1 = 0.860; \quad k_2 = 0.0028$ – for single-hull submarines.

We should note that coefficient δ in (9.21) and (9.22) is referred to the overall beam, length and draught on the design waterline while coefficient (is referred to the beam on the full-buoyancy waterline.

The centre of gravity offset is found in fractions of the hull midship height:

$$z_g = \mu H \qquad (9.23)$$

where $\mu = 0.43$ to 0.45 for double-hull submarines;
$\mu = 0.39$ to 0.41 for single-hull submarines.

If a preliminary load balance estimated from weights and moments is already available, the z_g offset should be taken from that table of loads.

Higher μ values for double-hull submarines are explained by the fact that in this case the pressure hull is elevated above the base plane by the height of the vertical keel.

Substituting values z_c, z_g and ρ into (9.19), we obtain the stability equation

$$h = k_1\left(1 - 0{,}415\frac{\delta}{\alpha}\right)T + k_2\frac{(2\alpha + 1)^3}{\delta}\frac{B_0^2}{T} - \mu H \qquad (9.24)$$

Considering that the beam on the full-buoyancy waterline is related to the overall beam by a rather stable (within one architectural type) coefficient

$$B_0 = k_3 B \qquad (9.25)$$

where $k_3 = 0.92$ to 0.95 for double-hull submarines;
$k_3 = 0.75$ to 0.80 for single-hull submarines;

equation (9.24) may be written as:

$$h = k_1\left(1 - 0{,}415\frac{\delta}{\alpha}\right)T + k_2 k_3^2 \frac{(2\alpha + 1)^3}{\delta}\frac{B_0^2}{T} - \mu H \qquad (9.26)$$

It should be emphasised that coefficients k_1, k_2, k_3, as well as μ, quoted in this book come from submarines with different reserve buoyancy values. Therefore, they are not directly applicable to practical submarine design and should always be corrected to the relevant prototype.

While we don't yet have any results of static calculations, formulae (9.25) and (9.26), the same as approximate formula (9.20), may be used for:

– approximate estimation of the initial surface metacentric height h_0;
– approximate estimation of h_0 based on tentative main dimensions B and T and coefficients α and δ the latter are then taken from the prototype or statistic data described in Chapter 8);

– estimation of the submarine beam required for the specified initial surface metacentric height.

For this purpose it is convenient to solve (9.24) for B. If the draught T has already been tentatively estimated, the expression for B is written as:

$$B = \sqrt{\frac{(h + \mu H)T - k_1(1 - 0{,}415\,\delta/\alpha)T^2}{k_2 k_3^2 \frac{(2\alpha + 1)^3}{\delta}}} \qquad (9.27)$$

When the draught is unknown, we can divide the right and the left terms of (9.27) by B and after relevant transformations arrive at:

$$B = \frac{h + \mu H}{k_1\left[1 - 0.415\,\frac{\delta}{\alpha}\right]\left[\frac{T}{B}\right] + k_2 k_3^2 \frac{(2\alpha + 1)^3}{\delta}\left[\frac{B}{T}\right]} \qquad (9.28)$$

To use (9.27) it is necessary to assume some coefficients α and δ, and for formula (9.28) we also need a value of $\left(\frac{B}{T}\right)$.

Having estimated the submarine beam we can find all other main dimensions.

PROVEN EXCELLENCE IN NAVAL AND COMMERCIAL SHIPBUILDING

State-owned enterprise Admiralty Shipyards is the oldest yard in Russia. The pedigree of Admiralty Shipyards stretches back to the earliest days of St.Petersburg. In 1704, a year after the foundation of the new capital of the new Russia, the genius of Peter the Great did more than simply established a shipyard, he created the cradle of the Russian Fleet thus laying the foundations of Russia's might and influence for centuries to come.

The yard has built more than 3000 ships and vessels of various classes including more than 300 submarines. In particular, state-owned enterprise Admiralty Shipyards is manufacturer of Kilo-class submarine which is designed by the CDB ME "RUBIN" and is acknowledged one of the most successful non-nuclear submarines recently built in the world. This submarine which have accumulated a 90-year experience in the construction of submarines has won world renown reputation for its low noise level, high sensitivity of the sonar antenna and high reliability. 21 submarines of Kilo-class are delivered in 6 countries of the world.

At present the Admiralty Shipyards manufactures two submarines of the new generation.

203, Fontanka Emb., St.Petersburg, 190121, Russia
Tel.: +7-812-114-88-63. Fax: +7-812-311-13-71
E-mail: azovceva@ashipyard.main.ru

10. SUBMARINE LINES DRAWING

10.1 General

The significance of the lines drawing is due to the fact that it is a fixed record of the hullform, which, as we have already stated, has a direct impact on characteristics of the submarine. Therefore, along with load balance calculations, constant buoyant volume calculations and general arrangement drawings, the lines drawing should be regarded as a major design document [14].

The lines drawing is generated based on two fundamental postulates:
- at fixed main dimensions of the submarine it is necessary to meet the specified volume and integral hullform parameters like area and volume coefficients;
- preferably, all other hullform parameters, like the parallel middlebody length and its longitudinal position, fore and aft end shapes, etc., should be close to what is optimum for the subject design.

The lines drawing serves for the following purposes:
- to give a comprehensive and clear idea about the external shape of the submarine hull;
- to make calculations associated with tasks aimed at finding design characteristics of the submarine;
- to manufacture models for towing tank and wind tunnel tests aimed at finding performance and manoeuvrability qualities of the intended submarine;
- to achieve the proper shape of the submarine hull during its construction by full-scale lofting of plates, frames, stringers and other parts at the yard;

- to help in service-associated tasks (hull repairs, docking, assistance to damaged submarine, etc.).

For the sake of these purposes the lines drawing is included in the technical document package issued for every submarine [31].

The submarine lines drawing is understood as the whole family of projections (to principal planes) of hull sectional lines formed by planes passed parallel to the respective principal ones, including:
- the horizontal plane passing through the upper edge of the middle straight portion of the keel is called the *base plane (BP)*;
- fore-and-aft vertical symmetry plane of the submarine hull passing perpendicular to the base plane is called the *centerplane (CP)*;
- transverse vertical plane (perpendicular to the above two planes) passing through the midship plane of the submarine length is called the *midship plane*.

Principal plane intersections are usually taken for grid axes of the submarine body-fixed co-ordinate system:
- the centerplane intersection with the base plane is the X (abscissa) axis positive towards the bow;
- the centerplane intersection with the midship plane is the Z (Z-direction) axis, positive upwards;
- the midship plane intersection with the base plane is the Y (ordinate) axis, positive towards the starboard.

The grid origin is the cross-point of principal planes and respective co-ordinate axes.

Planes parallel to the centerplane are called *bow and buttock planes*; planes parallel to the base plane are called *moulded waterplanes*; planes parallel to the midship plane are called *transverse sectional planes*.

The lines drawing normally shows the inner surface of the submarine hull external plating.

Submarine hull intersection lines on the bow and buttock planes are called *bow and buttock lines*. The centerplane intersection with the submarine hull surface - the centerline - forms the line of the keel, stempost and sternpost lines and the deck centerplane. Bow and buttock lines are counted off from the centerplane, separately to starboard and portside.

Intersections of moulded waterplanes with the submarine hull surface are called *moulded waterlines*. The base plane is taken as the zero waterline and all other moulded waterlines are counted off from it, upwards.

Intersections of transverse sectional planes with the submarine hull surface are called *frame stations*. Frame stations are counted from the bow sternwards starting with the zero station. Usually, the pressure hull length or the length between the foremost bulkhead of the bow MBT to the aftmost bulkhead of the stern MBT (the watertight length) is divided into 10 or 20 frame spacings. The end portions have their own station spacings assigned depending on their lengths and on the sophistication of their shapes.

Projections of frame stations, moulded waterlines, bow and buttock lines to the centerplane together form the *profile* plan, their projections to the base plane form the *half-breadth* plan and projections to the midstation plane form the *body* plan. Thus, the submarine lines drawing consists of three views: profile, half-breadth and body (Fig.10.1.)

The profile plan shows bow and buttock lines in their true shape while moulded waterlines and frame stations are indicated as horizontal and vertical lines respectively.

The half-breadth plan shows moulded waterlines in their true shape while bow and buttock lines and frame stations are represented by straight horizontal and vertical lines. Since moulded waterlines are symmetric with respect to the centerplane, the plan shows only their halves, usually portside from the centerplane, and hence comes the name of this projection.

For the sake of convenience they often draw two half-breadth plans and place them one under the other. In this case one view shows moulded waterlines from zero to full-buoyancy while the other one represents waterlines from full-buoyancy to submergence.

The body plan shows the true shape of the frame stations with moulded waterlines and bow and buttock lines marked by straight horizontal and vertical lines. Since bow and buttock lines are symmetric with respect to the centerplane, the plan shows the complete projection only for the midship section. The rest are presented as half-sections: fore stations to the right of the centerplane and aft stations to its left (Fig.10.2).

The lines drawing shows 10 or 20 station sections, 8 to 10 moulded waterlines and 2 or 3 bow and buttock lines for either side.

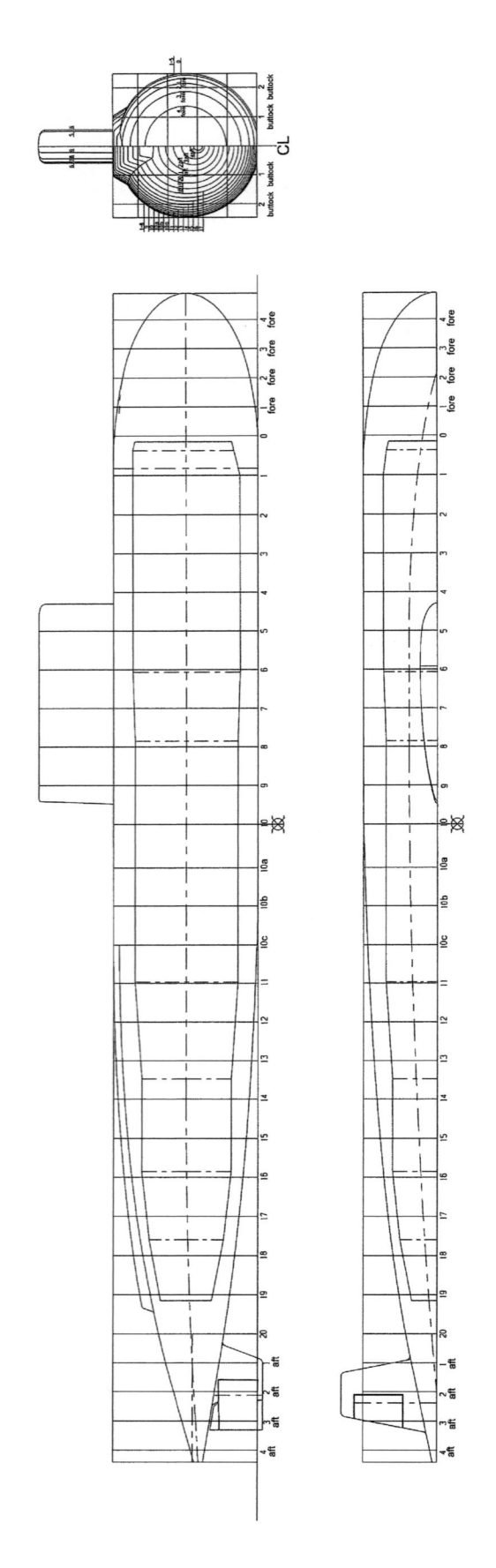

Fig.10.1 Submarine Lines Drawing

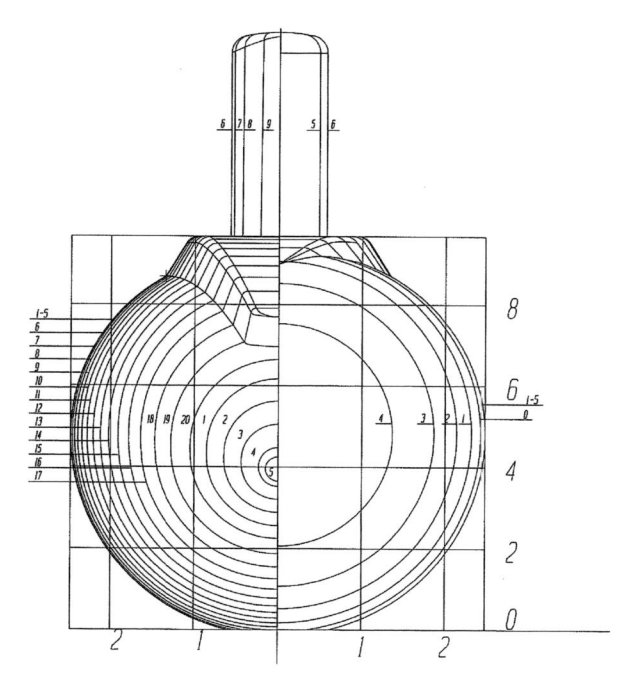

Fig. 10.2. Submarine Lines Drawing (Body Plan)

Additionally to submarine outer hull sectional lines, the lines drawing shows:
- the outline of the submarine pressure hull;
- shaftline axes, propeller disks, axes of torpedo tubes, etc.;
- basic particulars of the submarine, overall dimensions, etc.

The main task of the lines drawing designer is to achieve a faired, matched, in terms of all co-ordinates, surface of the hull that would satisfy the whole complex of requirements set to the submarine, as well as to find hull surface co-ordinates with an accuracy sufficient for final computations and for simplifying submarine construction work. These tasks are performed throughout the submarine design process, though with different accuracy and detail elaboration as suits the goals of each design stage.

10.2. Phase-Wise Specifics of Submarine Lines Drawing Design

The fact that the submarine has a pressure hull and a light (outer) hull, and the priority assigned to submerged characteristics and at the same time the need to ensure certain surface seakeeping parameters have forced quite specific approaches and methods for lines drawing design to evolve. These approaches and methods vary from one design stage to another because different phases pose different tasks that have to be considered with the help of the lines drawing. At the same time the submarine lines drawing design undoubtedly retains many common features with the surface ship lines drawing design, has a much longer background and has accumulated extensive experience in generating complex hull surfaces [60].

Lines drawing design is naturally related to work on the general arrangement, on the load balance, on stability, seakeeping, performance, manoeuvrability parameters, and therefore needs close cooperation with experts on those disciplines and obligatory checks for all possible consequences of any changes in the lines.

Computer-based procedures for lines drawing generation, which have been developing so rapidly in recent years, are actively overtaking traditional «drafting» techniques. However, the latter are still relevant in the global context of understanding the task of hull surface generation and relations among various design characteristics. At the same time one should never forget that the overall appearance of any ship should be beautiful and that there are no better tools to achieve this than the eye of the designer and his perception of the future submarine. That is why at the concluding stage of hullform design, the lines drawing needs to be finalised «visually», and actually after that the verification tests in the cavitation tunnel prove the high quality of the flow around the submarine hull.

Thus, lines drawing design is a rather wide and specific domain of engineering that requires special training in terms of complex surface generation and a very broad general background in naval architecture [49].

Let us review the tasks that have to be solved with the help of the lines drawing design at different phases of submarine design.

The most crucial aspect of conceptual design work is the comparative analysis of various options, and therefore the submarine lines drawing is often made as a combination of simplistic shapes that enables principal geometric characteristics in their finite analytical format to be obtained and considered as functions of variable parameters. The most commonly-used geometrical models are combinations of an ellipsoid bow, a cylindrical middlebody and a straight or a parabolic cone for the after end.

Such a simplistic model enables us to investigate feasibility and advisability of the specified main dimensions and their ratios, draught, displacement and integral hullform characteristics like area and volume coefficients. Parallel to these efforts, they optimise geometric ratios aiming at highest values of the chosen particular and general criteria (performance, manoeuvrability, noise, cost, efficiency). This is done using both direct and inverse methods of the design theory.

At this stage the lines drawing exists as a mathematical model describing variations in the total submerged displacement, main dimensions, wetted surface area, draught, etc. Its graphical format is limited to a basic schematic of hull lines [7].

It might be interesting to note that the rapid progress in computer hardware and numerical methods for surface generation already extends this approach to later design stages, so that the lines drawing starts losing its significance as a graphical document and is increasingly substituted by a sort of mathematical model that allows any geometric data to be obtained skipping graphic outputs.

At the technical proposal or tentative design stage the lines drawing is usually not submitted with the design package as a separate document. At that phase it is generated in a simplified form with only such a degree of detail elaboration that is required to find main dimensions, make basic calculations and prepare general design drawings.

Basically, at this design stage there is no special need to carefully elaborate the hullform and match lines drawing details. However, as long as the lines drawing is generated using modern plotting or computing hardware and methods this is in any case done automatically.

At the preliminary design phase the lines drawing is believed to be vitally important. This is associated with producing the whole set of other calculations and drawings covering various aspects of naval architecture and other relevant disciplines that need the lines drawing as a source of input data. Another important feature of this design phase is that at this stage they carry out a large amount of model tests in order to determine or verify performance, manoeuvrability and other design parameters. These exercises require models built to different scale ratios. In pre-computer days generation of model drawings and model manufacture used to be quite labour-intensive since some of the drawings had to be as long as 7 m. In this respect the situation improved when routing machines were suceeded by NC machines that work not from the drawings themselves but from numerical data about external surface offsets.

The lines are finalised based on the outcome of model tests for the structure of the flow around the submarine hull. This work includes optimisation of those areas that reveal unacceptably high velocity gradients, stimulate vortices, cause overflows among permeable structures, etc.

On the whole, hull lines together with the sail and control surfaces should fully satisfy all requirements set to the submarine and fixed in the approved design specifications. Therefore, at the engineering design stage the lines drawing is elaborated as a separate document that describes the final hullform of the submarine. Such a high significance of the lines drawing means that it should be produced as a high-quality and detailed document so that one could use it to scale and get any section or parameter associated with that section (perimeter, area, volume, etc.). Today, at the engineering design stage, lines drawings are generated exclusively by computers.

The detail design phase makes it necessary to adapt the lines drawing to construction facilities available at the yard. At this stage they need surface co-ordinates in the numerical format for hull detail nesting, for making templates, assembly jigs and other fixtures. The outcome of this work is a special document called the lofting table. Earlier (and sometimes still) they used to do this on the shipyard mould-loft floor where the hull was developed to the full size and they could make final checks for matching of surface points with respect to all co-ordinates and see whether the whole surface was properly faired.

In that classical arrangement the mould-loft floor was the tool to finalise sizes of all details to be fabricated at the yard.

Lofting has always been a very labour-intensive, expensive and long component of shipyard work, and therefore the history of submarine construction contains many attempts to somehow simplify this process. To make lofting easier, it is first of all necessary to know exact surface co-ordinates in terms of frame stations, waterlines and bow and buttock lines. Even when these co-ordinates are measured directly from a large-scale finalised lines drawing, discrepancies on the loft floor may eventually reach several tens of millimetres. This situation has made full-scale lofting an absolutely necessary stage of submarine construction. Obviously, only presenting the hullform in an analytical format enabling surface co-ordinates to be accurately computed, can offer a realistic way to faster and cheaper lofting, up to and including a complete changeover to computer simulation of hull surfaces and details instead of geometrical construction of structural components.

Such computer-aided design systems are already available and extensively utilised in shipbuilding all over the world. They are also applied to submarine design, though with certain modifications associated with specifics of this class of ships. Still, in the background of modern analytical methods rests the experience gained from classic graphical techniques of hullform design and their fundamentals remain to be vital elements of education for future naval architects.

10.3. Methods for Lines Design and Specification

All methods applied in lines drawing design may be grouped as follows:
 a) Methods applicable when the hullform is generated by sight;
 b) Methods based on reconstructions of prototype lines drawings;
 c) Methods based on geometric constructions;
 d) Analytical methods for lines drawing.

When designing and specifying complex surfaces, the designer has a wide choice of methods, every one of which has certain advantages and drawbacks. In order to review their comparative merits, let us apply them to generate one simple surface. This surface envelopes a body symmetric about the centerline and about the extreme half-breadth. The body is actually an analogue of the submarine hull. At the foremost point tangent lines are perpendicular to the centerplane and to the extreme half-breadth. The aft end has different angles of run with respect to aftlines and waterlines. The model has no cylindrical middlebody.

The Method of Orthogonal Sections

The eldest design technique for faired surfaces, used when drawing lines by sight, is the method of orthogonal sections otherwise known as the method of bow and buttock lines (fore-and-aft vertical sections) and contours (waterlines). This method is based on the laws of descriptive geometry. It appeared in the earliest days of shipbuilding and is still successfully used for some studies at initial design stages. The essence of this purely graphical technique boils down to generating a large number of transverse or longitudinal sections of the subject surface in three projections. The outlines of these sections are corrected against each other by projections of a large number of individual points in order to get fair lines. Intervals between bow/buttock planes and waterplanes are chosen depending on the size of the subject body and its surface curvature. The greater the number of sections, the more accurately specified the surface [7].

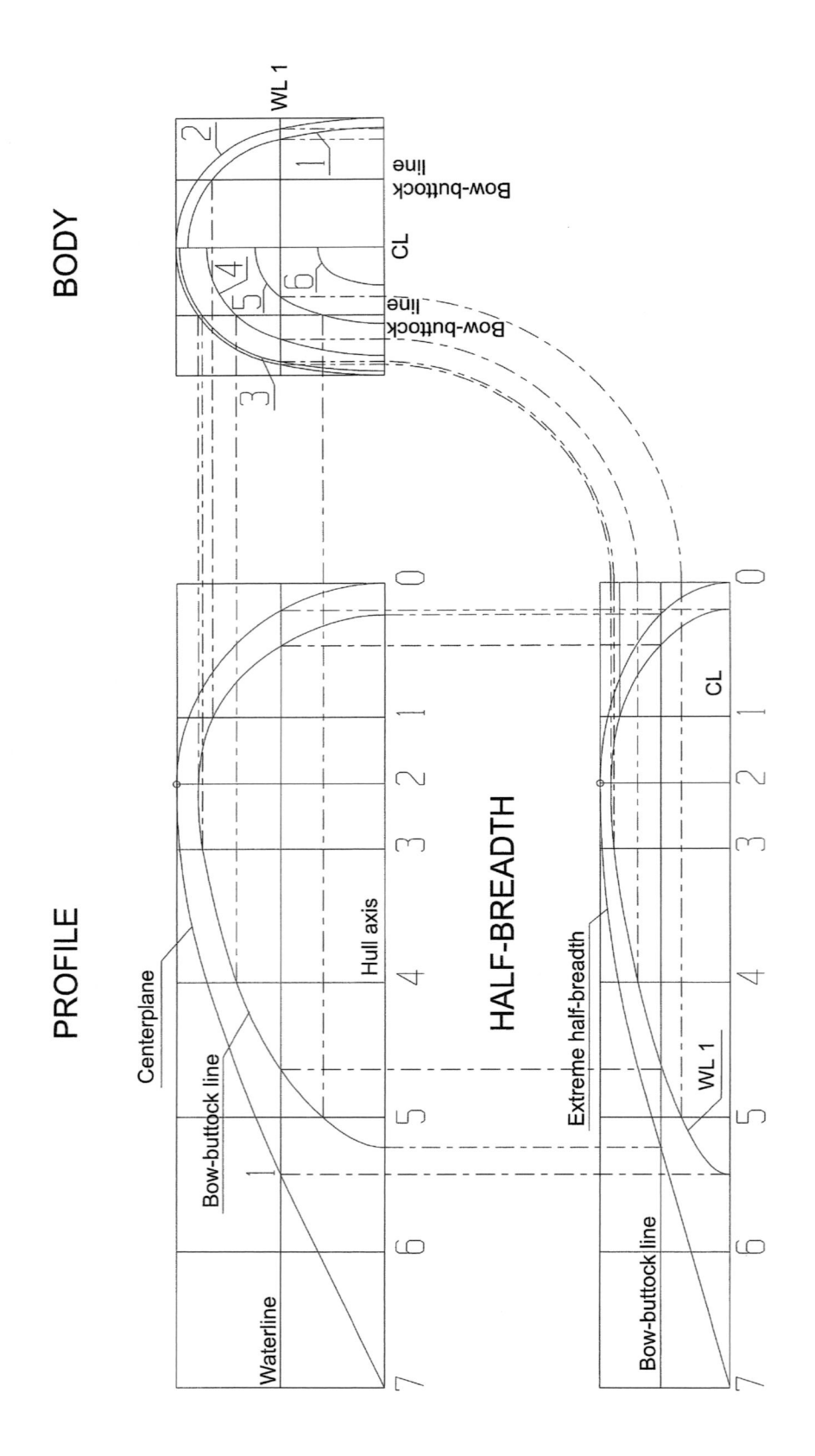

Fig. 10.3. The Orthogonal Section Method

The surface design and specification procedure is as follows (Fig.10.3.):

Using a draftsman's curve or a flexible bar, they draw specified principal longitudinal lines (in our example: bow and buttock lines - centerline and main contour - extreme half-breadth) through «critical» points where the hull line should pass (due to pressure hull and between-hull space equipment arrangements);

Then they plot co-ordinates of initial points (in terms of the centerline and the extreme half-breadth) for the chosen sections on the profile projection and tentatively trace these sections by freehand;

Then they mark traces of bow/buttock and waterline planes on projection plans;

Then from the hand-traced section lines they take co-ordinates of points to be transferred to the profile plan. The thus obtained points are connected with curves. Then they plot waterline curves on the horizontal projection in the same way;

After that co-ordinates of those individual points that fall out of fair curves of fore/aftlines and waterlines are corrected until all fore/aftlines, waterlines and station sections are properly faired.

Then with a draftsman's curve or a flexible bar they draw final curves of bow and buttock lines, waterlines and stations;

Finally, they measure point co-ordinates on the bow and buttock lines and waterline grid of respective sections and enter them in a table.

The method of orthogonal section gives a clear idea about the surface but fails to provide any unbiased evaluation of the surface since the only criterion of faired lines is their visual assessment.

The Method of Affine Reconstruction of Prototype Lines

In practical design they quite often apply various techniques of reconstructing lines from prototype drawings. This group of methods includes reconstruction by interpolations, reconstruction by the curve of sectional areas. The simplest way to re-draw prototype lines is to apply affine reconstructions. This reconstruction is possible only when the main dimensions are changed without changing the form factors [7]. The displacement, the centre of buoyancy, the areas of frames and waterplanes, the same as other particulars of the drawing, can be easily calculated with simple formulae based on partial similarity of the new submarine hullform and the prototype.

Considering a particular case of such a reconstruction when the only variable is the submarine length, we can find the change of the distance between frame stations due to the new length

$$\ell_a = \frac{L_1}{L_0} = \frac{\Delta L_1}{\Delta L_0} \qquad (10.1)$$

where index «0» refers to the prototype while «1» marks the new lines drawing.

In case the only changed parameter is the submarine hull height, to reconstruct the drawing we need only change the distance between the waterlines that may be found from:

$$t_a = \frac{T_1}{T_0} = \frac{\Delta T_1}{\Delta T_0} \qquad (10.2)$$

In case we are changing the beam, the offsets for the new drawing are obtained by multiplying the original ones by

$$b_a = \frac{B_1}{B_0} = \frac{\Delta B_1}{\Delta B_0} \qquad (10.3)$$

Thus, $y_1 = b_a y_0$. It is quite simple to plot such a drawing since faired lines of the prototype remain faired on the new drawing. Fig.10.4 shows a lines drawing reconstructed by length-wise affine enlargement of a drawing generated using the previous method. The enlargement factor was k=1.50.

The Radius Graphics Method

The radius graphics method for lines drawing generation in submarine design appeared comparatively recently, in the 1960s, but has quickly become quite popular. The greater part of lines drawings for post-WWII submarines, including «Typhoon», «Delta», «Yankee», «Kilo» and other classes, were produced using this very method.

The radius graphics method for complex surface design (introduced by engineer D.S.Kitainov) [37] is based on a geometric model that considers the surface of any voluminous body as a 3D evolvent curve of a certain evolute surface. Methods for representing and generating this evolute, which is called the centric evolute and is a family of 3D curves of centres, essentially constitute the radius graphics technique. For the sake of an example Fig.10.5 offers an assembled section parallel to the vertical plane ZY of an ellipsoid hull line and its evolute surface.

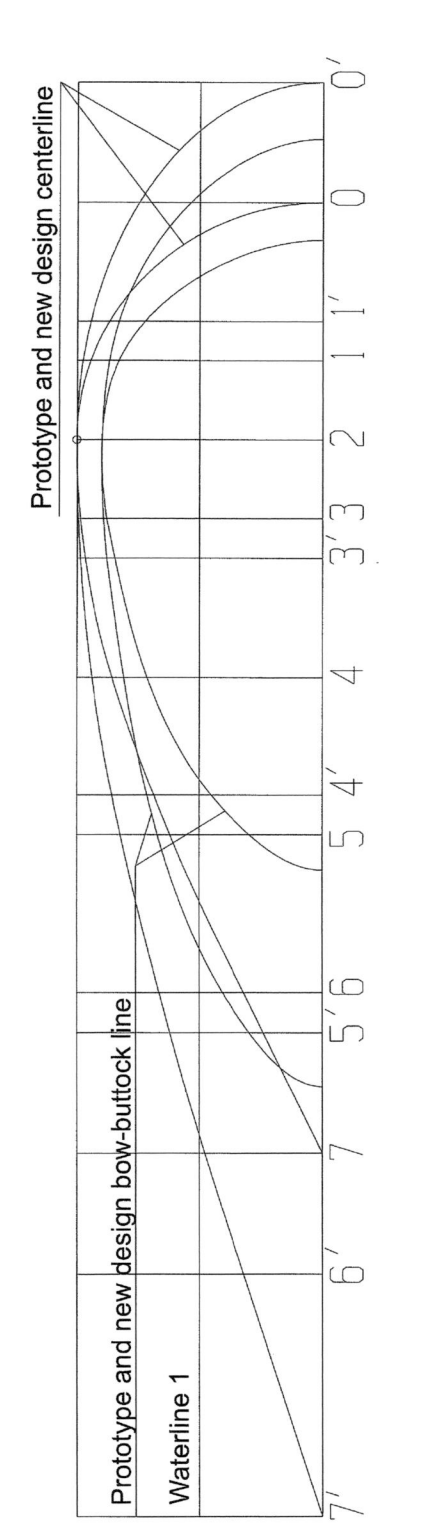

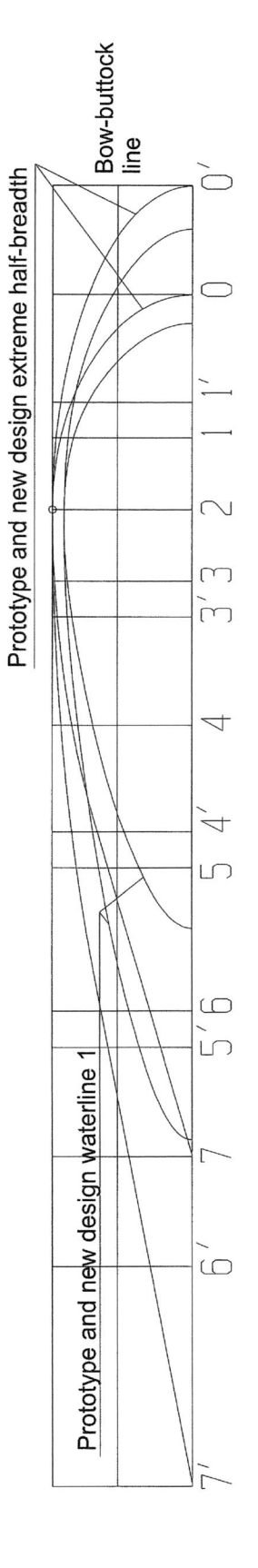

Fig.10.4 Lines Drawing Affine Reconstruction

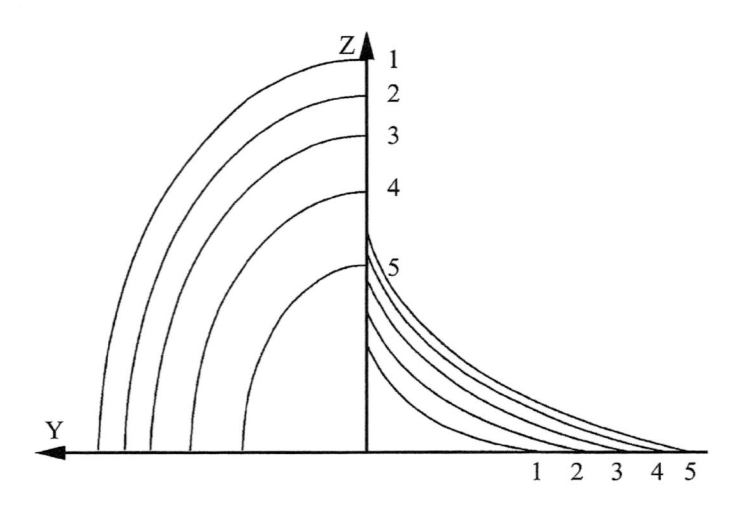

Fig. 10.5. Assembled Vertical Sections of an Ellipsoid Octant and its Evolute Surface

Practical experience shows that any comparatively complex evolute can always be substituted by several straight lines, i.e. quite significantly simplified. In this case the evolute may be presented as a sum of several segments of circles. The number of these arcs depends on the number of straight lengths substituting the evolute. Thus, from the subject curve, which in our case is an ellipse, we may go to its approximate equivalent: an oval consisting of several arcs. For a sufficiently experienced designer, substituting a curve by a polygonal line represents the hull curvature with a quite acceptable accuracy for practical applications. Since we have a simplified evolute, we can take it to the centre key (Fig.10.6). The centre key of a curve is a schematic notation of the initial radius and centre points that is written in the adopted system of co-ordinates and allows us to plot, analytically express and calculate the subject curve in Cartesian or polar co-ordinates using elementary equations of the circumference.

$$(y - y_0)^2 + (z - z_0)^2 = R^2 \qquad (10.4)$$

where y_0 and z_0 – co-ordinates of the centre of the circumference;
  y and z – current position co-ordinates.

Though circumference segments are second-order curves, the radius graphics method does not belong to quadratic curve procedures because it is based on absolutely different principles. It utilises both the purely graphic construction and the construction based on analytical calculations.

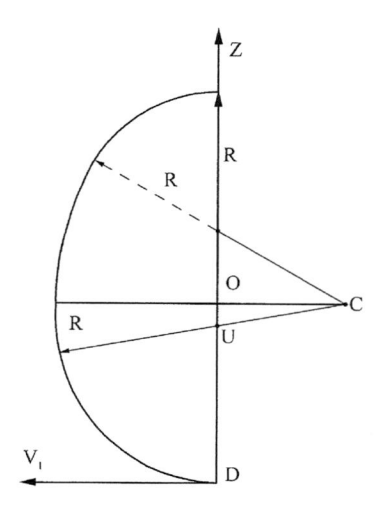

Fig.10.6. The Centre Key of a Flat Curve

It should be noted that it is quite difficult to take a very slight curve to the centre key. This comment concerns the longitudinal generating line: the centreline or the extreme breadth waterline. In this case the length of the submarine or of the subject end is compressed to $L_{com} = L/k$ where k is the compression factor found for the simplest centre key and convenient lines grid. Depending on specified conditions and on the involved type of longitudinal lines, compression factors for the fore and the aft ends may be different. This means that the radius graphics method utilises affine reconstructions as a part of its procedure. Let us apply the radius graphics method for generating a lines drawing from the same input data as was used for the orthogonal section exercise. Fig.10.7 shows the same surface as in Fig.10.3 but specified according to the radius graphics method.

The design and specification procedure is as follows:

Based on initial data, they draw principal fore-and-aft lines (in the subject case these are the centerline and the extreme half-breadth) and tentatively trace cross-section lines;

Depending on the type of cross-sections, with a pair of compasses they perform constructions to find the required number of conjugate arcs of different circumferences that will constitute cross-section lines (in our example it is enough to have two arcs with radii R_1 and R_2);

Then they tentatively mark position lines of the centres of R_1 arcs along the surface and this actually defines the method of section construction. The R_2 does not need any special specification since it is fully defined by the R_1 value and due to the fact that the circumference centre rests on the extreme half-breadth waterline (the axis of the subject body).

257

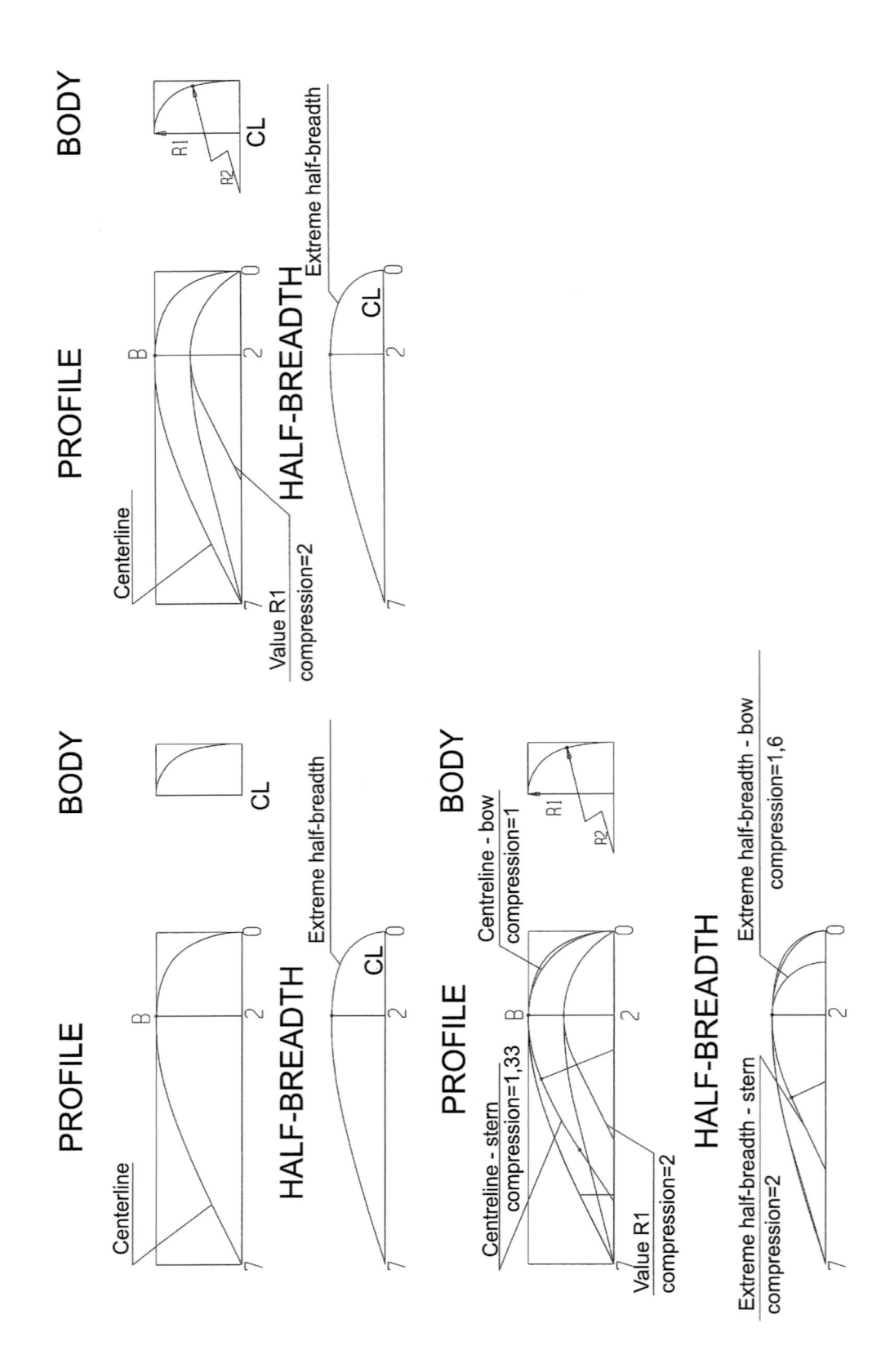

Fig.10.7. Radius Graphics Method for Lines Drawing Construction

Then they fit conjugate circumference arcs for principal and auxiliary fore-and-aft lines (the centerline, the extreme half-breadth, the R_1 radii plot). In order to cut down the size and the number of arc radii they apply scaled compression of longitudinal lines. In this example we have compressed the extreme half-breadth (k = 1.60), the aft portion of the centerline (k = 1.33) and the R1 radius plot (k = 2.00). The fore portion of the centerline may be adequately approximated by a single radius without any compression (k = 1.0). The compression and the subsequent representation of fore-and-aft lines by circumference arcs are performed in order to ensure obtaining faired curves that could be later calculated analytically. Faired longitudinal curves, combined with a single principle for generating all cross-sections, ensure that the whole surface is properly matched and faired. Basically, the fore-and-aft curves may be specified by any method that can warranty a fair surface, e.g., by quadratic curves.

Using the thus corrected initial points, they construct finalised cross-sections;

Then they establish dimensions for the lines of centres and arc radii and, if required, calculate lofting co-ordinates.

One of the major specific features of the radius graphics method is the warranted smooth variation of line curvatures (since the lines are specified by conjugate arcs of different circumferences and all of them are conjugated in terms of one and the same tangent line).

Hull lines graphic construction with the radius graphics method requires certain skill and experience because, to envelope fore-and-aft and cross-section lines, it is usually necessary to specify several curves of centres. On the other hand, with this method the lines drawing is generated mostly using only the simplest instruments: a pair of compasses and a ruler. Analytical calculations for co-ordinates are made, if required, after the lines drawing is already constructed. The whole family of centre keys together with compression factors forms the so-called radius graphics key of the lines drawing. With this key one can get any cross-section of the hull. And it is important to remember a valuable feature of this method: all cross-sections are sure to be matched and the hull has faired lines [49].

The radius graphics method has a number of advantages. The most important ones are:
 – the obtained hull lines are more correct and smooth than when constructed by drafting;

- if entered in a computer, the radius graphics key makes it possible to generate any hull section on a plotter;
- the method essentially renders lofting unnecessary since any hull section can be computed from the radius graphic key without going to the mould-loft floor;
- with the radius graphic method the lines drawing generation much faster since under this procedure there is no need to perform the most time-consuming step - matching projections;
- this method is rather well adapted for computer-aided design, and this feature significantly expands its capabilities.

This method is an optimum tool for circular hull lines typical for submarine architecture. The opportunity to bend hull details along several but constant radii makes construction technology much simpler. Surface co-ordinates can be calculated with any required accuracy.

Thanks to these merits, the radius graphics method has become very popular.

Analytical Methods for Lines Drawing Generation

Methods belonging to this group, whose founding father was a Swedish engineer F.G.Chapman, have been applied in international shipbuilding since 1760 and in Russia since 1831. These methods emerged at a time when there was a lack of other criteria and designers had to judge about the hullform by the beauty and smoothness of the drawn lines [7], [59], [60].

The best known analytical curves among all options suggested at different times for representing hullform sections are:

- parabolas

$$y^n = px \qquad \text{(F.G.Chapman)}$$
$$y = 1 - (ax^m - cx^n) \qquad \text{(D.W.Taylor)}$$
$$y = ax^4 + bx^3 + cx^2 + 1 \qquad \text{(L.M.Nogid)}$$
$$y = kx^m(1 - x^n)^p \qquad \text{(I.A.Jakovlev)}$$

- progression-based

$$y = \frac{1 - x^n}{1 - mx^n} \qquad \text{(A.A.Popov)}$$

$$y = \frac{(1-c)\sqrt{1-x^2}}{1-c\sqrt{1-x^2}} \qquad \text{(S. Artseulov)}$$

- plastic lines

$$y = (u^m - mu^{m/2}\ln u)^n \qquad \text{(V.I.Afanasiev)}$$

– elliptical inversion

$$y = \frac{a}{\sqrt{\text{tg}^2\theta + \left(\dfrac{a}{b}\right)^2}} \qquad \text{(I.P.Alymov)}$$

All these functions were originally intended for surface ship hull-forms. Nevertheless, I.G.Bubnov tried for the first time to apply them when designing his «Bars» submarine. This attempt to use such methods was made in the early days of submarine design but was not pursued further. Studies aimed to develop analytical methods for lines drawing construction were carried out by I.G.Bubnov, G.E.Pavlenko, A.A.Kurdiumov, K.B.Malinin.

It should be noted that practical application of analytical methods to lines drawings is rather difficult for a number of reasons. Firstly, it is difficult to find a suitable form of the equation for the generator curve because this equation should be sufficiently simple to be convenient for various mathematical manipulations and at the same time accurately enough describe the optimum hullform. Secondly, detailed analytical expressions should contain design characteristics found based on established main dimensions, form factors, etc.

Let us assume that the hullform curve may be generally described as

$$y = f(x, a, b, c) \qquad (10.5)$$

where a, b, c – coefficients found based on boundary or other conditions.

With a large number of coefficients the (10.5) equation enables any sort of a hullform curve to be reproduced. However, when we apply analytical curves we have to restrict the number of coefficients included into involved equations, and this leads to a more stringent specification of the curve. When there are too many coefficients, it becomes impossible to set any rational conditions for finding them.

It should be mentioned that the above-described radius graphic method is also analytical only its mathematical expression for the surface is very complex and there is no way to analyse it without computer assistance.

From these comments it follows that among analytical methods for lines drawing construction we should distinguish those that strictly define the hullform and those that allow one or two degrees of freedom to be retained. The majority of methods based on mathematical equations of curves belong to the first category. It should be noted that till recently the major obstacle hampering the progress of analytical methods was the lack of computation power. Naval architects tended to apply only the simplest curves that could be plotted with drafting tools. Today such difficulties are largely history and modern software packages manipulate with the most sophisticated functions, describing any sorts of surfaces. However, certain fundamental principles have never changed because they represent specific features of naval architecture, including submarine hullforms as a particular case.

Let us for the sake of certainty assume that equation (10.5) is a formula for the submarine waterline. Then it should satisfy at least four conditions mandatory for hullform curves. These generally known conditions are:
- at the foremost point ($x = 0$, $y = 0$) the angle between the tangent line to the generator curve and the OX grid axis should be $90°$;
- at the aftermost point ($x = 1$, $y = 0$) the slope angle of the tangent line to the OX axis should be, for the sake of achieving attached flow around the body of revolution stern and gaining the optimum propulsive coefficient, within 8 to $16°$;
- in order to minimise the submerged resistance, the largest cross-section of the body of revolution should be within $x = 0.30\sim0.40$ from the co-ordinate grid origin;
- at these X values the curve has a horizontal tangent line allows to put in the cylindrical middlebody.

These conditions can also be satisfied by the surface specification method of quadratic curves with variable coefficients. When the coefficients are specified by smooth differentiable functions, it is possible to get analytical hull surfaces.

Let us review lines drawing construction by the quadratic curve method extensively applied in shipbuilding and aircraft industries. The essence of this method is that quadratic curves are used to specify fore-and-aft lines, cross-section envelopes and auxiliary lines for surface generation. The lines themselves may be constructed graphically or calculated analytically. Both the analysis and the graphic construction techniques are based on the theory of quadratic curves which postulates that to define these curves it is enough to satisfy five conditions [8].

A quadratic line (curve) is a line described by an equation that in Cartesian co-ordinates is risen to the second power with respect to current co-ordinates.

The general 2nd – order (quadratic curve) equation with two unknown variables is formulated as:

$$Ax^2+Bxy+Cy^2+Dx+Ey+F=0 \qquad (10.6.)$$

Values of coefficients in the quadratic curve equation allow the type of the curve (elliptical, hyperbolic or parabolic) to be found. The relevant criterion is the value of $B^2 - 4AC$ called the analytical discriminant.

If $B^2 - 4AC < 0$ – the curve is elliptical;

$B^2 - 4AC > 0$ – hyperbolic curve;

$B^2 - 4AC = 0$ – parabolic curve.

As we may see from (10.6), a quadratic curve is defined by five coefficients (A, B, C, D, E), i.e. by five geometrical conditions. Indeed, let us assume we know co-ordinates of five points belonging to a quadratic curve.

Substituting their values into the general quadratic curve equation we obtain a set of five equations. Solving them jointly, we find such values for coefficients (A, B, C, D, E) at which co-ordinates of each of the five specified points satisfy (10.6) solution. As has been noted earlier, for submarine design it is better to specify a curve through angles of tangent lines to the hull line at the foremost (90°) and the aftermost (angle of run α =8~16°) points and three intermediate points. It is often more convenient to make graphic constructions with the help of the so-called projective discriminant, enabling us to plot a curve by specified tangent lines and a specified point [20].

Let us now review the design and specification procedure for the earlier-considered surface using quadratic curves based on three known points for the centerline and the extreme half-breadth, as well as two tangent lines for every curve (Fig.10.8).

First, we construct a triangle of three sides that are the hull length and tangents to the centerline at the foremost and the aftermost points. Inside this triangle we mark the maximum section point B.

Through this point B we draw half lines 1, 2 and 3. In this case half line 3 is the median of the triangle. The ratio of lengths AB and AP is called the projective discriminant: f = AB/AP. Together with the vertices of the triangle, the projective discriminant f uniquely defines the shape of the quadratic curve.

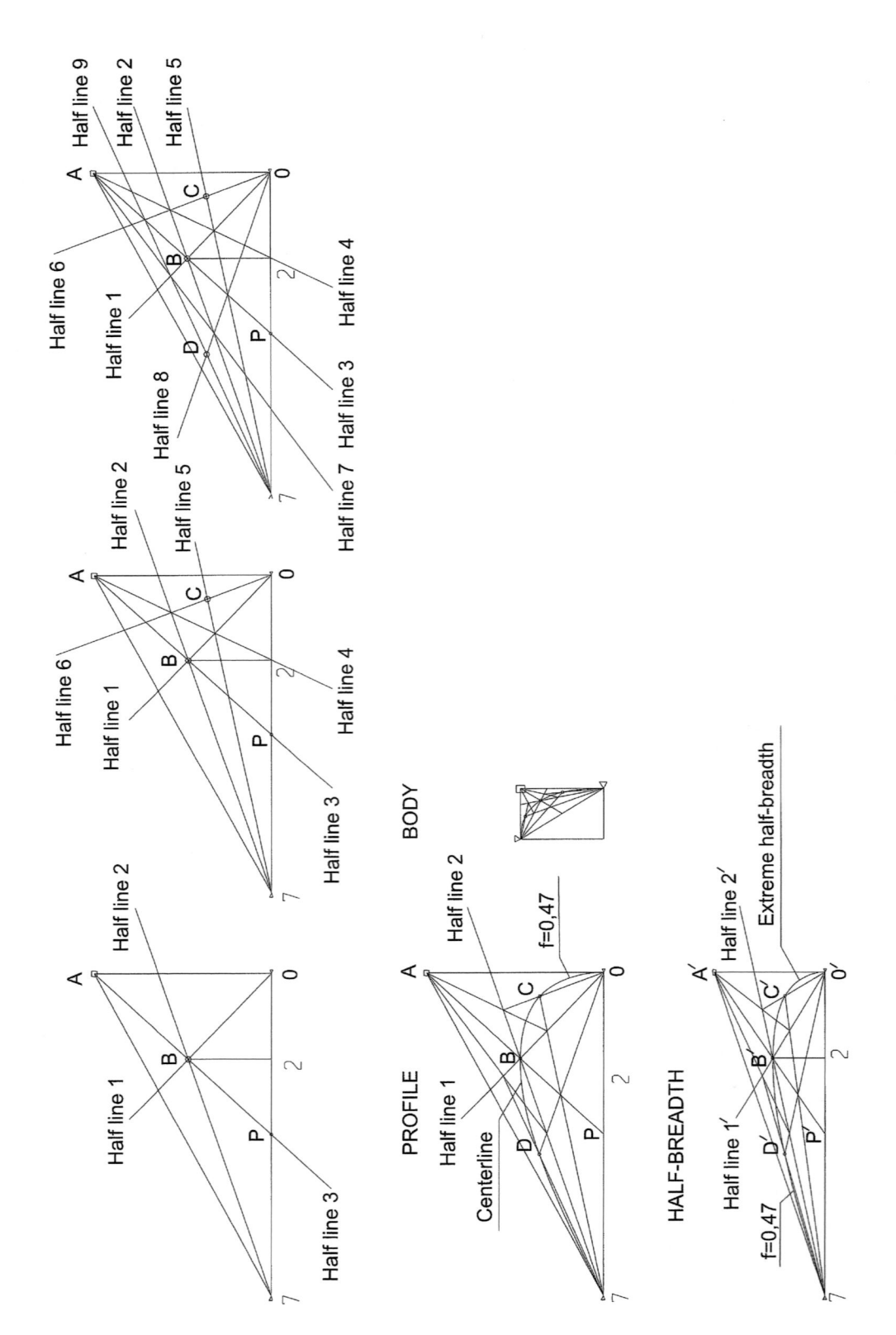

Fig.10.8 The quadratic curve method

To plot the second point of the curve, we arbitrarily draw half lines 4 and 5 intersecting on half line 1. Then we draw half line 6 through the intersection of half lines 4 and 2. The intersection of half lines 5 and 6 is point C that belongs to the sought curve. Similarly, with the help of half lines 7, 8 and 9 we construct the curve point D and, if necessary, other additional points. Then we check whether the thus generated curve fits «critical» dimensions.

In a similar way we perform constructions for the extreme half-breadth (points B, C and D) and the transverse sections.

In order to get regular smooth curves of longitudinal lines and station sections we assign a pattern for projective discriminant variations along the submarine.

Quite often the transverse sections (stations) are plotted with the help of additional bow and buttock sections that do not coincide with either waterlines or fore/aftlines. Such sections inclined to the base plane and centerline sections are called ribband lines.

A valuable advantage of this method is that it allows fast and simple construction of hull lines for any number of transverse sections by specifying just several longitudinal lines connecting the initial points. Thanks to this, the drawing gives a clear impression of the designed hullform.

The fact that quadratic curves can be subjected to affine transformations does not allow us to draw fore-and-aft lines of big bodies to the full size. Instead, it is possible to construct a «compressed contour» with all fore-and-aft lines specified by quadratic curves where straight lines are scaled down while the offsets are shown in true dimensions. The «compressed contour» enables us to reduce the size of rooms necessary for lines construction.

Thanks to the fact that quadratic curves are suitable for graphic constructions and analytical calculations of geometric parameters, it is possible to comparatively easily and quickly plot or compute them with sufficient accuracy.

It is also possible to combine the quadratic curve method with the radius graphics method. Then transverse sections (frames) are generated with the radius graphic method while centre key curves and principal side fore-and-aft lines are specified by quadratic curves.

The Stringer Plan

The stringer plan is a lines drawing (body plan projection) which, instead of frame stations, shows transverse sections corresponding to bulkheads of main and auxiliary ballast tanks and various bays arranged in the between-hull space. The plan represents longitudinals that limit tank and bay volumes, i.e. the stringers after which they call such design documents. Geometrical and digital information from the stringer plan enables us to mark these transverse sections on the mould-loft, to match them with longitudinal and transverse framing details, to develop expansions of plates and shaped materials and to fabricate construction fixtures.

The stringer plan is generated at the final submarine design stage, the engineering and detail design phases. Nowadays they sometimes successfully substitute this schematic with mathematical models in order to build submarines without the loft floor constructions. Then the stringer plan may serve as an illustration providing a clear idea of the between-hull architecture (see Fig.10.9).

10.4. Submarine Sail and Fin Lines Drawings

In present design routines, outlines and main dimensions of the mast sail and control surfaces are usually shown on the hull lines drawing in order to illustrate submarine overall dimensions defined by these components. At the same time, the sail and the fins are complex enough to require their own lines drawings. Special attention has to be paid to finding stock positions and graphical verification of rudder/hydroplane freedom to tilt through specified angles. As we have already mentioned, hull appendages should be described by dedicated lines drawings. The basic philosophy in this case is the same as for constructing the hull lines, though the grid is different and should be chosen so that subject lines would be represented in proper detail. Naturally, before we start making these drawings, it is necessary to determine sail sizes and the type of its lines, as well as aspect ratios of control surfaces. On the sail lines drawing they specify the centerline, several transverse sections and the half-breadth. Then they show transverse sections and waterlines on the drawing. The bow and buttock lines are usually omitted. The designer may choose any method to construct the drawing. Nowadays they do it on computers with whatever CAD system is available at the design office (Fig.10.10).

In fin lines drawings they make special emphasis on section design for control surfaces as low-aspect-ratio wings. For this purpose they apply catalogues of various wing sections. An example of a lines drawing for submarine horizontal control surfaces may be seen in Fig.10.11.

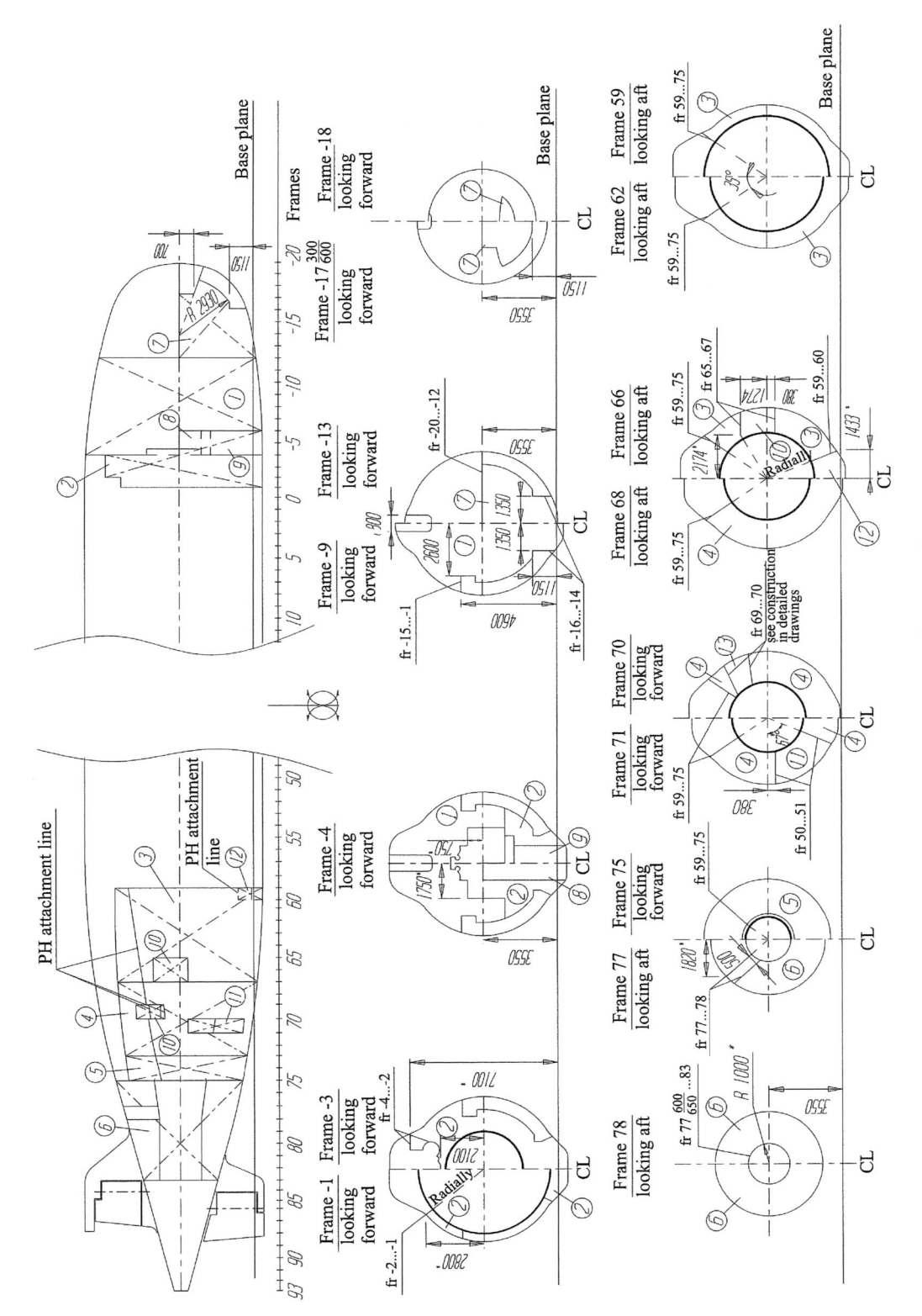

Fig. 10.9 A stringer plan

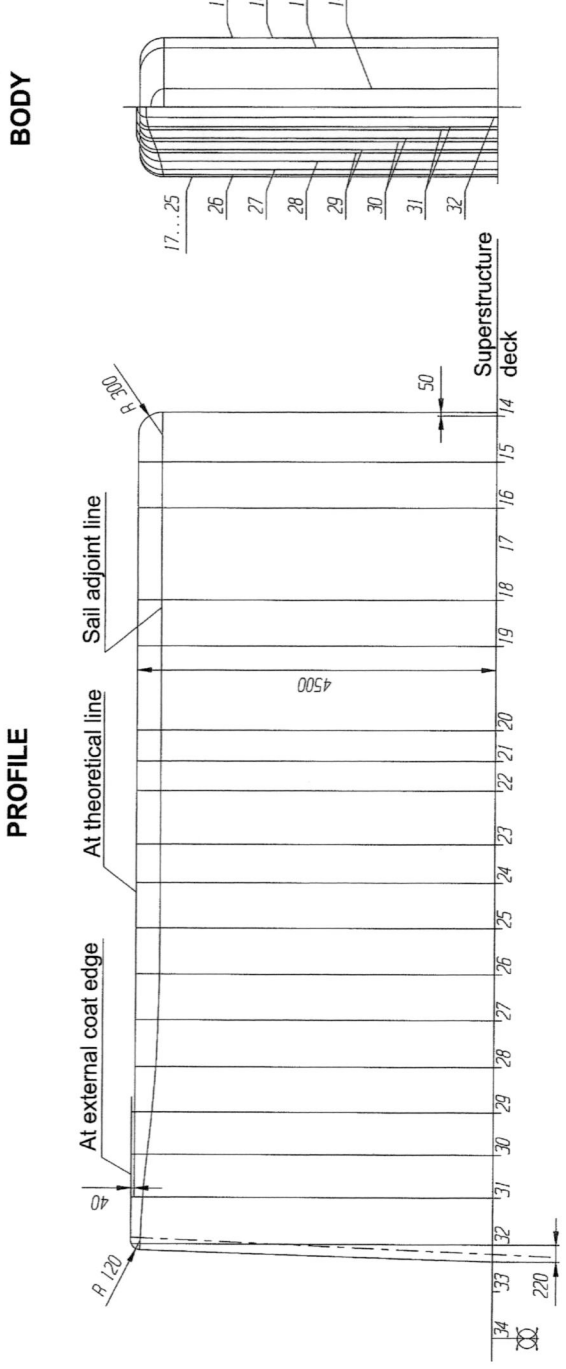

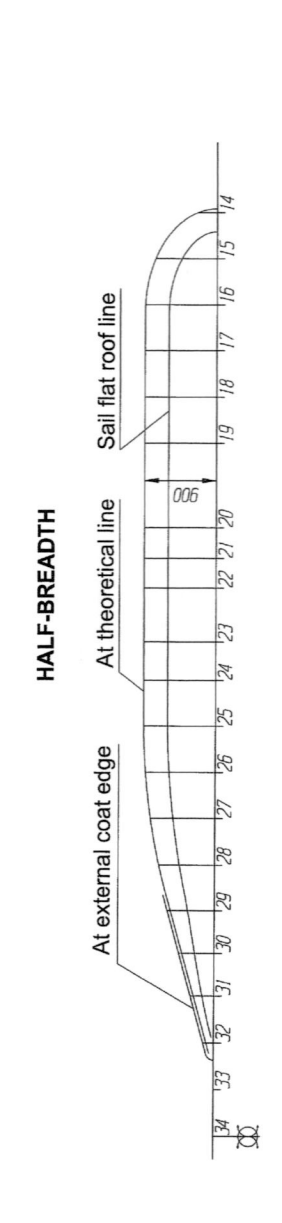

Fig.10.10 A Sail Lines Drawing

AFT HORIZONTAL PLANES (AHP)

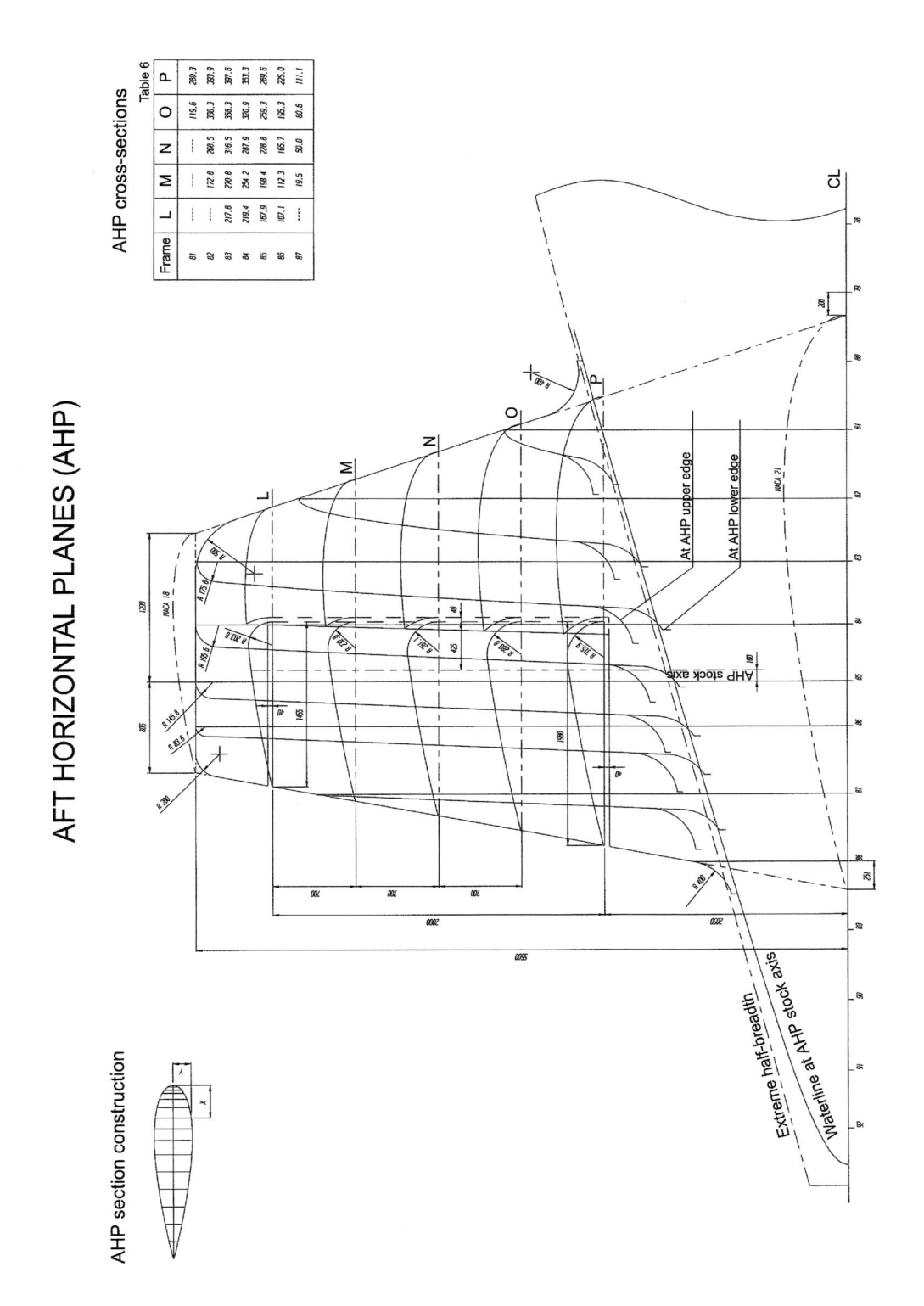

AHP cross-sections

Table 6

Frame	L	M	N	O	P
81	----	----	----	119.6	260.3
82	----	172.8	269.5	336.3	393.9
83	217.8	270.8	316.5	358.3	397.6
84	218.4	254.2	287.9	320.9	353.3
85	167.9	198.4	228.8	259.3	289.6
86	107.1	112.3	165.7	195.3	225.0
87	----	19.5	50.0	80.6	111.1

AHP section construction

Fig.10.11 A Fin Lines Drawing

10.5. Modern Computer Technologies for Lines Drawing Generation

Today there are many various CAD/CAM systems for the design of hull surfaces, structures and equipment. Each system has certain merits and disadvantages.

In submarine design they use CADDS5, AutoCAD, CADKEY and APIRS. AutoCAD and CADKEY are mostly applied for producing drawings and developing simple parts and assemblies. With the help of the CADDS5 package they engineer hull structures and pipelines, and build 3D computer models of compartments. The hullform is designed with a special package APIRS (Russian abbreviation for «computer-aided design and computation system»).

APIRS is a computer-aided system for shipbuilding applications that includes a package for geometrical modelling, a product-oriented database, computation and interface modules.

Principal features of the APIRS system include dialogue-mode generation of the hull surface, definition of decks, superstructures, floors, bulkheads, general arrangement, pipe and cable routing based on a single mathematical arsenal of bispline curves and bispline surfaces. All constructions, both for the hullform and for the general arrangement and hull structures, utilise a common graphic interface and a common database. This unified set of mathematical tools enables at all design phases to manipulate with an analytically continuous hull surface, and therefore to avoid accumulation of mistakes in hull part geometry.

The APIRS package incorporates an integral parametrisation system and enables changes coming up in the course of design work to be monitored. The unified database contains all relations set by the user among different components and makes it possible to access these components from networked user stations, monitoring all alterations entered by different users through their computers. All information in the database is hierarchically structured and the structure of this hierarchy can be chosen by the user to suit the adopted design work model and the number of involved user stations.

The APIRS system makes it easy to generate hull surfaces for any type of ships, from submarines to large ocean liners, high-speed boats, air cushion vehicles, hydrofoils and wing-in-ground craft. The hull surface mathematical model is defined by a set of sections of analytically continuous bispline surfaces.

Hull surface generation starts by specifying side lines with subsequent definition of surface portions. The obtained surfaces can be corrected and refined at later design stages, and these corrections are reflected in hull configuration, general arrangement, etc. It is also possible to generate a library of typical hullforms and thus to utilise their typical elements in new designs.

The APIRS system has capabilities for generating hull structure components, general arrangements, pipelines. All components are topologically related to the hull surface and to each other, and all of them respond to any variations introduced at any design stage.

The system includes a subroutine for finding key statics particulars of the designed submarine. Thanks to this, simultaneously with hullform design it is possible to get all relevant information on displacement, wetted surface area, waterline area and inertia moment, submarine trim and other particulars. Therefore, while constructing the lines the designer can find engineering solutions that would, in an optimum way, satisfy requirements and criteria specified for the submarine. If necessary, data are relayed to dedicated computation packages capable of high-accuracy calculations on submarine statics or dynamics and for outputting design documents. Modern systems can be expanded by additional routines for computations on hydrodynamic characteristics directly based on naval hydrodynamics theory equations. This enables hull lines to be optimised while designing them without resorting to expensive model tests. This technique is known as «pre-tank optimisation». It does not allow us to completely do without any model testing but the scope of model tests and the time required for them can be reduced quite sizeably.

MORTEPLOTEKHNIKA

Torpedo Weapons and Underwater Propulsion Systems

Since foundation in 1948 Morteplotekhnika has been developing torpedo weapons and torpedo thermal propulsion plants.

Morteplotechnika boats unique experience in the creation of underwater power plants featuring open, closed and combined working cycles based on the own powerful gas-turbine and piston engines using mono- and binary liquid, solid or paste fuels, including fuels reacted with sea water.

Due to the large number of various ground and sea tests the weapons feature high characteristics and the required degree of reliability. Morteplotekhnika creates and successfully operates specialized test equipment.

At present Morteplotekhnika completes the development of new versatile torpedo, designated UGST to hold a firm place on the world's market.

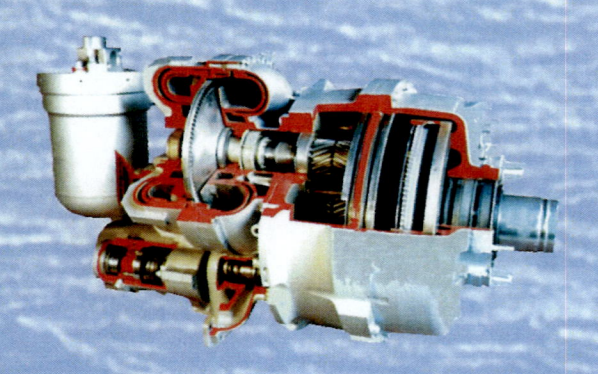

Morteplotekhnika is the world's leading developer of underwater thermal power plants and is ready to co-operate with the Russian and foreign companies to accomplish a whole range of work, including research, design and delivery of all types of 324, 533 and 650 mm diameter underwater systems weapons as well as equipment for full-scale ground and sea tests of torpedoes and power plants.

Federal State Unitary Enterprise
MORTEPLOTEKHNIKA Research & Design Institute
1 Verkhni Park, Lomonosov, St. Petersburg 189510, Russia

11. DESIGN OF SUBMARINE CONTROL SURFACES

11.1. Submarine Control Surfaces: Types and Purposes

All surface ships move, unless we consider the wave effects, in the horizontal plane only. A submerged submarine exercises three-dimensional motions. Submarine control surfaces are indented to ensure the specified level of controllability for both 2D and 3D manoeuvres.

It is generally known that controllability is a complex notion encompassing ship motion and behaviour at different speeds both with and without controls. It is also known that controllability may be generally described through course stability under the effects of disturbing forces and through manoeuvrability under the effects of applied controls.

Both the course stability and the manoeuvrability are characterised by traditional criteria while values of these criteria depend on the submarine designation and are stipulated in design specifications.

Thus, submarine control surfaces are intended to provide course stability and manoeuvrability in all cases of defined service conditions.

All control surfaces installed on submarines can be divided into fixed (stabilisers) and tiltable (planes/rudders). Both groups can be further subdivided into vertical and horizontal.

A special place belongs to stern control surfaces. Their existence is dictated exclusively by hydrodynamics. Should a modern submarine have no stern stabilisers, any disturbance in the infinite liquid would cause a non-asymptotic change of the submarine motion path, i.e. the submarine motion would never be stable [72].

To prevent non-authorised path variations, it is necessary to ensure submarine dynamic stability. For design purposes this is achieved by fitting the submarine aft with surfaces (stabilisers) to damp the overturning moment. Stabilisers are shaped as fins with a span that provides a satisfactory load-bearing capacity. Lengthwise, the stabilisers are placed as close as possible to the stern end in order to gain highest characteristics in terms of the moment.

There are certain considerations restricting the desire to shift stabilisers further afterwards:
- first, noise control requirements prescribe that trailing edges of stabilisers should be removed rather far from the propeller;
- second, we need space to arrange steering actuators and to fix stabilisers to the stern.

There are three identifiable basic configurations of stern control surfaces – cruciform, X-type and H-type (Fig.11.1).

Cruciform control surfaces – horizontal and vertical stabilisers with planes/rudders fitted within them.

Cruciform control surfaces are most commonly used on modern single-shaft submarines. This arrangement allows for separate controls in vertical and horizontal planes. Actually, due to the asymmetry of the hull with appendages with respect to the overall beam waterplane and, respectively, due to 3D hydrodynamic forces associated with course changes, horizontal and vertical plane controls are not exactly independent. Nevertheless, areas of horizontal and vertical stabilisers and planes/rudders for cruciform control surfaces are selected separately.

To achieve normal controllability, the dynamic stability criteria is usually chosen to be:
- over 1.0 in the vertical plane;
- about 1.0 in the horizontal plane.

Thus, a submarine is dynamically stable in the vertical plane and dynamically neutral in the horizontal plane.

These criteria are achievable when relative areas of the stern control surfaces $S_{ST}/V_{FS}^{2/3}$ are:
- 0.12 to 0.16 for horizontal stabilisers,
- 0.075 to 0.110 for vertical stabilisers.
- The stabiliser elongation, i.e. the ratio of the squared extreme span on one side of the hull to the stabiliser area, varies within 0.8 to 1.5.
- It should be kept in mind that if stern horizontal stabilisers project beyond the fullest cross-section of the submarine hull, their efficiency is better but this creates considerable difficulties in the submarine operation as it becomes necessary to be especially careful when mooring. Therefore, to save stabiliser efficien-

cy when the span has to be decreased they sometimes install so-called tip plates. Such control surfaces were chosen for Project 641B II-and- generation submarine.

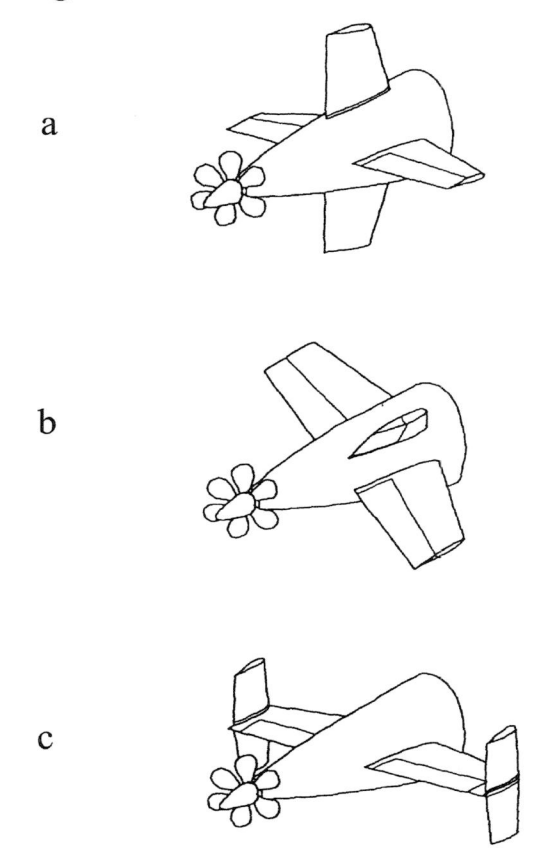

Fig 11.1. Typical Arrangements of Stern Control Surfaces
a) Cruciform, b) X-type, c) H-type

X-type control surfaces may be regarded as the same cruciform arrangement turned 45° around the hull axis. Such control surfaces are used on a number of Swedish, Dutch, Australian and German diesel-electric submarines.

With the X-type arrangement, forces arising on all four stabilisers and planes/rudders include both vertical and horizontal components. When one pair of planes/rudders is used for control, these components force the submarine to enter a 3D motion. In this case, to enable a single-plane manoeuvre, i.e. to change only the course or only the depth, it is necessary to simultaneously apply both pairs of planes/rudders. Therefore, areas of plane/rudder pairs should be equal. In real life all four fins have equal areas. And as all four fins simultaneously work in all planes of submarine motion, their area, the same as that of the stabilisers, can be reduced compared to the cruciform arrangement by about 25%. When areas of all four planes/rudders, and hence of all stabilisers, are equal, as soon as we achieve the required stability in one plane, e.g., vertical, the submarine acquires excessive stability in the horizontal plane. Otherwise, when the design is driven towards the horizontal-plane stability, the submarine stability in the vertical plane becomes absolutely insufficient.

3D forces arising on X-type planes/rudders and generating both horizontal- and vertical-plane components make the submarine control more complicated. Therefore, a submarine with X-type control surfaces for all practical purposes can be controlled only by an automatic system and a failure of one plane/rudder pair may result in a serious accident, especially at a high speed. Perhaps this is why X-type control surfaces are chosen only for diesel-electric submarines with their limited submerged speeds.

The H-type arrangement is different because the rudders are placed on the tips of horizontal stabilisers. Such control surfaces were chosen for the German Project 201 submarine and they appear to be quite promising for submarines with full-lined sterns.

Hydrodynamic conditions for cruciform and H-type control surface arrangements may be regarded as identical. However, the latter option is more complicated in terms of design. Besides, when vertical fins are shadowed by the hull they may lose some of their efficiency.

A submarine is controlled by tiltable control surfaces, i.e. vertical rudders and hydroplanes, as well as forward (or fin) hydroplanes.

Rudders can be made either as stabilizer flaps or as separate fins. The first option is preferable from the point of view of hydrodynamic efficiency as then rudders work jointly with stabilisers and this improves their efficiency by 1.3 to 1.5 times. With the second arrangement, the rudder efficiency is lower but without vertical stabilisers it is possible to reduce the horizontal-plane stability that may sometimes be excessive.

The area of rudders is defined by the specified minimum submerged turning circle diameter. Usually, the relative rudder area $S_{VR}/V_{FS}^{2/3}$ on attack submarines is:
  – about 0.050 for flap rudders on vertical stabilisers;
  – about 0.075 for rudder fins.

It is desirable for lower vertical control surfaces not to project beyond the submarine keel plane because otherwise there are additional difficulties with docking or mooring when the depth near the pier is limited.

Aft hydroplanes are usually designed with flaps of horizontal stabilisers that should ensure considerably higher stability in the vertical plane. The area of aft hydroplanes is selected so as to achieve the specified vertical speed for depth manoeuvres with trim angles of 15 to 25°. This speed may reach 4 m/s. The relative area of hydroplanes $S_{AFRL}/V_{FS}^{2/3}$ is about 0.05.

A special feature associated with aft hydroplanes is the inversion speed, i.e. a speed at which hydroplane tilting does not cause any change in the submarine path. As a rule, this inversion speed rests within the range of low submarine speeds (less than 3 knots). At speed below the inversion speed the submarine has inverted controllability. The presence of the inversion speed at aft hydroplanes is the reason for installing a second pair of hydroplanes which should be located forward of the centre of hydrodynamic pressure. This enables inversion speed effects to be avoided when the submarine is controlled by the fore hydroplanes and to ensure that the submarine is controllable at low speeds [72].

Besides, due to their position forward of the centre of pressure, fore hydroplanes generate a trimming moment directed opposite to that of the aft hydroplanes. When both pairs of hydroplanes are applied together, the submarine can bodily sink or ascent without any trim.

Practical experience shows that the relative area of fore hydroplanes $S_{FPL}/V_{FS}^{2/3}$ is 0.03 to 0.04.

Fore hydroplanes may be shifted to the midship section and placed at the base of the mast sail. In this case they are called midship planes. They may also be mounted on the mast sail (usually, if it is a foil-type sail). In this case they are made fixed (non-retractable) and called fin planes.

Both options (midship and fin planes) are applied to reduce hydroplane effect, especially pronounced while tilting, on the performance of the bow sonar.

Fixed fin planes increase submarine resistance. Besides, because of their location in the sail, they sit much closer to the centre of pressure and have a shorter longitudinal lever than fore planes. Therefore, to generate an equal trimming moment, their area should be 1.5 to 2.0 times larger than that of the fore hydroplanes. From the point of view of hydrodynamics, fore planes are more effective than fin planes. Advantages of fore planes have been confirmed by the experience gained from Russian submarines and, apparently, those of the US Navy. At any rate the US Navy rejected fin planes for the latest modified series of «Los Angeles» attack submarines and for

«Sea Wolf» submarines of the new generation and returned to the fore hydroplanes. These examples highlight the necessity further research on this issue considering all aspects associated with the position of fore hydroplanes [50].

From the design point of view, each pair of hydroplanes can have either a continuous or divided stock. In the latter case starboard and portside planes/rudders can be tilted to different angles, one - to surface and another - to dive. This may be helpful in special cases of submarine control, e.g., to correct the heel. In some cases submarines are fitted with two pairs of aft hydroplanes: large and small (Fig.11.2). Small planes are used at high speeds to simplify submarine control.

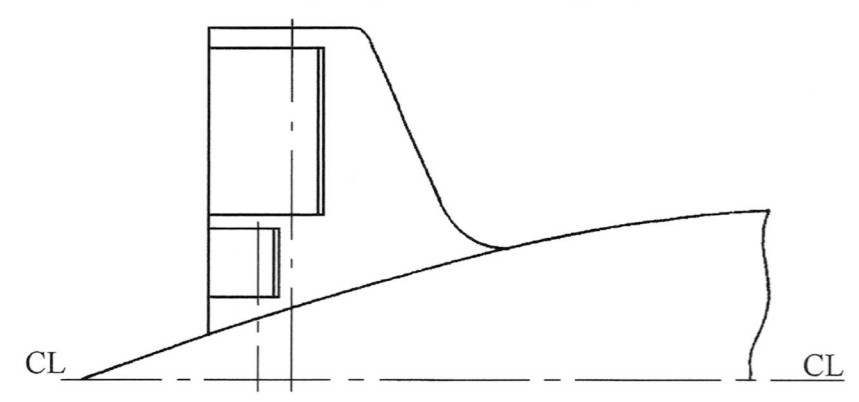

Fig.11.2. Aft Hydroplanes of a Submarine

Concluding this section it should be noted that all submarine planes/rudders have been until recently designed as balanced planes/rudders with balancing factors of k ≈ 0.25 to 0.30. Recently, however, there appeared a trend towards reducing the balancing factor (especially for hydroplanes). This is explained, mostly, by attempts to reduce the flow component of noise associated with hydroplane tilting, as well as to improve damage control conditions.

11.2. Determination of «Hull + Control Surfaces» Hydrodynamic Characteristics

For design purposes all submarine control surfaces are regarded as wings with low aspect ratios. Therefore, at initial design stages hydrodynamic characteristics of control surfaces are found using the tools of low-aspect-ratio wing theory. If any control surface element has a trapezoidal shape, in calculations it is substituted by an equivalent rectangular wing of equal area (S_0) and aspect ratio ℓ_i (Fig.11.3).

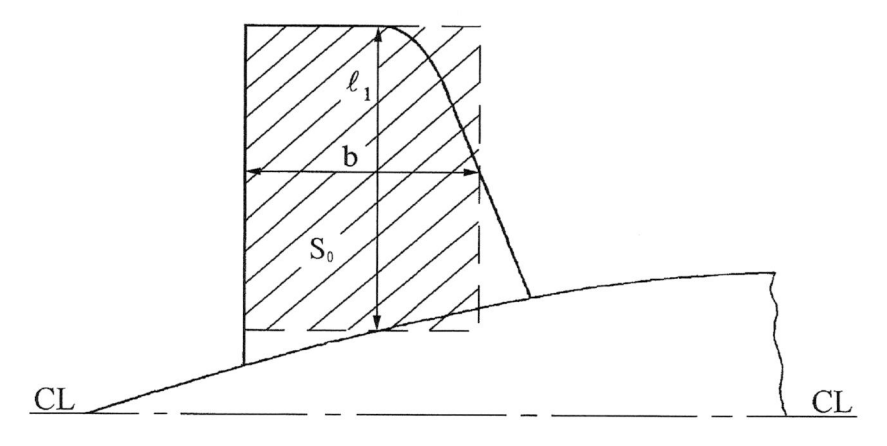

Fig.11.3. An Equivalent Wing

Aspect ratios of submarine control surface elements are: $\lambda_i = \dfrac{\ell_i^2}{S_i}$

For a rectangular wing $\lambda_i = \dfrac{\ell_i}{b_i}$

where b_i – wing chord that usually does not exceed 2.0 to 2.5.

Generally speaking, the submarine hull itself may also be considered as a low aspect-ratio wing ($\lambda_k \approx 0.1$).

To solve the task of submarine control surfaces design, it is necessary to be able to find by calculations, before model tests, hydrodynamic characteristics of the «hull + control surfaces» complex with operating propellers.

This task is quite complicated and it is difficult to get strict solution at the early design stages. Therefore, designers apply approximate methods and certain assumptions.

First, when selecting submarine control surfaces at the early design stages, hydrodynamic characteristics of the complex are estimated ignoring the propellers.

Secondly, surfaces are selected based on exclusively single-plane motion considerations, separately for vertical and horizontal planes. 3D motions are, at this design stage, ignored. This significantly simplifies the estimation of hydrodynamic characteristics of the designed complex since they become a function of only one angle: of attack or of drift.

All hydrodynamic characteristics of the subject motion – translational and rotary – are conventionally estimated by adding together the relevant elements. The procedure is often called the design or the «dry» method. This means that any hydrodynamic coefficient of the subject complex C_i is presented as a sum of such coefficients for the hull, the fore hydroplanes, the aft hydroplanes, etc. Thus, we may formulate any hydrodynamic coefficient as

$$C_i(\alpha) = \Sigma C_i(\alpha) = C_{iH}(\alpha) + C_{iFPL}(\alpha) + C_{iAFPL}(\alpha) \qquad (11.1)$$

Then, when submarine control surfaces are selected and other controllability issues are considered at early design stages, all hydrodynamic coefficients are linearised, i.e. presented as

$$C_{yi}(\alpha) = C_{yi}^{\alpha}\alpha \qquad (11.2)$$

where C_{yi}^{α} is a derivative of coefficient C_i with respect to the angle of attack α and describes the slope of the curve $C_i(\alpha)$ (Fig.11.4).

Thus, any hydrodynamic coefficient may be formulated as

$$C_i(\alpha) = \Sigma C_i^{\alpha}\alpha = C_{iH}^{\alpha}\alpha + C_{iFPL}^{\alpha}\alpha + ... C_{iAFPL}^{\alpha}\alpha \qquad (11.3)$$

Similar expressions can be written for those hydrodynamic coefficients that are functions of the drift angle.

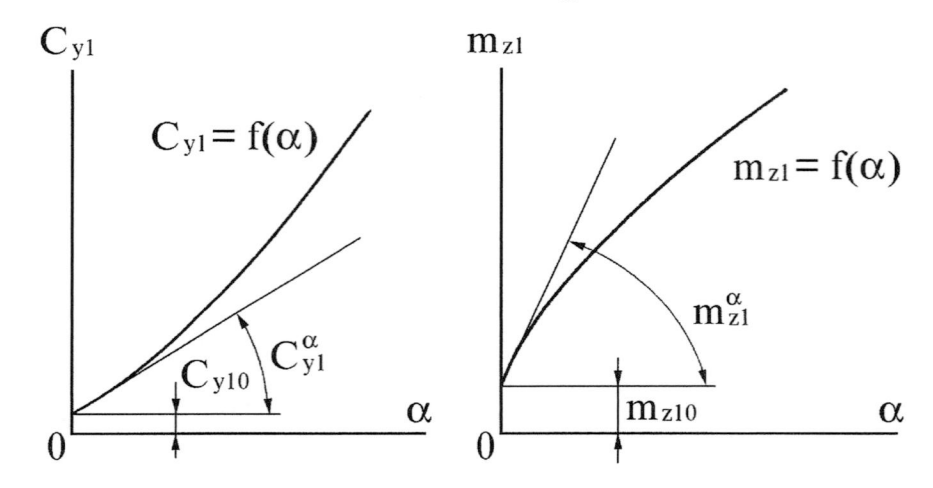

Fig. 11.4. Hydrodynamic Characteristics of Control Surfaces

Hydrodynamic Coefficients of Control Surfaces

As has been mentioned above, all elements of submarine control surfaces may be regarded as low-aspect-ratio wings.

Let us consider, first, any element of submarine control surfaces separate from of the hull, i.e. as an isolated wing. For an isolated wing with a known low aspect ratio λ we can calculate the derivative of the normal wing lift with respect to the angle of attack α. As a rule, for this purpose they apply the formula derived by G.F.Burago. According to this formula

$$C_{y1INS}^{\alpha} = k(C_{y1}^{\alpha})_{\infty} \cdot \frac{2{,}75\lambda}{2{,}75\lambda + (C_{y1}^{\alpha})_{\infty}} \overline{S} \qquad (11.4)$$

where k = 0.92 — K.K.Fyedyaevsky's coefficient for the solidity of the wing;

$(C_{y1}^{\alpha})_{\infty}$ = 5.6 — derivative of the normal force with respect to the angle of attack at an infinite wing span;

$\overline{S} = S/V_{FS}^{2/3}$ — relative area of the wing;

V_{FS} — total submerged displacement of the submarine.

If we exclude multipliers $\overline{S}$ from (11.1), we get the derivative with respect to the wing area S proper. Provided the wing area is constant, formula (11.4) has only one variable λ

$$(C_{y1}^{\alpha})_{INS} = A\lambda S \qquad (11.5)$$

From the obtained expression we may see that at a given wing area the efficiency with respect to the normal force increases together with the aspect ratio λ, i.e. C_{y1}^{α} grows with λ (Fig.11.5)

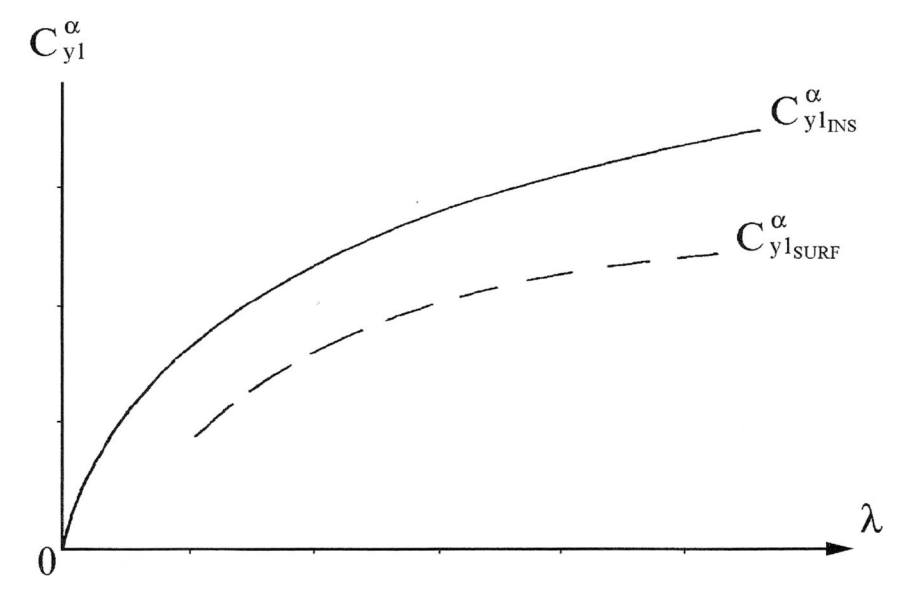

Fig.11.5. C_{y1}^{α} Versus Wing Aspect Ratio λ.

Hence, when submarine control surfaces are designed and their elements are selected, it is advisable to increase the span of control surfaces, though keeping in mind service requirements.

Flow conditions in way of the wing attachment to the hull are changed for both the wing and the hull. A part of the wing is within the boundary layer of the hull while the hull flow respectively distorts the wing flow and changes angles of attack on the wing. The hull in way of control surface attachment acts as a wall and changes the wing aspect ratio. The wing affects the hull pressure profile, etc. All these effects change hydrodynamic characteristics of both the wing and the hull. To simplify the situation, all variations in hydrodynamic characteristics are, for the sake of convenience, referred to the wing assuming that they remain unchanged for the hull.

The actual reciprocal effects are taken into account by introducing a coupling factor into hydrodynamic characteristics of the isolated wing that represents the subject element of control surfaces. Thus, the formula for the normal force derivative C_{y1} with respect to the angle of attack $\alpha - C_{y1}^{\alpha}$ of the wing (control surface) installed on the hull is

$$C_{y1_{SURF}}^{\alpha} = \mu_i (C_{y1_{SURF}}^{\alpha})_{INS}, \qquad (11.6)$$

where μ_i – coupling factor.

Special attention should be paid to the fact that numerical values of μ_i for the submarine hull and for control surfaces are different for different control surfaces and different hydrodynamic coefficients (forces, moments, translational and rotary). They are also different for various shapes of the light hull: stem-type and airship-type. Numerically, coupling factors may either be less than 1 or more than 1.

Unfortunately, it is impossible to find values of these coefficients analytically. They are determined only experimentally in wind tunnels.

By now there is a sizeable accumulated bulk of experimental data on coupling factors for different submarine hulls, control surfaces and hydrodynamic characteristics [72].

As we have already mentioned, to design submarine control surfaces it is necessary to calculate translational and rotary hydrodynamic coefficients, as well as their translational and rotary derivatives.

Thus, to select particulars of horizontal control surfaces we have to find C_{y1}^{α}; m_{z1}^{α}; $m_{y1}^{\omega z}$ and $C_{y1}^{\omega z}$ for each of them.

At early design stages these derivatives are expressed through derivatives of the normal force C_{y1}^{α}. Fig.11.6 shows one of the control surface elements: aft hydroplanes. If we designate their area (for both sides) as S_{AFPL}, the aspect ratio as λ_{AFPL}, the lever from the centre of buoyancy of the total submerged displacement to the hydroplane stock axis as ℓ_{AFPL}, the sought values will be, taking into account the coupling factor presented as:

$$C^{\alpha}_{y1_{AFPL}} = \mu_1(C^{\alpha}_{y1_{AFPL}})_{INS} = \mu_1 A(\lambda_{AFPL})\overline{S}_{AFPL}$$

$$m^{\alpha}_{z1_{AFPL}} = \mu_2(m^{\alpha}_{z1_{AFPL}})_{INS} = \mu_2 A(\lambda_{AFPL})\overline{S}_{AFPL}\overline{\ell}_{AFPL}$$

$$C^{\omega z}_{y1_{AFPL}} = -\mu_3 \overline{\ell}_{AFPL}(C^{\omega z}_{y1_{AFPL}})_{INS} = -\mu_3 \overline{\ell}_{AFPL} A(\lambda_{AFPL})\overline{S}_{AFPL} \qquad (11.7)$$

$$m^{\omega z}_{z1_{AFPL}} = \mu_4(m^{\omega z}_{z1_{AFPL}})_{INS} = \mu_4(C^{\alpha}_{y1_{AFPL}})_C \overline{\ell}^2_{AFPL} = \mu_4 \overline{\ell}_{AFPL} A(\lambda_{AFPL})\overline{S}_{AFPL}$$

In addition to the above explained symbols:

$\overline{\ell}_{AFPL} = \ell_{AFPL}/V^{1/3}$ – relative lever of aft hydroplanes.

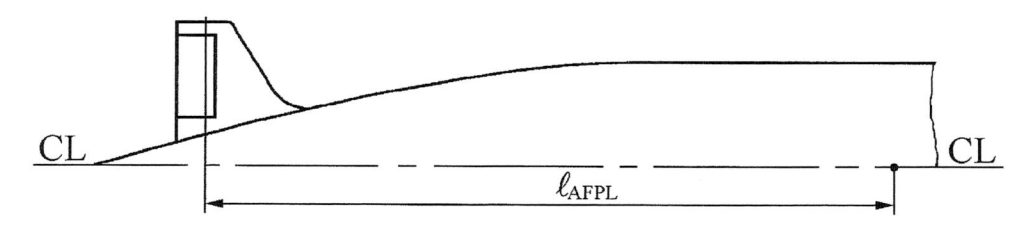

Fig.11.6. A Submarine Aft Hydroplane

Derivatives of forces and moments of planes with respect to their tilting angles (δ) are found assuming that $C^{\delta}_{y1} \cong C^{\alpha}_{y1}$. Strictly speaking, this is not absolutely true but fits the accuracy of other assumptions.

$$C^{\delta k}_{y1_{AFPL}} = \xi_1(C^{\alpha}_{y1_{AFPL}})_{INS} = \xi_1 A(\lambda_{AFPL})\overline{S}_{AFPL}$$

$$m^{\delta k}_{z1_{AFPL}} = \xi_2(C^{\alpha}_{y1_{AFPL}})_{INS}\overline{\ell}_{AFPL} = \xi_2 A \overline{\ell}_{AFPL}(\lambda_{AFPL})\overline{S}_{AFPL}. \qquad (11.8)$$

where ξ_1 and ξ_2 – coupling factors of the hull and the aft hydroplanes when they are tilted.

Hydrodynamic Coefficients of the Bare Hull

When designing submarine control surfaces, for either plane (vertical and horizontal) we need four derivatives: two translational and two rotary. Thus, when considering the vertical-plane motion it is necessary to find derivatives: $C^{\alpha}_{y1_H}$; $m^{\alpha}_{z1_H}$ and $C^{\omega z}_{y1_H}$; $m^{\omega z}_{y1_H}$.

Translational and rotary derivatives of the bare hull can be determined in two ways: by calculations or by experiments.

For calculations, the submarine hull is considered as a wing with the minimum span and as suggested by K.K.Fyedyaevsky substituted by a horse-shoe vortex equivalent to a triaxial ellipsoid. For the sake of convenience the required coefficients of added masses are calculated assuming that the hull shape is equivalent to the triaxial ellipsoid [83].

By now there is a large amount of experimental data on hydrodynamic characteristics of submarine hulls of two architectures: stem-type and bodies of revolution with different aspect ratios and cross-section ovals.

This enables a prototype model hull to be selected close to the subject one in shape and geometry parameters.

Nevertheless, it is difficult to imagine that it will ever be possible to find a truly identical model.

Usually they take the most suitable hull model of the subject type and scale its hydrodynamic characteristics to the new hull. E.g., the scaling formula for $C_{y1_H}^{\alpha}$ is

$$C_{y1_H}^{\alpha} = \chi_1 \chi_2 \chi_3 \chi_4 (C_{y1_H}^{\alpha})_0 \qquad (11.9)$$

where $(C_{y1_H}^{\alpha})_0$ – value of the derivative for the chosen prototype model;
χ_1 – coefficient accounting for the effect of the new L/B ratio,
χ_2 – coefficient accounting for the new H/B ratio, etc.

Numerical values of coupling factors of hullform parameters are, for every subject hydrodynamic coefficient, found from plots like those shown in Fig.11.7 made based on test series results available from catalogues and project reports. Other hydrodynamic characteristics are scaled in a similar way. It should be noted that thus obtained derivatives of submarine hull hydrodynamic coefficients are referred to the bare hull volume V_{BH} with a corresponding exponent.

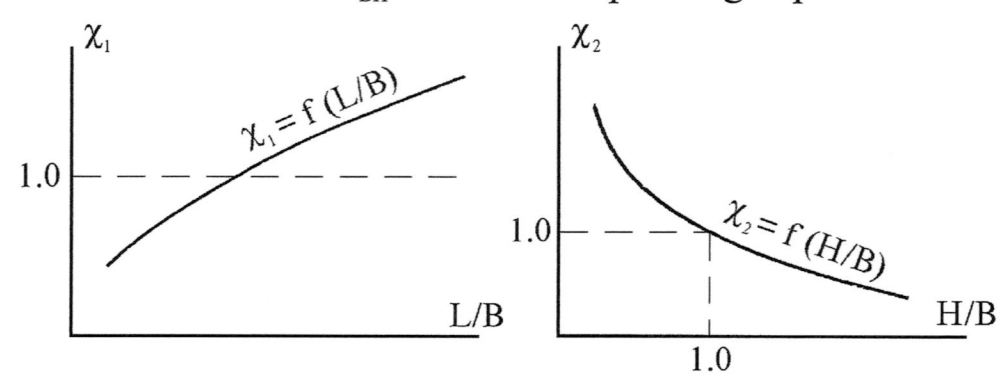

Fig. 11.7. Hull Parameter Coupling Factors

However, for further calculations we need derivatives referred to the total submerged displacement V_{FS}. This means that the previous results have to be scaled, which is done by multiplying relevant derivatives by V_{BH}/V_{FS} taken to a corresponding power.

After obtaining derivatives of hydrodynamic forces for the bare hull, it is necessary to consider their changes due to the mast sail. The sail itself brings changes to all hydrodynamic characteristics and at the same time re-distributes hull pressures and thus changes hull characteristics. Both these factors are usually accounted for by introducing corrections to earlier obtained hull characteristics. Values of these corrections ($\Delta C^{\alpha}_{y1_{SAIL}}$; $\Delta m^{\alpha}_{z1_{SAIL}}$; $\Delta m^{\omega z}_{z1_{SAIL}}$; $\Delta C^{\omega z}_{y1_{SAIL}}$) are calculated with the help of functions derived from serial tests.

Correction values depend on geometric parameters of the sail and of the hull, as well as on their mutual arrangement.

Thus, taking into account the mast sail, translational and rotary derivatives of hull geometric forces are formulated as:

$$C^{\alpha}_{y1_H} = C^{\alpha}_{y1_{BH}} + \Delta C^{\alpha}_{y1_{SAIL}}$$
$$m^{\alpha}_{z1_H} = m^{\alpha}_{z1_{BH}} + \Delta m^{\alpha}_{z1_{SAIL}}$$
(11.10)

11.3. Criteria Used in the Control Surface Design

The task of submarine control surface design is, as has been pointed out earlier, to elaborate all control surfaces, i.e. both the fixed and the tiltable parts. It is generally known that neither Russian nor any other World War II submarines had any stabilisers. Their control surfaces consisted of planes/rudders only. There was no need for stabilisers due to submarine low speeds. The plane/rudder areas were found from a prototype with good manoeuvring qualities. Thus, the rudder area was found as

$$S_{VR} = k_1 LH$$

The area of aft hydroplanes S_{AFPL} was estimated with a similar formula

$$S_{AFPL} = k_2 LB$$

where k_1; k_2 – coefficients taken from the prototype;
L, B, H – main dimensions of the hull.

However, there are several reasons why this method for plane/rudder area estimations should be considered rather unreliable.

First, LB and LH products do not at all define the submarine hull-form, and therefore tell nothing about its hydrodynamic characteristics. Secondly, due to different designations, different submarines should have different course stability and manoeuvrability characteristics. This should be somehow taken into account by coefficients k_1 and k_2. Naturally, this sort of plane/rudder area calculations is not used today.

Nowadays, at initial stages of submarine control surface design they apply criteria of steady horizontal motion and steady motion manoeuvrability. Consideration of the steady conditions makes the task much simpler. There are two stability criteria: static instability $\bar{b}$ or $\bar{b}_L$ and dynamic instability k. It may be appropriate to remind readers that these submarine stability criteria are expressed through submarine hydrodynamic characteristics as follows:

$$\bar{b} = \frac{b}{V^{1/3}} = \frac{m_{z1}^{\alpha}}{C_{y1}^{\alpha}}$$

$$\bar{b}_L = \frac{b}{L} = \bar{b}\frac{1}{L/V^{1/3}} = \frac{m_{z1}^{\alpha}}{C_{y1}^{\alpha}} \cdot \frac{1}{L/V^{1/3}}$$

(11.11)

$$k = \frac{\bar{b}^{\omega}}{\bar{b}} = \frac{\dfrac{\left|m_{z1}^{\omega z}\right|}{2(1+k_{11})-C_{y1}^{\omega z}}}{\dfrac{m_{z1}^{\alpha}}{C_{y1}^{\alpha}}}$$

(11.12)

where k_{11} – coefficient of the added mass for motion along the OX_1 axis. In these formulae all derivatives of hydrodynamic coefficients are applicable to the entire moving hydrodynamic complex: hull+sail+control surfaces.

The static stability criterion $\bar{b}$ or $\bar{b}_L$ represents the relative lever of resultant hydrodynamic forces acting on a submarine steadily moving along a straight path (Fig.11.8). It is counted from the origin of co-ordinates coinciding with the centre of mass of the total submerged displacement. On all submarines the resultant R application point is located forward of the grid origin. This means that in steady-state motion all submarines are statically unstable and the force R is not righting but capsizing the submarine. This is why we need to design and install aft control surfaces.

Thus, criterion $\bar{b}$ expresses the static instability as the arm based on $V^{1/3}$ or L.

The dynamic stability criterion k in accordance with (11.12) represents the non-dimensional value describing the relationship of the damping force lever $\bar{b}^{\omega}$ to the lever of the resultant of hydrodynamic forces $\overline{b}$. This relationship is always positive when submarines are moving in the vertical plane, and thus if k>1 they are dynamically stable in this plane. Criteria $\overline{b}$ and k are used for the selection of the total area of control surfaces, fixed and tiltable, as both of them influence the submarine motion stability.

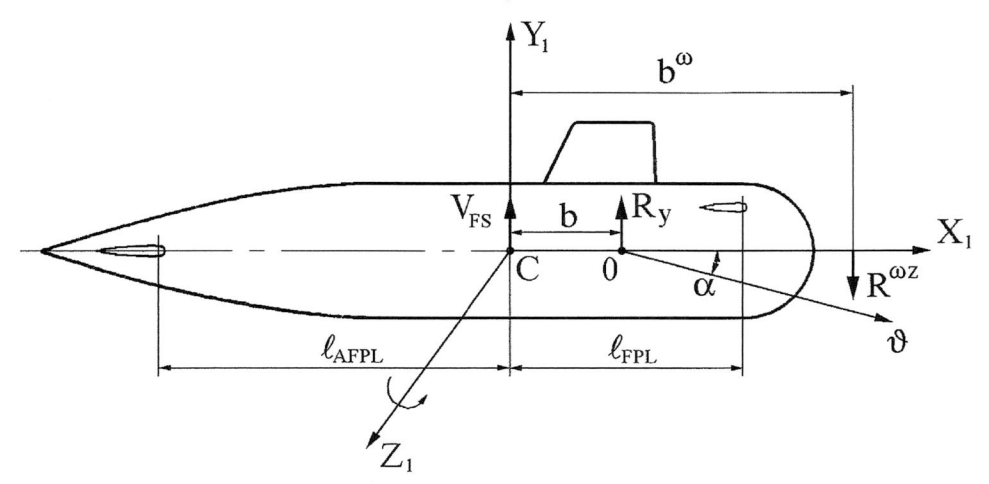

Fig. 11.8. Forces Acting on a Submarine

Tiltable parts of control surfaces are selected by applying manoeuvrability criteria. One such criteria, which is among the simplest and therefore applied at early design stages, is the steady-state rectilinear rate of climb. For the vertical plane motion, and hence for the selection of horizontal tiltable surfaces (aft and fore hydroplanes) these criteria are formulated as

$$\frac{\partial \vartheta_\eta}{\partial \delta_{PL}} = \frac{\vartheta^3}{2g\overline{h}} \cdot C_{y1}^{\delta_{PL}} \left(\frac{m_{z1}^{\delta_{PL}}}{C_{y1}^{\delta_{PL}}} - \frac{m_{z1}^{\alpha}}{C_{y1}^{\alpha}} \right) \qquad (11.13)$$

or

$$\frac{\partial \vartheta_\eta}{\partial \delta_{PL}} = \frac{\vartheta^3}{2g\overline{h}} \cdot C_{y1}^{\delta_{PL}} (\overline{\ell}_{PL} - \overline{b}) \qquad (11.14)$$

Here, in addition to earlier-used symbols:

ϑ_η – vertical component of the submarine speed;

δ_{PL} – diving-plane angle (fore or aft hydroplanes);

$\overline{\ell}_{PL} = \ell_{PL}/V^{1/3}$ – relative lever of the hydroplanes from the origin of coordinates;

$\overline{h}$ – initial metacentric height referred to the total submerged displacement and the squared speed.

The rate of climb criterion may be expressed in physical terms as the vertical speed of the submarine centre of gravity in the steady-state rectilinear vertical-plane motion with the subject hydroplanes tilted through one degree.

From (11.13) and (11.14) it follows that the rate of climb criterion depends on the submarine speed. Numerical values of the above-mentioned criteria of motion stability and manoeuvrability for submarines of different designations are specified in their SDSs. Thus, the static instability, which is defined as the ratio of the horizontal force lever to the submarine hull length L, is usually within $\bar{b}_L = \frac{b}{L} = 0.15$ to 0.25, but never beyond 0.35.

Naturally, the less of $\bar{b}_L$, the larger the area of aft control surface should be.

Dynamic stability coefficients of submarines in service today are within k=2.0 to 4.0.

The larger the value of k, the larger the dynamic stability of the submarine.

One should pay attention to the fact that stability criteria $\bar{b}$ and k are in a certain way related to each other. This is evident from formula (11.12). Design rate of climb criteria $\frac{\partial \vartheta_\eta}{\partial \delta_{PL}}$ for fore and aft hydroplanes of the submarine are regulated for a certain speed (ϑ_s=10 kts). As a rule, for this speed with hydroplanes tilted through 1 degree the criteria are $\frac{\partial \vartheta_\eta}{\partial \delta_{FPL}} \geq 0.03$m/sec for aft hydroplanes and $\frac{\partial \vartheta_\eta}{\partial \delta_{AFPL}} \geq 0.08$m/s for fore hydroplanes.

The fact that the submarine rate of climb due to aft hydroplanes is considerably larger than due to fore ones is a feature of special interest.

Similarly to the above described criteria of submarine vertical-plane motion stability and manoeuvrability, one can formulate horizontal-plane motion criteria. However, these criteria should be different from the former ones in both the form and the content. When a submarine is moving in the horizontal plane there is no righting moment. For the vertical plane it is represented by the metacentric height, and therefore for the horizontal plane there is no notion of a static stability. For the horizontal plane we can use only the dynamic stability criterion. There is no notion of the rate of climb for the horizontal plane as well, since when the rudder is tilted to any angle the submarine enters the turning circle. Therefore, submarine horizontal-plane manoeuvrability is evaluated through the turning circle diameter D_{CIRC} as a function of the rudder tilt angle:

$$D_{CIRC} = f(\delta_{VR})$$

Requirements of the dynamic stability in the horizontal plane are much lower than those in the vertical plane and usually amount to $k_g \approx 1.0$. The horizontal-plane manoeuvrability is usually specified by the relative diameter of the steady turning circle D_{CIRC}/L, where L – submarine length, when the rudder is tilted to the maximum angle. The value of D_{CIRC}/L varies from 3 to 6 and depends on the submarine designation [83].

It is also a common practice to specify the heel in turn angle for a steady turning circle at a certain speed with the fully applied rudder.

11.4. Methods Used in Submarine Control Surface Design

The task of submarine control surface design is to solve a number of issues that govern the boat's controllability. This includes selecting control surface configuration necessary to satisfy all requirements on the controllability of the designed submarine, arranging these control surfaces on the submarine, finding the area of every surface, etc. This task involves many unknown values, and therefore it is impossible to solve it with only analytical methods.

There are two design methods for control surfaces applicable to initial design stages. The first method is as follows. The designer, based on similar submarines and statistic data, chooses an arrangement of control surfaces including their elements and calculates stability and manoeuvrability criteria for this configuration of control surfaces. The found criteria are compared to specified values. If the obtained criteria do not meet the requirements, control surfaces are modified and criteria are checked again. Essentially, this is again the convergence method.

The second method is basically a calculation procedure, though not a «pure» one. In this case the designer assumes a configuration of control surfaces and numerical values of course stability and manoeuvrability criteria. Then, assuming some part of control surface particulars to be known (e.g., aspect ratios, longitudinal levers), calculates areas of major elements. This approach does not need iterations and allows individual parameters to be varied in the process of calculations: levers $\overline{\ell_i}$, aspect ratios λ_i, etc.

Let us deliberate more on the second method, especially as this will, at the same time, clarify what should be done when using the first one. Besides, the second method, to our mind, enables to identify the influence of each particular element.

First of all, it is necessary to assume a configuration of control surfaces. This means that we need to establish the composition of the control surface package and approximate locations of individual fins.

Today, practical submarine design has already mastered control surface configurations for single-shaft and double-shaft submarines. Some examples may be seen in Figs.11.9, 11.10 and 11.11.

It is evident that on single-shaft submarines the aft control surfaces have the cruciform arrangement and planes/rudders are usually flaps of the stabilisers. From the point of view of higher efficiency of control surface elements, it would be reasonable to shift them as far aft as possible because then the levers of control surfaces ℓ_i would increase, and hence the moments of control surfaces due to the lever would be greater. However, when selecting a suitable configuration it is necessary to consider a number of other critical factors, e.g., the position of the aft control surfaces with respect to the propeller. The more control surfaces are removed from the propeller, the less the nonuniformity of the flow in way of the propeller and the less the noise. It is also important to consider the feasibility of fitting actuating gears of planes/rudders, as well as to check for the possibility that control surfaces would go beyond the overall hull dimensions (e.g., the beam and the height) which would be undesirable.

Let us for example consider designing the horizontal control surfaces for a single-shaft submarine.

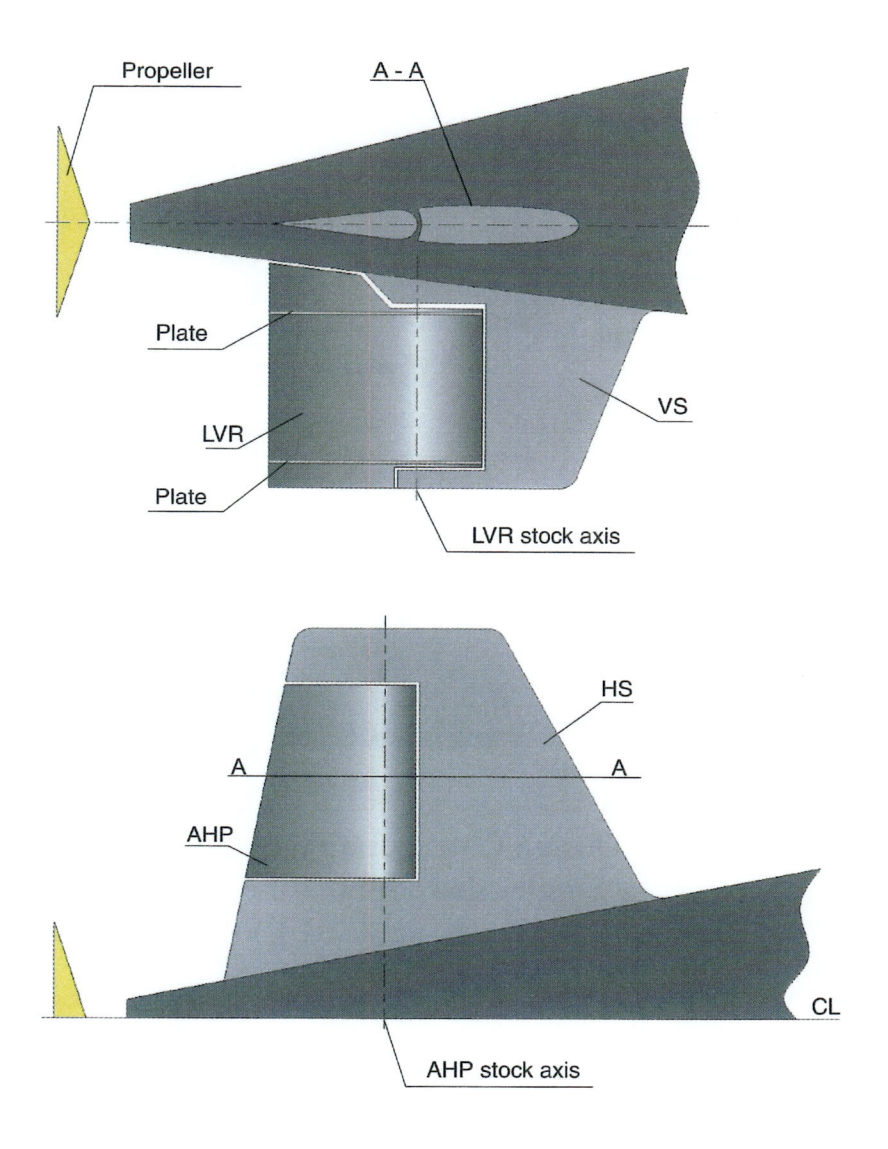

Fig.11.9. Design Configuration of Aft Control Surfaces
on a Single-Shaft Submarine
VS – vertical stabiliser, LVR – lower vertical rudder.
HS – horizontal stabiliser. AHP – aft hydroplane.

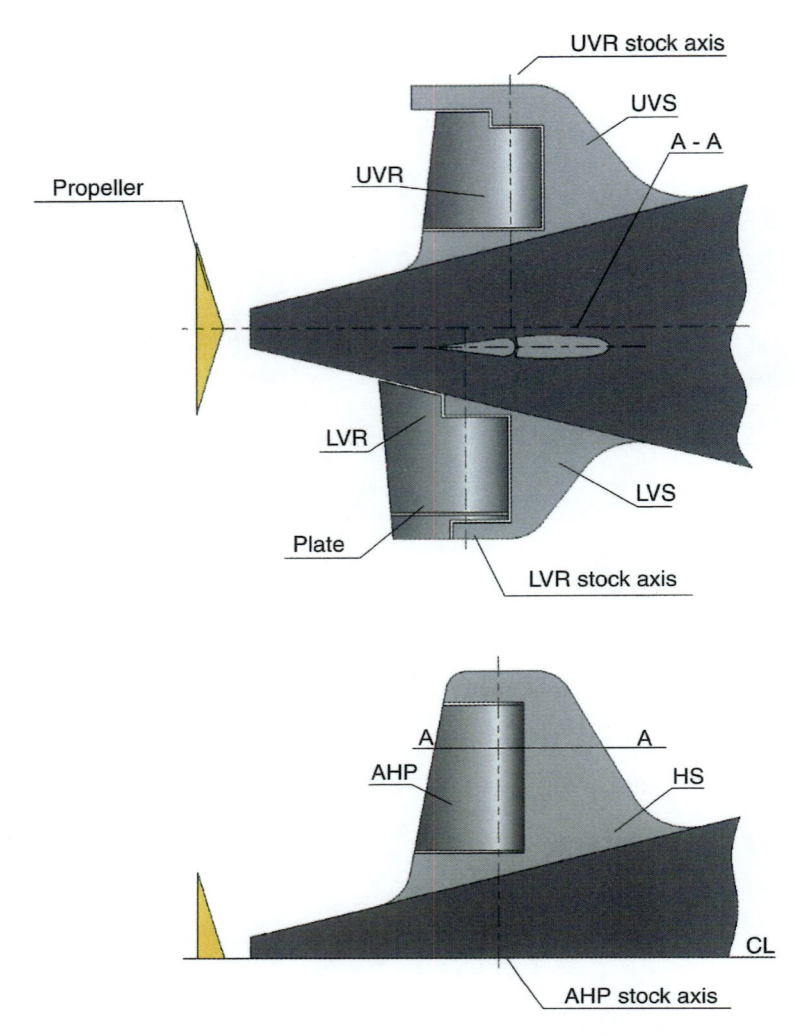

Fig.11.10. An Option of Design Configuration of Aft Control Surfaces on a Single-Shaft Submarine

UVS and LVS – upper and lower vertical stabilisers. UVR – upper vertical rudder

Fig.11.11. Design Configuration of Control Surfaces on a Double-Shaft Submarine

Let us assume that the intended package of horizontal control surfaces includes fore hydroplanes (retractable), aft stabilisers and planes/rudders designed as stabiliser flaps. For every such element we assume a lever from the origin of coordinates ℓ_i, which is located in the centre of buoyancy of the total submerged volume of the submarine, and an aspect ratio λ_i.

After that, using the earlier described methods, we calculate derivatives of hydrodynamic characteristics for the submarine hull: translational $C^\alpha_{y1_H}$ and $m^\alpha_{z1_H}$ and rotary $C^{\omega z}_{y1_H}$ and $m^{\omega z}_{z1_H}$ with the account for the sail and scaled for the total submerged displacement V_{FS}.

Then, based on the requirements the designed submarine, we assume a coefficient value or the relative lever of statical instability $\overline{b}$ or $\overline{b}_L$. Let us express the relative lever $\overline{b}$ through derivatives and present them as hull and control surface components.

$$\overline{b} = \frac{m^\alpha_{z1}}{C^\alpha_{y1}} = \frac{m^\alpha_{z1_H} + m^\alpha_{z1_{PL}}}{C^\alpha_{y1_H} + C^\alpha_{y1_{PL}}} \qquad (11.15)$$

Let us first calculate $\overline{b}$ for higher speeds, i.e. with fore hydroplanes retracted. For the chosen configuration, when calculating $\overline{b}$, horizontal stabilisers and aft planes may be regarded as a single wing with an aspect ratio λ_{PL} and a relative lever ℓ_{PL}.

Using 11.7, we may write:

$$\overline{b} = \frac{m^\alpha_{z1_H} + \xi A(\lambda_{PL}) \overline{S}_{PL} \overline{\ell}_{PL}}{C^\alpha_{y1_H} + \mu_1 A(\lambda_{PL}) \overline{S}_{PL}} \qquad (11.16)$$

where ξ and μ_1 – coupling factors for control surfaces and the hull for the force and moment derivatives.

There is one unknown $\overline{S}_{PL}$ in formula (11.16). Thus, we obtain the aft control surface area S_{PL} satisfying $\overline{b}$ and $\overline{b}_L$ as

$$\overline{b}_L = \overline{b} \frac{1}{L/V^{1/3}} \qquad (11.17)$$

Based on S_{PL} and λ_{PL} values, the aft control surfaces are drawn and their absolute span is found. Further, for the same motion mode we can determine the dynamic stability coefficient k. All derivatives of hydrodynamic characteristics included into the formula are presented as sums of components. Thus, C_{y1}^α and m_{z1}^α are presented similarly to (11.10) and rotational derivatives are presented as:

$$C_{y1}^{\omega z} = C_{y1_H}^{\omega z} + C_{y1_{ST}}^{\omega z} = C_{y1_H}^{\omega z} + \mu_3 \overline{\ell} A(\lambda_{ST}) \overline{S}_{ST}$$
$$m_{z1}^{\omega z} = m_{z1_H}^{\omega z} + m_{z1_{ST}}^{\omega z} = m_{z1_H}^{\omega z} + \mu_4 A(\lambda_{ST}) \overline{\ell}_{ST}^2 \overline{S}_{ST} \qquad (11.18)$$

The coefficient of added mass while moving along axis $OX_i - k_{1i}$ is calculated either as a triaxial ellipsoid equivalent to the bare hull or using the strip (plane section) method. The resulting value k is compared to the requirements. If it meets the requirements, the calculation can be continued. If it fails to do that, it is necessary to specify a value of k and to establish the stabiliser area $\overline{S}_{ST}$ and a new value of $\overline{b}$ based on it.

Then it is possible to determine the required area of aft hydroplanes.

For this purpose let us specify the rate of climb while the submarine is controlled by aft hydroplanes (See Para 11.3). Let us present derivatives included into this formula in a form similar to the above-described one and derivatives of forces and moments of the plane tilt angle (δ_H) will be written as:

$$C_{y1_{AFPL}}^{\delta k} = C_{y1_{AFPL}} = \xi_1(C_{y1_{AFPL}}^\alpha) = \xi_1 A(\lambda_{AFPL}) \overline{S}_{AFPL}$$
$$m_{z1_{AFPL}}^{\delta k} = C_{y1_{AFPL}}^{\delta k} \cdot \overline{\ell}_{AFPL} = \xi_2(m_{z1_{AFPL}}^\alpha) = \xi_2 \overline{\ell}_{AFPL} A(\lambda_{AFPL}) \overline{S}_{AFPL_T} \qquad (11.19)$$

From expression $\dfrac{\partial \vartheta_\eta}{\partial \delta_{AFPL}}$ presented in the extended form we can determine the sought value $\overline{S}_{AFPL}$ and, hence, S_{AFPL} as well.

This is the way the submarine aft control surfaces are selected based on the stability and the manoeuvrability S_{ST} and S_{AFPL} because $S_{ST} = S_{PL} - S_{AFPL}$.

Knowing the specified value λ_{AFPL}, we can sketch the aft control surfaces.

Now we need to select the fore hydroplanes. Let us specify their area S_{FPL} and aspect ratio λ_{FPL}. When the submarine is running with FPLs rigged out, the motion stability is changed and stability criteria $\overline{b}$ and k have to be recalculated accounting for FPL effects. This is made similarly to the procedure described above but the formula for derivatives of hydrodynamic characteristics should include FPL components. Obtained values $\overline{b}$ and k are again compared to the required ones. If they are within permissible limits, the rate of climb for the fore hydroplanes control is determined. In case the calculation results do not meet the requirements, the problem is solved in the second approximation varying $\overline{S}_{FPL}$.

In spite of the fact that the above-described method of the control surface design is comparatively sophisticated, at subsequent submarine design work the results have to be updated.

For this purpose they carry out wind tunnel model tests for attack and drift angles during which the hydrodynamic parameters of the complex (the hull, the control surfaces, the sail) are refined. Additionally, they find hydrodynamic parameters depending on the plane/rudder tilting angles in respective planes.

Tests are also carried out in cavitation tunnels and model test tanks. All above-listed tests allow discrepancies in the complex to be revealed and recommendations on their elimination given.

Further, based on model test results and existing calculation methods, the submarine controllability is estimated with higher accuracy. In order to do this, they consider submarine transient motions in horizontal and vertical planes, as well as the 3D manoeuvres.

The propeller effect on hydrodynamic parameters and on the submarine stability and manoeuvring parameters is estimated using ship theory methods [72]. Changes in the above-described submarine controllability parameters can also be estimated for cruising at the periscope depth or near the sea floor. In such cases they apply the notion of free surface or solid wall effects.

Shapes of all control surface components are thoroughly elaborated in the submarine control surface lines drawing. Special attention is paid to the compatibility of control surface components and the hull, as well as to operational aspects related to the arrangement of control surfaces.

After the selection of control surfaces the designer has a natural desire to compare the obtained data to those of similar types of submarines.

It is obvious that such a comparison should be made in relative rather than absolute values. For such a non-dimensional parameter, one may use the relative control surface area (see Para 11.1).

ИНТЕЛТЕХ INTELTECH

INTELTECH RIGHT COURSE IN THE WORLD OF TELECOMMUNICATIONS

Inteltech J.S.C. (Information and Telecommunication Technologies Joint Stock Company) is one of the leading and eldest companies in Russia in the field of design and delivery of networks, systems, hardware developments and software tools for telecommunications, including those for naval applications-data exchange with submarines and ships, to be supplied in the frame of military and technical co-operation with foreign countries.

Information and Telecommunication Technologies Joint Stock Company
197342 Russia St. Petersburg
Kantemirovskaya st., 8
Tel. +7-812-245-50-69

State Unitary Enterprise Designers' Office

State unitary enterprise design bureau SVIAZMORPROEKT was founded in 1932 and at present is the only company of Russia's shipbuilding industry which designs, develops and manufactures unique systems and device of marine radio electronic equipment for special and general purposes.

Design bureau SVIAZMORPROEKT invites to cooperation and offers:

- Design, manufacture and delivery of various antenna feeders and switch communications, retrieval towed antenna vehicles, retrieval autonomous information devices, radio communications control boards, equipment for the emergency inside-the-ship communications, as well as special radio communication systems for equipping submarines, deep-water Navy's devices, commercial ships and coastal communications installations belonging to different governmental agencies;
- Delivering equipment including GMDSS in sets in accordance with developed projects;
- Design, manufacture, testing, installation and maintenance of the technical devices intended for protect of the information;
- Realization of electrical, mechanical, environmental tests, ensurance of electrosafety and EMC, in particular for the purposes of the radio communication and common industrial production certification at the in-house Testing Center accredited by the State Committee of standards, Russian Marine and River Register of navigation and Ministry of Transport of Russian Federation.

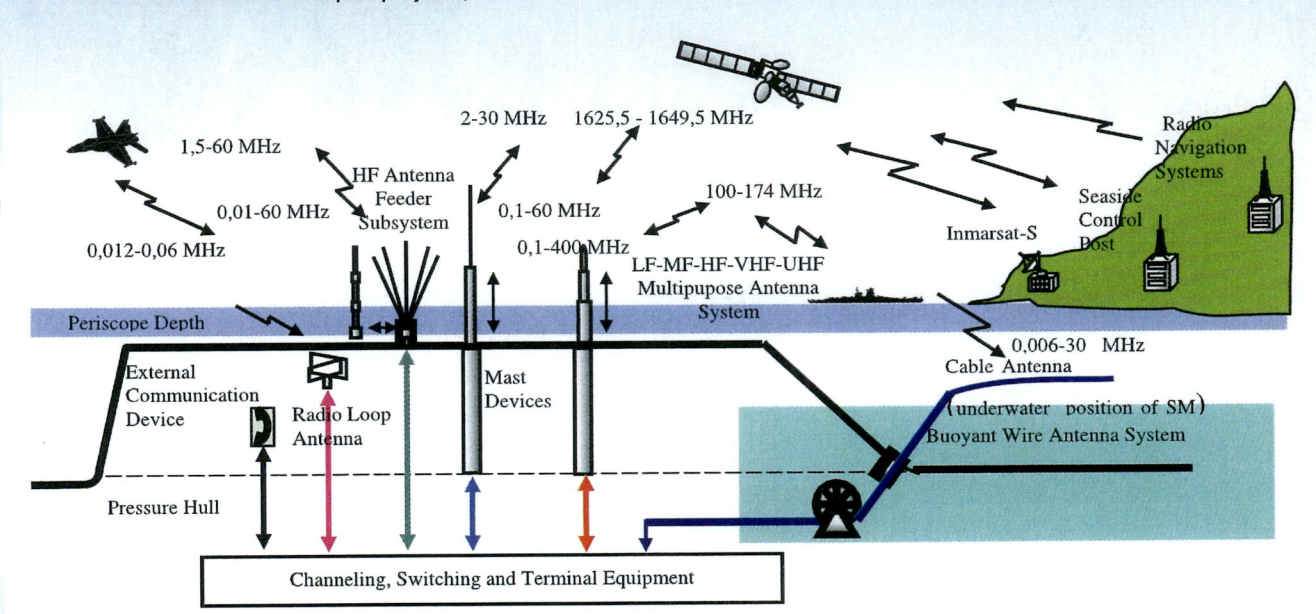

Antenna feeder devices of the Communication Complex «Distance-E» for submarines (project «Amur-1650»)

TOP-CLASS SHIP MACHINERY for NAVAL and MARINE APPLICATION

All Russian and Soviet-made submarines are equipped with steering gears, hydraulic mechanisms, devices for installation and retrieval of towed sonar arrays, winches for installation and retrieval of underwater radar aerials developed by the Central Research Institute of Marine Engineering and manufactured by JSC Proletarsky Zavod.

JSC Proletarsky Zavod and Central Research Institute of Marine Engineering also offer for export specially designed

- take-off and landing system for aircraft-carriers;
- dry and liquid cargoes transfer-at-sea gears;
- ship stabilizers and steering gears;
- deck electrohydraulic cranes and winches;
- axial-piston hydraulic motors and hydraulic pumps;
- water treatment equipment;
- underwater manipulators.

Our products and services have been highly appreciated by many customers all over the world, including U.K., Germany, Finland, India, China, Singapore, Vietnam, Iran, Turkey, and many others.

JSC Proletarsky Zavod

Central Research Institute of Marine Engineering

Russia, 193171, St.-Petersburg, ul. Dudko, 3.
Tel.: +7-812-5673230, 5673480.
Fax: +7-812-5673733, 5673260.
E-mail: cnii-sm@peterlink.ru
htpp://www.bestrussia.com/proletar/index.htm

12. MILITARY-AND-ECONOMIC ANALYSIS AND ITS ROLE IN SUBMARINE DESIGN

12.1 The Subject of Military-and-Economic Analysis

In submarine design, like in many other fields of human activities, a designer continuously has to make choices as he is always faced with the inevitable question «What is better?» Two additional knots of speed or 50m more diving depth? Ten extra days of endurance or 4 additional reload torpedoes? He discards the worst options trying to find the best solution for the specified task. Practical experience shows that it is impossible to choose the best alternative without considering limitations, and limitations are always there. They may include restrictions on weights or dimensions (it is impossible to go on increasing the displacement without some detrimental effect to the submarine), limitations due to shipyard facilities (construction sites dimensions) and certain economical constraints, including limited resources available for submarine development and operation [12]. These «resources» do not always mean financing – they may refer to materials, capabilities of the industry, etc., though it is more convenient to express all resources allocated for submarine construction and upkeep in costs. Without taking resource limitations, i.e. the economic aspect, into account the task of choosing a variant becomes meaningless: with unlimited resources one can build any number of submarines of every kind, i.e. accomplish any conceivable mission.

These circumstances made it necessary to introduce and develop military-and-economic analysis (MEA) as an independent research discipline which has by now become an integral part of any engineering project, especially when it comes to such sophisticated things as submarines. One can name three reasons for this process.

The first one is the expanding choice of options available to the designer. Indeed, the scope of potential design solutions has over the past few decades become considerably wider. New types of power plants, various new materials for submarine pressure and light hulls have appeared, and an especially rich selection is offered in terms of weapons and sensors. Multiple alternatives available within every class of weapons and equipment allow us to conclude that the number of their possible combinations defining the submarine configuration is also quite sizeable.

These developments have entailed a drastically swift rise in ship-associated expenses. It may be, with a comparatively minor error, claimed that every 20 years the cost of submarines within one class on average escalates by an order of magnitude. The cost of a submarine series becomes commensurable with entire military budgets of major countries, even those as big as the USA (Table 12.1). The soaring costs may be explained by two factors:
1. Inflation, which is on average 3 to 5% per year;
2. Ever-increasing sophistication of modern military hardware incorporating huge numbers of highly-expensive instrumentation and electronic components.

Table 12.1

Post-WWII Climb of Series-Built Submarine Costs [12], [76]

Years	Submarine	Cost, mln. $
1945	Ocean-going SS «Balao»	5~7
1965	SSN 2nd generation «Sturgeon»	70~75
	SSBN 2nd generation «Lafayette»	120~130
1978	SSN 3rd generation «Los Angeles»	400
1980	SSBN 3rd generation «Ohio»	1200
1986	SS Type 209/1500 (Germany)	450
1998	SS «Agosta-90» (France)	235

A special role is associated with the third factor: the increased innovation period, i.e. the interval between the emergence of a technical concept and its implementation in series-built products - constructed and mastered in terms of service operation submarines. It takes some 8 to 10 years from starting work on the technical proposal to commissioning the submarine and sometimes, when it comes to a fundamentally new development, this period becomes even longer. Thus, submarines become outdated while they are still on slipways. This means that when choosing a design option it is vital to consider how well this or another decision can resist moral depreciation due to emerging new technical solutions in both the national and foreign navies.

Due to these reasons, the cost of an error in the adopted decision becomes much greater. The process of submarine design involves a large amount of research and development efforts. They prepare grounds for implementing new engineering solutions in practical projects. Therefore, an erroneous decision results not only in unjustified expenses for submarine construction and operation but also in the choice of false research targets, i.e. in wasted R&D budget and time [11].

The object of military-and-economic analysis is the development and implementation of methods of sharing the part of the gross national product allocated for defence in such a way so as to reach the required level of national security with the least expenses. Obviously, at the top – the Armed Forces Command – level alone it is impossible to fully achieve this goal, i.e. not only to establish the required number of every kind of weapon system (ships, missiles, tanks, etc.) but also to choose their individual tactical and technical characteristics. Therefore, military-and-economic analysis has to be carried out at several levels.

At the Armed Forces Command level they aim to determine the total strength of every Arm of Forces required to fulfil a strategic task or a certain multitude of such tasks taking into account joint operations with other Arms.

At the Arm Command level they aim to determine a balanced composition of the Arm required to fulfil a certain multitude of operational tasks taking into account co-operation both with other Arms and among various components of the Arm itself. Particularly, this is a task to be resolved when determining a balanced composition of the Navy.

In the process of combat system (e.g., submarine) conceptual design, they have to choose the optimum option of the subject system taking into account its possible interaction with other forces.

In the process of combat subsystems (e.g., submarine power plant or sensor package) conceptual design, they have to achieve the specified level of performance.

Comparative evaluation of options and selection of the optimum alternative out of the multitude of possible solutions is the primary, though not sole, aim of military-and-economic analysis at the conceptual design stage. Other routine MEA tasks include: submarine tentative and approximate cost estimations during the design, in order to forecast the total required resource budget for the project, comparison of the designed submarine against the best foreign examples, estimation of the submarine optimum service life, the anticipated moral life of the project, and the advisable upgrading schedule.

At the same time it should be pointed out that military-and-economic analysis (called «cost-efficiency analysis» in the USA) does not determine the decision. It can only provide the person in charge, e.g., the chief designer of a submarine, with some additional data to help better study and understand relevant technical solutions (actually, there is no other way to get such information). The process of decision-making remains the prerogative of the designer who makes choices based not only on the outcome of military-and-economic analysis but also on other factors outside the scope of this analysis.

Under the present conditions it would be at least unreasonable to refuse to make use of additional information available from military-and-economic analysis of design solutions. However, it would be just as unreasonable to blindly rely on MEA conclusions and apply them without any critical assessment [58].

12.2. Basic Notions of Military-and-Economic Analysis

Military-and-economic analysis has appeared at the nexus of several disciplines: the combat efficiency theory, the design theory, economics, and a comparatively young science – operation analysis. As with any scientific discipline, MEA has its own system of notions or, as they often call it, «terminology» [12], [87]

A significant contribution to the development of military-and-economic analysis was made by Prof., Dr.Eng.Sciences L.B.Breslav. It was he who suggested the function of tactical degradation, and developed models of military-and-economic analysis dynamic tasks, formulated the cost matrix model (Breslav's matrix).

The basic notions in MEA terminology are: operations, combat and military-and-economic efficiencies, tactical and technical elements, parameters of technical solutions.

«Operation» means an action or a certain totality of actions consolidated by a common concept and an aim to be achieved as a result of these actions. «Operation» is a rather broad notion, which may mean destruction of an enemy object, a troops landing, a submarine crew rotation, a reactor refuelling, etc. An operation always has some result. The degree of correlation between the intended and the achieved results in terms of achieving the target is called the operation efficiency.

Submarines are constructed for combat. Their most important characteristic is the combat efficiency which is estimated as their capability to fulfil assigned missions in operations to be carried out with their participation. Thus, the combat efficiency of a submarine is understood as the degree of correlation between the achieved result and the assigned target in one operation.

In order to get an unambiguous characteristic of submarine combat efficiency, it is necessary to specify three values:
– the amount of damage inflicted or prevented during an operation by a certain number of newly designed submarines;
– the probability of inflicting or preventing this damage during one operation;
– the number of submarines necessary to inflict or prevent the specified amount of damage with the required probability.

These values are called the indices of combat efficiency. The combination of the former two describes the combat efficiency level. Obviously, there is no way to compare several options of one submarine design or different submarine designs of the same designation under simultaneous variations of all three indices. Therefore, two of them are fixed at some values equal for all considered options and the comparison is made in terms of the third index which becomes the criterion for comparison and selection. The value of this index is then used to judge the submarine combat efficiency, the ability to achieve mission targets.

In some cases it may be justifiable to compare submarine design options exclusively in terms of combat efficiency, but this is not always so. Actually, the three combat efficiency indices indirectly contain a certain measure of economics – a characteristic of the spent resources. It is reflected through the number of submarines necessary to fulfil the assigned task. It is acceptable to compare submarines only by their combat efficiency indices when subject submarines (design options) are approximately similar, i.e. have similar displacements, weapon and equipment packages, complement strengths. However, more often comparisons have to be made for submarine design options and submarines in service with different kinds of weapons and equipment. Such submarines require different amounts of resources for their development and upkeep, and therefore it is insufficient to compare them only by their combat efficiencies.

The system of indices, which take into account both the submarine combat efficiency and the resources spent for developing and upkeeping this submarine, characterises the submarine military-and-economic efficiency.

Military-and-economic efficiency is understood as a measure of correlation of the expected results of using the submarine for a certain mission, the target of this mission as specified before the operation, and the resources spent to achieve these results.

Military-and-economic efficiency of a submarine is uniquely described by a set of three indices: the amount of inflicted or prevented damage during one operation, the probability of inflicting or preventing this amount of damage, and an index characterising total resources spent to achieve these results with the required probability.

The above comments make it evident that for evaluating both the combat and the military-and-economic efficiencies of a submarine, it is necessary to clearly understand the nature of the operation: the process of combat utilisation of the submarine. Besides, it is necessary to know the accumulation pattern of involved expenditures. Since it is impossible to make any accurate prediction of all particular ways the designed submarine will ever be employed, such tasks are considered with the help of models.

A model is in this application understood as the subject process simulation or description that covers only its major features. With regard to military-and-economic analysis, a model is understood as a combination of assumptions and algorithms enabling us to find values of military-and-economic efficiency indices as functions of parameters varied in the process of parameter analysis.

The military-and-economic model consists of two parts and it is generated in two phases.

The first phase: formulation of a descriptive model of submarine development, service and combat employment that is written as a text. In particular, this description indicates the place of construction (this defines the price) and the submarine deployment area (this affects service expenses). The part of the descriptive model that covers combat employment deals with the anticipated area of combat missions, characteristics of hostile forces against which the designed submarine may be used, the mode of combat employment, etc.

Based on the descriptive model, they develop the second component: the formalised model. In this model they derive formulae for estimating the combat efficiency indices (criteria).

As has already been mentioned, the main MEA task at the submarine conceptual design stage is to choose an option, particularly – to choose a set of tactical and technical elements (TTEs) and parameters of technical solutions that define the submarine configuration.

Tactical and technical elements of a submarine are understood as the totality of characteristics that determine the combat efficiency. Submarine tactical and technical elements include:
– number of weapons and reloads, and their characteristics;
– signature levels;
– sonar package characteristics;
– speeds (full speed, search speed, etc.);
– displacement;
– maximum diving depth, etc.

Technical solution parameters (TSPs) are understood as technical characteristics of the submarine as a whole and of her subsystems that determine the submarine development and upkeep costs at specified TTEs. The key TSPs for a submarine are:
– architecture and design configuration;
– pressure hull material grade, its physical and mechanical characteristics;
– power plant type and parameters characterising weight to output ratio;
– extent of automation, etc.

It may easily be noticed that it is impossible to specify all TTEs at fixed TSPs because their combination may become incompatible. E.g., at specified full speed, displacement and performance qualities of the submarine, no-one can just arbitrarily assign the power plant capacity: it is dictated by specified parameters and found from well known equations of the design theory [11].

However, out of the multitude of TTEs it is possible to choose a necessary and sufficient family for specifying all other ones. The elements belonging to this family are called basic elements. All other TTEs may be found in the process of design from the specified basic elements and the TSPs. These elements are called design elements. The choice of the basic TTE set is not a simple task. If, e.g., the speed is specified and thus becomes a basic element, the power plant capacity has to be calculated, i.e. becomes a design element and vice versa.

In the military-and-economic model, indices of military-and-economic efficiency are regarded as functions of basic TTEs and TSPs. MEA suggests the possibility of solving optimisation problems. When any two military-and-economic efficiency indices are fixed, the third one becomes a criterion which is a function of the chosen set of basic TTEs and TSPs. An optimisation problem means searching for such a set of basic TTE values at which the military-and-economic efficiency criterion would reach its extremum. These deliberations have an important geometric interpretation. Basic TTEs are linearly independent. Therefore, every set may be considered as an n-dimensional vector where n is the number of basic TTEs. Since every option of the submarine with specified TSP values is uniquely described by a set of basic TTE values, it is convenient to regard the design option itself as an n-dimensional vector or a point in the n-dimensional vector space and to consider the criterion of the military-and-economic efficiency as a function of the TTE vector in (n+1)-dimensional space.

Beside TTEs and TSPs, submarines have certain economic characteristics. They are called technical-and-economic indices and describe expenses for the development and upkeep of submarines during various periods of their life. Technical-and-economic indices serve to form the military-and-economic efficiency index that characterises the total cost of resources spent in achieving the specified level of combat efficiency. Resource expenditures are classified by their nature (single non-recurrent, lumpsum or capital, operating or current) and purpose (direct, i.e. directly related to the submarine, and associated, particularly – those related to submarine base infrastructure development).

The major technical-and-economical indices include:
- development costs (design and R&D work to support the project), C_1;
- construction costs, C_2;
- average annual operation costs, C_3;
- weapon replenishment costs (the first package is included in construction costs), C_4;
- shared costs of setting-up or expanding shore or floating base deployment system for the subject submarine, C_5;
- shared costs of the base system, C_6.

Each technical-and-economical index of the submarine can be considered as a function of TTEs and TSPs. E.g., the construction cost or, more exactly, the firm wholesale price of the submarine, is usually considered to be a function of major design elements: displacement, power plant capacity and some basic TTEs, in particular, the number of missiles. On the other hand, the submarine cost depends considerably on some TTEs, e.g., on the type and mechanical characteristics of the pressure hull material. Since technical-and-economical indices are different in their nature, they are involved differently in the index of military-and-economic efficiency. The latter, being the measure of resources spent to resolve the task, in its physical sense should be linear with regard to technical-and-economical indices. The resource index of the military-and-economic efficiency may be formulated as

$$b = \sum_{i=1}^{6} \alpha_i C_i \qquad (12.1)$$

where b — resource index of military-and-economic efficiency;
C_i — technical-and-economical index of the submarine;
$\alpha_1 - \alpha_6$ — weighting coefficients defining the structure of the resource index of military-and-economic efficiency.

Weighting coefficients α_i are found based on the military-and-economic model, i.e. from the involved set of assumptions. This should be clarified by an example. Let us assume that the intended series includes m submarines and the number of submarines required for accomplishing the mission is M. Then the share of expenses for the development of this project based on mission accomplishment is found as follows: shared expenses for one submarine development are C_1/M and for m submarines employed for accomplishing a combat mission are $C_2 m/M$.

Hence, the ratio of weighting coefficients α_1 and α_2 must be equal to m/M. In a similar way they find the ratio of α_2 and α_3. Construction costs are paid only once, while operational expenses continue throughout the entire service life of the submarine T. Therefore, the ratio of α_2 and α_3 must be equal to 1/T. Thus, the ratio of these coefficients is an analogue of the depreciation deductions standards.

Ratios of other coefficients are found in a similar way based on physical factors. Absolute values of these coefficients are related to the number of submarines assigned to accomplish the mission.

12.3. The Military-and-Economic Efficiency Evaluation Model

As has already been mentioned, the model is understood as a totality of assumptions and algorithms based on these assumptions that, in a formalised way describe the subject process or a system and contain the sought value as a function of variable parameters and their restrictions.

Models can be divided into analytical and statistical ones. The former contain the sought value as a function of the arguments in analytical format and are very convenient because they are easy and simple to work with. In particular, such models are convenient for investigating effects of various parameters upon the value of the military-and-economic efficiency criterion. A considerable disadvantage, however, is the fact that the simplicity of analytical functions is achieved by building a simplified, rough model neglecting some potentially relevant specific features of the process.

Models of the second category, the statistical ones, are free of this drawback. In this case the approach is as follows. They consider one option of the subject process development at a random set of all input parameters in order to obtain some result. The total number of thus examined options is quite large and all input parameters, which are external with regard to the designed object, are produced by a random number generator according to their distribution patterns. Then they find the mean result of all simulated time histories and take it for the input parameter expectation.

Statistical models are more complicated than analytical ones but this is fully compensated by the fact that they are more trustworthy and enable oversimplifications to be avoided.

Model generation starts with the text description that introduces all involved assumptions.

The military-and-economic model defines the military-and-economic efficiency index, which is chosen as a criterion, of tactical and technical elements of the submarine and indices describing the mode of submarine combat employment. The system of notions explained in the previous paragraph allows the military-and-economic model structure to be shown graphically (Fig.12.1)[11].

Symbol → means that the right-hand parameter is determined according to the set (vector) of parameters located to the left of the symbol.

Reviewing the above-shown military-and-economic model structure one may notice that it can be divided into four blocks.

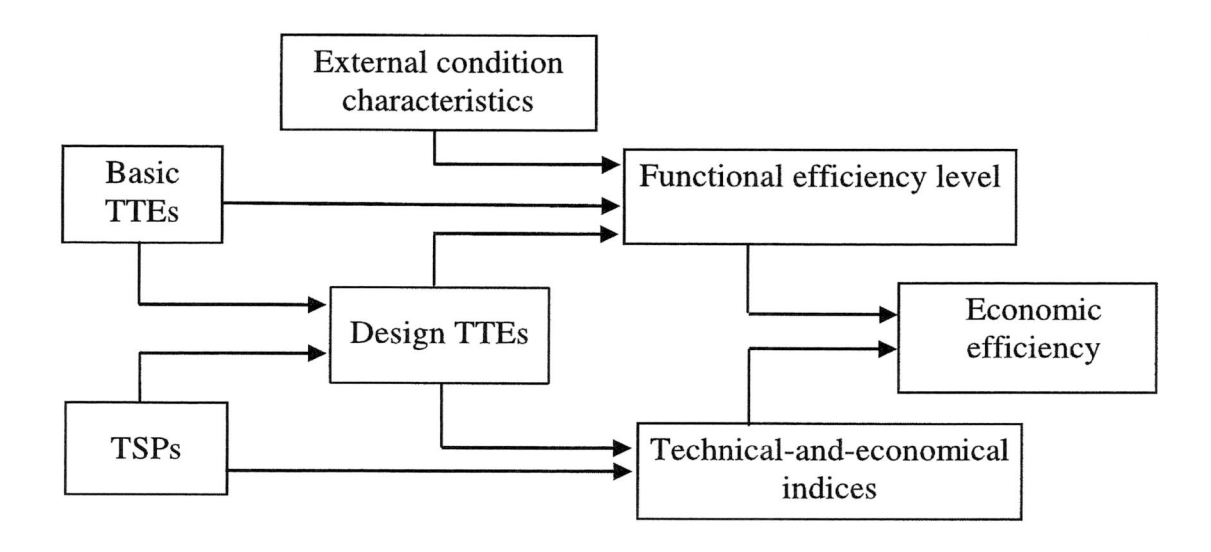

Fig.12.1. The structure of the Military-and-Economic Model

The technical block covers determination of design TTEs based on specified basic TTEs and TSPs. Solutions involved in this block are traditional design tasks.

The tactical block covers determination of the combat efficiency level index based on the complete set of specified TTEs and external condition characteristics. It sets correlations between submarine TTEs and the efficiency level, and its description includes operational and tactical environments in which the designed submarines will have to operate, including:
- description of the potential combat action area;
- characteristics of hostile forces, their composition, goals they may pursue, probable modes of employment, TTEs of enemy submarines, main data on weapons;
- intended employment modes of friendly forces.

The technical-and-economical block covers determination of submarine technical-and-economical indices (development, construction and per-year operation costs) based on known TTEs and TSPs. The technical-and-economical block description should contain data necessary to find these costs.

The military-and-economic block covers determination, based on known technical-and-economical indices and the combat efficiency level, of the resource efficiency index: total expenditures for accomplishing a certain mission. This index is usually taken for the criterion of military-and-economic efficiency at the specified level of combat efficiency. The military-and-economic block

description should contain data necessary for establishing this criterion:
- expected peacetime duration;
- number of submarines to be constructed under the subject project;
- expected submarine service life, as well as some other data characterising peace and wartime operation of subject submarines.

12.4. The Submarine Combat Employment Model

To form the model of submarine combat employment (the tactical block), it is necessary to make several assumptions regarding the character of combat actions, to describe the process of submarine combat employment based on these assumptions and to formalise this description.

As an example, let us consider a possible combat employment model for an attack submarine. The detection and track model may be taken from I.J.Diner [103]. It is assumed that the submarine mission is to reach a certain area (a square of S m^2), find the enemy unit deployed there, track it and kill it with the outbreak of combat actions announced by a special signal transmitted at some prearranged moment. Since the aim of the first stage of hostilities is to detect the enemy, the search efficiency index is the probability of detecting the enemy. The attack submarine («scout») detects the enemy unit («object») in a certain time «t» with a probability P(t). To find the detection probability, let us assume that the flow of objects appearing in the scout's field of vision is ordinary. This means that during a negligibly small time interval the scout will be unable to detect more than one object, i.e. there are only two possible scenarios: the number of contacts in an elementary time interval dt is either 1 or 0. Based on these assumptions, an elementary detection probability (t)dt is formulated as:

$$\gamma(t)dt = \frac{dP(t)}{1 - P(t)} \quad (12.2)$$

Integrating expression (12.2) assuming that the detection probability at the beginning of the search is 0, we arrive to:

$$P = 1 - e^{\int_0^t \gamma(t)dt} \quad (12.3)$$

The exponent in the detection probability formula (12.3) represents the expectation of the number of contacts during the search time t and is called the search potential

$$u(t) = \int_0^t \gamma(t)dt \qquad (12.4)$$

In contrast to the elementary probability $\gamma(t)dt$, probability $P(t)$ is called the accumulated or integral detection probability.

Depending on whether the search of unlimited duration ends with detecting an object or not, the process of searching is called converging or diverging. Let us consider the simplest case of a converging searching process: searching with a constant intensity γ = const. In this case the searching potential is a linear function of time $u(t) = t$ while the accumulated probability is an exponential time function

$$P(t) = 1 - e^{-\gamma(t)} \qquad (12.5)$$

Let us find the search time expectation:

$$T_{SEAR} = \int_0^\infty e^{-\gamma t} = \frac{1}{\gamma}(1 - 0) = \frac{1}{\gamma} \qquad (12.6)$$

Thus, when the scout is searching the enemy with a constant intensity, the search time expectation is inversely proportional to the search intensity. In order to determine the search intensity, let us assume that the location of the object in the area is subject to the law of uniform density, i.e. it can be at any point of the area with equal probability. Let us for the time being assume that the object is immobile though later its travel will have to be taken into account. The search is assumed to be conducted in a chaotic path pattern, i.e. both already inspected and not yet inspected areas are covered with an equal probability. Under such assumptions the elementary probability of object detection during a time unit is equal to the ratio of this site size and the total area [11].

In its turn, the elementary area covered by the search during a time unit is equal to the product of the scout speed multiplied by the double range of its detection facilities (Fig.12.2).

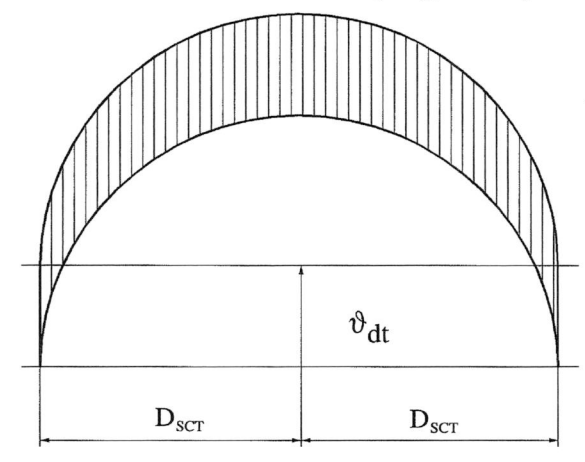

Fig.12.2. Elementary Site Covered by the Search

$$\gamma(t)dt = \frac{2D_{SCT}\vartheta_{SCT}}{S_0}dt \cdot P_{CON} \qquad (12.7)$$

where P_{CON} – probability of contact;

D_{SCT} – range of the scout's detection facilities;

ϑ_{SCT} – scout's search-mode speed;

Now it is necessary to account for the travel of the object. To do that, it is sufficient to consider not the absolute but the relative speed of the scout (Fig. 12.3.).

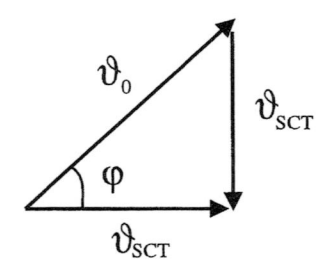

Fig. 12.3. To Relative Speed Estimations

The relative speed formula is:

$$\overline{\vartheta}_{SCT} = \sqrt{\vartheta_0^2 + \vartheta_{SCT}^2 - 2\vartheta_0\vartheta_{SCT}\cos\varphi} \qquad (12.8.)$$

The search, however, is only the initial stage of submarine employment. Having detected an enemy unit prior to the outbreak of hostilities, the scout must keep on continuously tracking the potential target.

Let us examine the target tracking process in the light of the following considerations. The tracking process starts immediately after the target is detected. Though the aim is to maintain uninterrupted lock on the target, the contact is periodically lost due to manoeuvres of the target. In case the contact is lost, the scout resumes searching and continues it till the contact is restored. One can easily guess that the less the time elapsed from the moment of losing the contact, the less will be the area where the target that has managed to break contact could be staying, and the easier it will be to re-gain the contact. It is obvious that the intensity of the resumed search would not be less than that of the first search $\lambda > \gamma$. At the same time, since the search area will be increasing with time, the intensity of the resumed search is not an increasing function of time. As a rule, two cases of tracking are considered:

a) the resumed search permanently prevails over the initial one: $\lambda > \gamma$. This happens when the tracked target escapes into a certain part of the search square $S' \ll S_0$;

b) the resumed search prevails over the initial one only for a short time $\lambda = \gamma$ This is true when the area within which the target may re-appear is growing rather quickly to $S' = S_0$.

Let us consider the first case. From the above deliberations it follows that the subject system can be in one of three possible states (Fig. 12.4.):

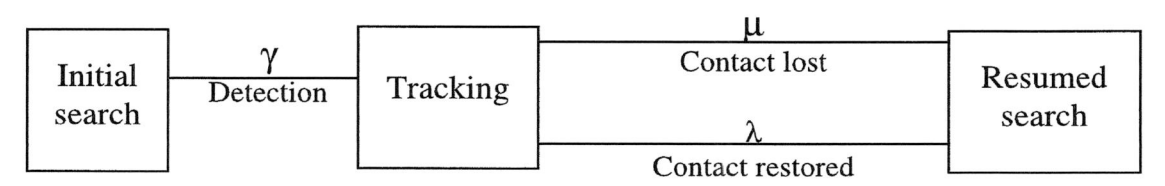

```
┌─────────┐   γ      ┌──────────┐    μ          ┌─────────┐
│ Initial │─────────│          │ Contact lost  │ Resumed │
│ search  │Detection│ Tracking │───────────────│ search  │
│         │         │          │      λ        │         │
└─────────┘         └──────────┘Contact restored└─────────┘
```

Fig. 12.4. The Model of Tracking with Contact Loss and Restoration

initial search: the scout searches an object throughout the whole area. This state continues till an object is detected;

tracking: the state the «scout-object» system has from the moment of target detection during the initial search till the target breaks contact, and from detecting the target during the resumed search till the contact is lost again;

resumed search – from breaking the contact in the process of tracking till restoring it in the process of the resumed search.

In accordance with the above-said, the tracking probability is generally described by:

$$P_{TRAC}(t) = \frac{\lambda}{\mu + 2} - \frac{\lambda - \gamma}{\mu + \lambda - \gamma} \cdot e^{-\gamma t} - \frac{\mu \gamma}{(\mu + \lambda)(\mu + \lambda - \gamma)} e^{-(\mu + \lambda)t} \quad (12.9)$$

where γ – detection intensity during the initial search;
λ – contact restoration intensity;
μ – contact loss intensity.

Let us consider some relevant particular cases:

1. Let $\gamma = 0$; it is obvious that if the initial detection is impossible, then the tracking is also impossible. Actually, substituting into (12.9) $\gamma = 0$, we get :

$$P_{TRAC}(t) = \frac{\lambda}{\mu + \lambda} - \frac{\lambda}{\mu + \lambda} \cdot 1^{-0} = 0 \quad (12.10)$$

2. $\mu = 0$; the target does not break contact. In this case the probability of tracking is equal to the probability of detection:

$$P_{TRAC}(t) = 1 - e^{-\gamma t} \qquad (12.11)$$

3. The resumed search does not have any advantages over the initial one:

$$P_{TRAC}(t) = \frac{\gamma}{\mu + \gamma}[1 - e^{-(\mu + \lambda)t}] \qquad (12.12)$$

Provided the necessary input data are specified these formulae enable the probability of tracking at any time within the submarine endurance limit to be determined.

The next phase of hostilities is the duel engagement, which in our example is modelled in a simplified manner. The analysis of this phase of submarine combat employment is in this case based on the following assumptions. After receiving the message about the outbreak of hostilities, the scout which is currently in contact with the target launches a weapon in order to kill the enemy. The salvo includes as many torpedoes as there are torpedo tubes on the submarine. The number of salvoes the scout, unless killed itself, can launch is found as:

$$n_3 = \frac{N_{TOR}}{D_{TOR}} \qquad (12.13)$$

where D_{TOR} – number of torpedoes in one salvo;
N_{TOR} – total number of torpedoes of the specified type.

Since the engagement may be started not only by the scout but also by the enemy, let us assume that the probability of the first salvo launched by the scout is $P = 1$. The probability of hitting the enemy by one salvo is found as:

$$P = 1 - (1 - P_v)^{v_{TOR}} \qquad (12.14)$$

where P_v – probability of hitting the target by a single torpedo.

Now let us consider the further development of the duel engagement scenario: the adversaries exchange salvoes till the scout's ammunition is exhausted (for the sake of simplicity we assume here that the enemy has an unlimited stock of ammunition). This will not change the probability of accomplishing the mission. The probability of hitting the target by each salvo, taking into account the probability that the scout itself will not be hit by enemy fire, may be represented as follows:

Salvo	Probability of hitting the target with a salvo	Probability of the scout been hit by the enemy salvo
1	P	(1-P)q
2	(1-P)(1-q)P	(1-P)2(1-q)q
3	(1-P)2(1-q)2P	(1-P)2(1-q)2q
i-th	[(1-P)(1-q)]i-1P	[(1-P)(1-q)]i-1(1-P)q
m-th	[(1-P)(1-q)]m-1P	[(1-P)(1-q)]m-1(1-P)q

It is easy to see that the general expression for the probability of hitting the enemy by the i-th salvo is:

in case the first salvo is made by the scout:

$$P_i^{(1)} = [(1-P)(1-q)]^{i-1} \cdot P \qquad (12.15)$$

in case the first salvo is made by the enemy:

$$P_i^{(1)} = [(1-P)(1-q)]^{i-1} \cdot (1-P)q \qquad (12.16)$$

and the total hit probability by all salvoes n_s possible with available reloads:

in case the first salvo is made by the scout:

$$P_{HIT}^{(1)} = P \frac{1 - [(1-P)(1-q)]^{n_s}}{P + q - Pq} \qquad (12.17)$$

in case the first salvo is made by the enemy:

$$P_{HIT}^{(1)} = P(1-q) \frac{1 - [(1-P)(1-q)]^{n_s}}{P + q - Pq} \qquad (12.18)$$

On the assumption that there exists an equal probability that either side engages the other, we arrive to the following formula of the total target hit probability:

$$P_{HIT} = 0.5 \left\{ (2P - Pq) \frac{1 - [(1-P)(1-q)]^{n_s}}{P + q - Pq} \right\} \qquad (12.19)$$

Having determined the probability of tracking the enemy by the outbreak of combat operations, and the tentative target hit probability calculated based on the assumption that by the outbreak of hostilities the scout was in contact with the target, let us now determine the total target kill probability taking into account both the tracking phase and the outcome of the duel engagement:

$$P_{KILL} = P_{TRAC} \cdot P_{HIT} \qquad (12.20)$$

where P_{TRAC} is determined from one of formula (12.9) to (12.12).

12.5. Determination of the Submarine Task Force Strength. Operational Strain and Operational Utilisation Coefficients

One of the combat efficiency indices is the number of submarines involved in accomplishing a mission: the task force. There are two kinds of task forces to be considered: the field task force and the full task force. By the field task force we understand the number of submarines that should be available in the combat zone in order to accomplish the mission. However, to ensure constant readiness for the mission it is not enough to have just the field task force. In real life, while some submarines are patrolling the area of potential combat operations, some others are on their way there to release those that are close to the limit of their endurance, and some are on their way back to the base. Besides, some submarines will always be found in the base undergoing between-sorties maintenance or some kind of repairs. Therefore, the number of submarines required to provide a permanent field task force in the combat zone is always sizeably larger than the field force strength. This greater number of submarines is what we call the full task force. The ratio between the full and the field task forces is characterised by operational strain and operational utilisation coefficients.

The operational strain coefficient is the ratio of the number of submarines at sea in the combat zone and in transit to and from it, to the total number of submarines at sea and in the base, i.e. to the full task force.

The operational utilisation coefficient is the ratio of the number of submarines in the combat zone to the total number of submarines at sea and in the base, i.e. the ratio of the field task force strength to the full task force strength.

The operational utilisation coefficient is found with the help of the operation cycle flow chart. This flow chart is a graphic schedule of submarine employment within one cycle (Fig.12.5).

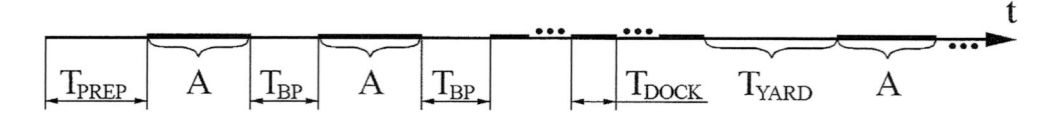

Fig.12.5. A Simplified Submarine Operation Cycle Flow Chart

«Cycle» in this context means time within which a submarine performs assigned functions and restores its combat capabilities.

For a diesel-electric submarine the cycle may include the following components:

T_{PREP} – preparation for campaign deployment;
A – sorties of duration determined by the design endurance;
T_{BP} – between-patrol intervals, including between-sorties repairs;
T_{YARD} – overhauls due to life expiration of major equipment;
T_{DCCK} – scheduled dry-docking.

The duration of every cycle component is found from empirical formulae as a function of the subject submarine particulars (TTEs) while the number of these components depends on the submarine designation. After calculating all these elements, they plot the operation cycle chart. It should be noted that compiling the operation cycle flow chart is always a creative design process. Having prepared the operation cycle flow chart and determined the cycle length, we can find the operational strain coefficient:

$$K_{OSH} = \frac{nA}{T_{CYC}} \qquad (12.21)$$

where A – design endurance or patrol duration;
T_{CYC} – total duration of the cycle.

The next stage is to find the operational utilisation coefficient. It is determined based on the following consideration. The submarine total endurance is used up in accordance with the adopted model of combat zone patrol and transits from and to the base. In this case the total sortie length is:

$$A = A_0 + 2T_{TRAN} = A_0 + \frac{2R}{\vartheta_{TRAN}} \qquad (12.22)$$

where A – submarine endurance;
A_0 – combat zone patrol duration;
T – time spent in transit from the base to the combat zone;
R – distance between the combat zone and the base;
ϑ_{TRAN} – average speed in transit.

Then the ratio of the time spent in the combat zone to the total duration of the submarine sortie:

$$\frac{A_0}{A} = 1 - \frac{2R}{\vartheta_{TRAN} A} \qquad (12.23)$$

Thus, the operational utilisation coefficient may be formulated as:

$$K_{OS} = K_{OS}\left(1 - \frac{2R}{\vartheta_{TRAN} \cdot A}\right) \quad (12.24)$$

Calculations for the tactical model are completed by determining the field task force strength required to accomplish the assigned mission with the specified probability:

$$N_{RNQE} = \frac{\ln(1 - P_1)}{\ln(1 - P_2)} \quad (12.25)$$

where P_1 – required probability of mission accomplishment;
P_2 – probability of mission accomplishment by at least one submarine.

The full task force strength required to accomplish the mission with the specified probability is, taking into account operational utilisation standards, found as:

$$N = \frac{N_{RNQE}}{K_{OPER}} \quad (12.26)$$

where K_{OPER} – operational utilisation coefficient.

12.6. The Concept of Military-and-Economic Analysis Dynamic Problems

Naval ships, primarily all modern submarines, are sophisticated and very expensive products with rather long service lives [75], [93]. The combat efficiency of any ship never remains constant during the whole service life, it changes due to physical, and a specific kind of moral, wear: tactical degradation. Considering combat efficiency time variation as a continuous process, it is possible to write the following expression:

$$\Delta E(t) = \int_0^t \left\{ \frac{\partial E}{\partial x_1} \cdot \frac{dx_1}{dt} + \frac{\partial E}{\partial x_2} \cdot \frac{dx_2}{dt} + \ldots \frac{\partial E}{\partial x_n} \cdot \frac{dx_n}{dt} + \right.$$

$$\left. + \frac{\partial E}{\partial z_1} \cdot \frac{dz_1}{dt} + \frac{\partial E}{\partial z_2} \cdot \frac{dz_2}{dt} + \frac{\partial E}{\partial z_n} \cdot \frac{dz_n}{dt} \right\} dt = \quad (12.27)$$

$$= \int_0^t \operatorname{grad}_x E \cdot \frac{d\overline{x}}{dt} dt + \int_0^t \operatorname{grad}_z E \cdot \frac{d\overline{z}}{dt} dt$$

where $\dfrac{\partial E}{\partial x_i}$ – partial derivative of the efficiency with respect to submarine TTEs (speed, diving depth, etc.) that describes changes in the combat efficiency due to changes in these TTEs;

$\dfrac{dx_i}{dt}$ – time variation of the i-th TTE due to physical wear;

$\dfrac{\partial E}{\partial z_i}$ – partial derivative of the submarine combat efficiency with respect to the j-th TTE of enemy naval and other forces.

The meaning of the first integral in expression (12.27) is quite simple, it is just the physical wear. Actually, with time submarine TTEs deteriorate, albeit slightly, and due to this process the combat efficiency also deteriorates. The first multiplier in the integrand is the gradient along the TTE vector and the second one is the time derivative of the TTE vector that describes the TTE deterioration rate for the subject submarine.

The physical meaning of the second integral is more complicated. Adversary forces are not going to remain unchanged during the whole service life of the subject submarine. Obsolete components of enemy ASW elements (ships, aircraft, etc.) will be substituted by new ones or upgraded. Therefore, for the subject submarine which does not change in technical terms (upgrading is a separate issue), it becomes more difficult to oppose upgraded hostile forces. In other words, with unchanging TTEs of the submarine, its combat capabilities decrease due to technical progress on the adversary side. This process, which is a specific kind of moral wear characteristic of only military hardware, is called tactical degradation. The first multiplier in the integrand is the submarine combat efficiency gradient due to improvements in TTEs of enemy forces and the second one is the rate of scientific and technological progress on the enemy side.

The subject submarine current combat efficiency, i.e. its value at the i-th year of service may be written as:

$$E(t) = E_0 - \Delta E(t) = E_0[1 - \dfrac{\Delta E(t)}{E_0}] = E_0 J(t) \qquad (12.28)$$

where E_0 – submarine combat efficiency at the time of her design;

$J(t)$ – tactical degradation function (Fig.12.6)

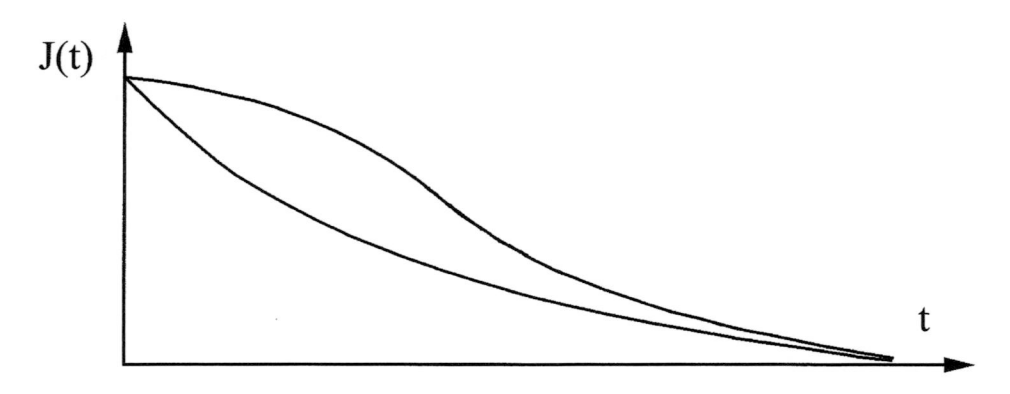

Fig.12.6. The Tactical Degradation Function Pattern

Tactical degradation functions are usually described with a formula satisfying the following conditions:

$$J(t) = 1 \text{ at } t = 0$$

$$J(t) = 0 \text{ at } t = \infty$$

$$\frac{\partial J}{\partial t}\bigg| \to 0 \; t \to 0$$

$$\int_0^\infty J(t)dt < M \qquad (12.29)$$

where M – is a certain limited positive number.

Integral $\int_0^\infty J(t)dt$ has a special meaning. Armed Forces are created to ensure national security, to deter potential aggressors, etc. In other words – to prevent war.

In this sense the higher the efficiency of combat elements constituting the Armed Forces, the better the overall efficiency of the Armed Forces. Therefore, the contribution of every combat element, every submarine included, to the accomplishment of this mission is defined not only by its combat efficiency on the day it is commissioned. This contribution will also depend on how many years a submarine will stay in service and what sort of combat efficiency she will have during every year of her service life. It is the ratio of the life-long contribution of the submarine into the national security to the initial combat efficiency of this submarine that is described by the integral in formula (12.29). In accordance with such understanding this integral multiplied by the initial combat efficiency is called the combat potential.

$$L = E_0 \int_0^\infty J(t)dt \qquad (12.30)$$

Let us consider a simplistic problem involving the category of combat potential: how the length of design work affects the military-and-economic efficiency index of the submarine. Let us assume the tactical degradation function in the form of an exponent

$$J(t) = e^{-vt} \qquad (12.31)$$

Let us determine the parameter of the tactical degradation function v assuming that in 10 years the subject submarine combat efficiency reduces by 25%. Then $e^{-10v} = 0.75$ and $v = 0.029$.

As in real life, let us assume that tactical degradation starts from the day the Submarine Design Specifications are approved, i.e. from the moment they fix submarine TTEs. Later, during the design process, these particulars remain unchanged and they do not improve in the process of service in the Navy. Let us assume that the scheduled design period of 4 years was somehow reduced to 2 years (Fig.12.7.)

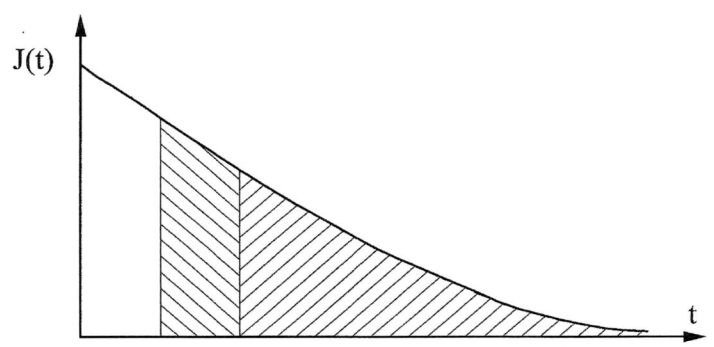

Fig.12.7. To the Calculation of the Design Time Reduction Effect

The relative increase in the combat potential is found from:

$$\frac{L_1}{L_2} = \frac{E_0 \int_2^\infty e^{-0,75t} dt}{E_0 \int_4^\infty e^{-0,029t} dt} = 1.36 \qquad (12.32)$$

Assuming that design costs and other technical-and-economical indices remain unchanged, we may see that under such theoretical conditions the design time reduction from 4 to 2 years will reduce the cost of the equally-efficient service of the submarine by 36%.

Based on similar considerations, it is possible to tackle more complicated problems of the military-and-economic analysis:
 – determination of optimum service life;
 – determination of optimum schedules for upgrading and other work.

12.7. Estimation of Submarine Technical-and-Economical Indices

Submarine development and upkeep expenditures may be considered to be the main technical-and-economical indices. The cost of developing a submarine, in a general case, includes the design cost and the construction cost.

The development cost is understood as project-related expenses of the major design contractor and involved subcontractors, as well as the cost of research and development work in support of the project.

The construction cost includes shipyard expenses for constructing and selling the submarine, as well as the profit of the shipyard.

The submarine operation cost consists of main components associated with repairs and scheduled replacements of equipment, logistic support and the crew. Total expenses for achieving a certain efficiency may then be presented as:

$$C = a_1 C_{DEV} + a_2 C_{BIL} + a_3 C_{OP} \qquad (12.33)$$

where C – index of expenses, C_{DEV} – development cost; C_{BIL} – construction cost; C_{OP} – operation cost; a_1, a_2, a_3 – weighting coefficients for ratios among various expenses. These factors are found based on the model of the intended mode of submarine construction and employment.

It is already a long-established procedure that the construction cost is taken as the baseline figure for submarine development cost estimations. The reason is, firstly, that this characteristic is more steady than the design cost and can be more easily identified and monitored as a function of major TTEs. Secondly, the design cost should be distributed over the entire series of submarines built under the subject project, but at early design stages the eventual number of submarines in the series is still unknown [13]. On the other hand, in the process of feasibility studies it is not as important to exactly estimate the cost of the future submarine and her operation expenses, as to get a correct idea about ratios between technical-and-economical indices of different design options of the submarine and to know how these indices are related to TTEs of the designed boat [4].

Therefore, in technical-and-economical analysis they mostly use fixed-parameter methods of cost estimations. These methods are based on the principle of finding how technical-and-economical indices of the designed submarine, first of all her cost, depend on TTEs and external (with respect to the boat) parameters characterising construction and operation conditions. Such dependencies are either established on the basis of experimental and statistical information about already-built submarines or derived from relevant regulatory documents. It is necessary to bear in mind that statistic methods can tell us what the submarine cost will be, unless the production undergoes some major changes, and regulatory documents tell what this cost should be if the production process management is at its best.

The submarine construction cost can be found as:

$$C = (1 + k)(C_{YARD} + C_M + C_{SCON} + C_{DIR}) \qquad (12.34)$$

where C – submarine construction cost;

$C_{YARD}, C_M, C_{SCON}, C_{DIR}$ – costs of shipyard work, materials, subcontractor supplies and other direct expenses;

k – coefficient taking into account commercial expenditures [87].

It is advisable to break down the submarine cost into components utilising both the classification of expenses adopted for budgetary estimations and the subdivision of the submarine into subsystems (elements). Such an «element-wise» submarine cost model allows us to apply comparatively simple formulae and make the model more «vigorous», which means that a variation in an element forces only one function, not the entire model to change. Cost items are in this case presented as a matrix where lines correspond to subsystems and columns refer to budget articles. Such cost models are called matrix models (Table 12.2)

Table 12.2.

The Submarine Cost Matrix Model (Breslav's matrix)

Subsystem	Shipyard work	Materials	Subcontractor supplies	Sub system cost
Pressure hull and pressure structures	C_{11}	C_{12}	C_{13}	$\sum_i 1$
Light hull	C_{21}	C_{22}	C_{23}	$\sum_i 2$
Gears	C_{31}	C_{32}	C_{33}	$\sum_i 3$
Power plant	C_{41}	C_{42}	C_{43}	$\sum_i 4$
Electric equipment	C_{51}	C_{52}	C_{53}	$\sum_i 5$
Equipment for compartments	C_{61}	C_{62}	C_{63}	$\sum_i 6$
Weapons	C_{71}	C_{72}	C_{73}	$\sum_i 7$
Radio electronic equipment	C_{81}	C_{82}	C_{83}	$\sum_i 8$
Provisions for weighing and trimming	C_{91}	C_{92}	C_{93}	$\sum_i 9$
Articles of expenses	$\sum_i 1$	$\sum_i 2$	$\sum_i 3$	$\sum\sum$

Value $C = \sum\sum$ represents the production cost of the submarine.

The item «shipyard work» includes the following budget articles: basic and additional wages of production workers, indirect workshop and shipyard expenses, expenses for production preparation and other direct expenses. This value is estimated based on technology-defined labour input requirements of every subsystem. Patterns of labour consumption functions are derived from prototype data or statistics.

Then the shipyard work cost is

$$C_{YARD} = C_{RATE} T (1 + k_{SHOP} + k_{YARD})(1 + k_{BON})(1 + k_{BIL}) \quad (12.35)$$

where C_{RATE} – average wage rate per hour; T – labour intensity for a subsystem, k_{SHOP}, k_{YARD} – coefficients for all indirect expenses; k_{BON} – coefficient taking into account additional wages, k_{BIL} – coefficient for special expenses [11].

The cost of materials required for submarine construction is calculated separately for each subsystem. The governing parameter for the «materials» item of costs is expressed through weight indices:

$$C_{MUT} = \frac{C_i P_M}{k_i k_{MUT}} \quad (12.36)$$

where P_M – subsystem weight; C_i – unit cost of materials; k_i – coefficient for transportation and purchasing expenses; k_{MUT} – material utilisation factor.

The most difficult task at the early design stages is to determine the cost of subcontractor supplies which constitutes a considerable amount of the submarine construction cost. This is due to the fact that the submarine development involves many branches of industry and each has its own specific features associated with the specifics of their products and production technologies. Submarine designers are not always aware of these specific features and sometimes it is impossible to fully account for them. Under such conditions it may be recommended to estimate costs by weight indices. Then the cost of subcontractor supplies and services for a subject subsystem is found from:

$$C_{SCON} = \sum_i C_i P_i \quad (12.37)$$

where P_i – weight of the subcontractor article of the subsystem;
C_i – specific cost (cost of a weight unit) of the article.

The values are derived from prototype data taking into account specific features of the subject article. E.g., costs of various types of power plants (nuclear, diesel-engine, air-independent) can be determined separately, proportional to their power outputs instead of weights.

$$C_{PP} = k_i \cdot N_{PP} \qquad (12.38)$$

where k_i – coefficient depending on the plant type;
N_{PP} – power plant output.

An idea on the cost of subcontractor supplies and services may be obtained based on information about US nuclear submarines (Table 12.3.)

Table 12.3

Components (%) of US Nuclear Submarine Costs (based on data from [16], [87])

Submarine type	Shipyard work	Materials	Subcontractor supplies and services
Torpedo	43 ~ 46	14 ~ 15	40 ~ 42
Missile	33 ~ 38	15 ~ 16	48 ~ 50

When we have determined the submarine construction cost C, it is possible to determine the cost of the series of submarines intended to be built to this design:

$$C_{SER} = \left[aN + b(\ln N + \varepsilon) \right] \cdot C \qquad (12.39)$$

where N – full task force required to accomplish the mission with the specified probability; $\varepsilon = 0.577$ – Eulerian constant.

An important factor in submarine combat efficiency estimations is the average annual operation cost index as such expenses during the service life are comparable with submarine construction costs and, in some cases, even exceed them. As has already been mentioned, all expenses for submarine operation during one year are divided into several items corresponding to their purposes:

$$C_{PER} = C_{PER} + C_{LOG} + C_{REPL} \qquad (12.40)$$

where C_{PER} – crew costs;
C_{LOG} – logistic support costs;
C_{REPL} – equipment repair and scheduled replacement costs.

Expenses for the crew are determined mainly by the complement strength and described as:

$$C_{PER} = n_{PER} \cdot C_1 m_1 + n_{TCR} \cdot C_2/m_2 \qquad (12.41)$$

where C_1 and C_2 — average annual costs of one submarine crew member and one shore-based technical crew member, respectively;
n_{PER} and n_{TCR} — the submarine complement and shore-based technical crew strength;
m_1 — number of rotating complements for one submarine;
m_2 — number of submarines supported by one shore-based technical crew;

The logistic support cost is determined mainly by the submarine displacement:

$$C_{LOG} = \left(kD^{2/3} + \frac{C_{SB}}{T_{SB}} \cdot 10^3 \right) \qquad (12.42)$$

where D — normal displacement;
C_{SB} — cost of storage batteries;
T_{SB} — storage battery service life;
k — numerical coefficient.

Cost of repairs and scheduled replacements includes expenses for hull repairs, electromechanical equipment and scheduled equipment replacements associated with repairs:

$$C_{YARD} = \frac{1}{T_{SHOP}} \left\{ [0{,}2(D \cdot 10^{-3} + Ne \cdot 10^{-4})] \sum_{i=1}^{3} t_i n_i C + (C_{CORE} + t_{CORE} \cdot C) n_{CORE} \right\} \cdot 10^{-3} \qquad (12.43)$$

where: t_i — the i-th repair length;
C — average labour quota hourly cost for shipyard repair jobs;
n_i — number of the i-th type repairs;
C — labour quota hourly cost for reactor refuelling jobs;
C_{CORE} — reactor fuel core cost;
t_{CORE} — reactor refuelling time.

After the above-described calculations it becomes possible to derive the military-and-economic efficiency criterion: the cost of the full task force and its upkeep for 10 years. This index, as has been mentioned, is applied to compare different design options of the future submarine.

$$C = C_{SER}^{(N)} + C_{OP} \cdot 10N \qquad (12.44)$$

And finally, let us for the sake of an example, consider comparative estimations of combat and military-and-economic efficiencies of employing submarines with nuclear, conventional diesel-electric and non-nuclear air-independent power plants at different theatres of naval operations. E.g., results of investigations into potential capabilities of task forces consisting of submarines of subject types in terms of the efficiency-cost criterion [57] show that in the ocean and faraway sea areas nuclear submarines have obvious advantages. This is due to the fact that because of their higher underwater speed, longer endurance, better conditions for sonars and other characteristics affecting efficiency, the nuclear submarines are 5 to 15 times more efficient than non-nuclear ones when employed in missions against hostile submarines and surface ships in remote sea and ocean areas.

At the same time the cost of nuclear submarine development and operation is only 2.5 to 5.0 times higher than for non-nuclear submarines (Fig.12.8). It should be noted, however, that index $W = E/C$ is only an approximate indicator.

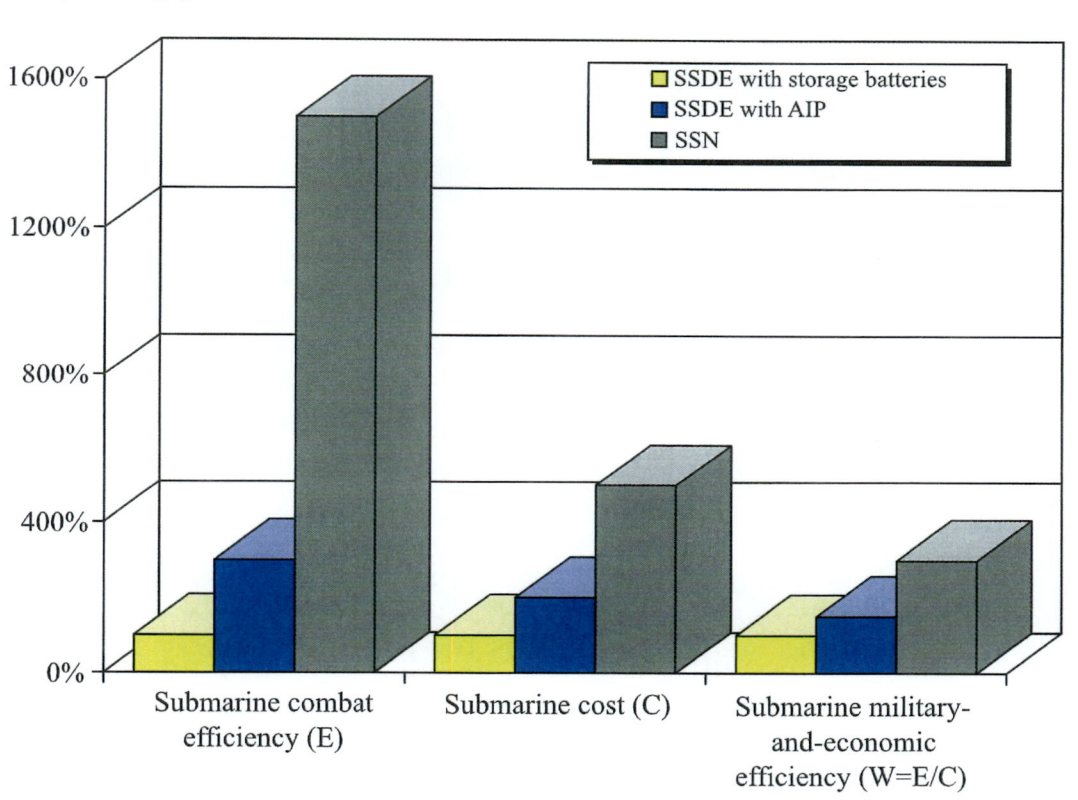

Fig.12.8. Ocean-Zone Military-and-Economic Efficiency of Submarines

When we estimate military-and-economic efficiency of submarines in littoral waters, where both the nature of assigned missions and the conditions under which they have to be accomplished are different (mostly due to the presence of supporting forces and to the shallow water hydrology) and, consequently, the submarine combat efficiency changes, we may see that non-nuclear submarines with air-independent power plants have an advantage over nuclear ones (Fig.12.9).

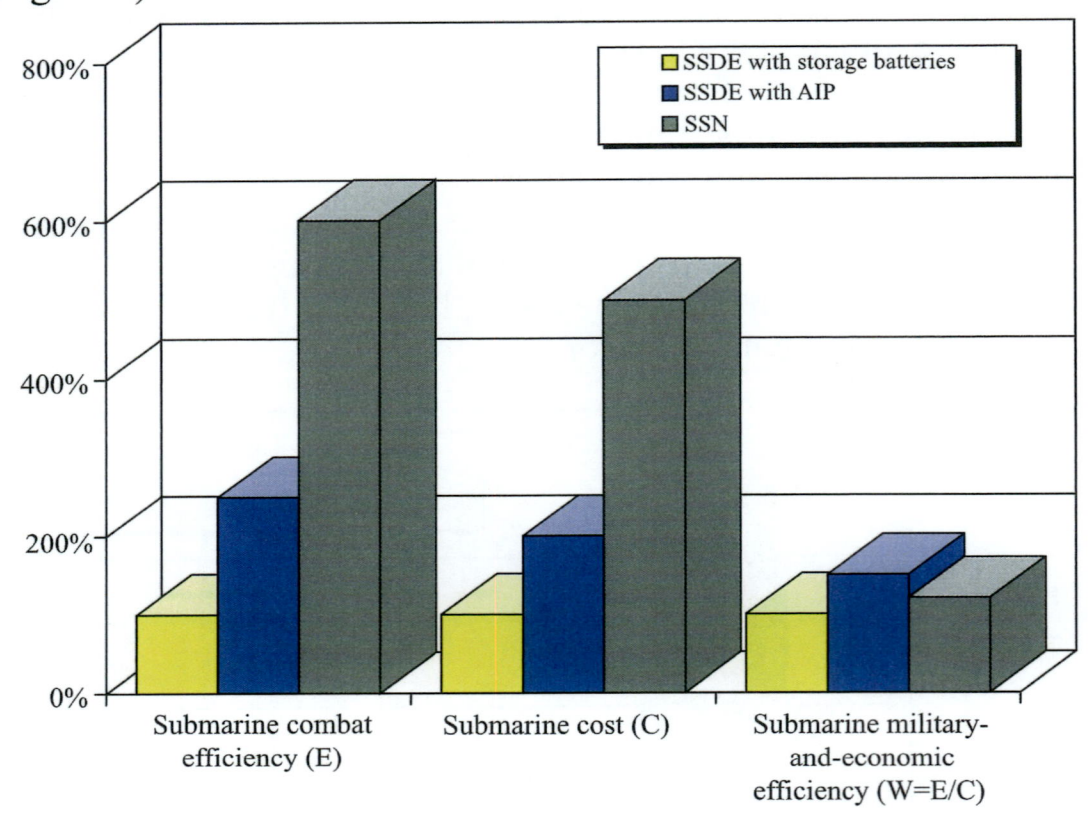

Fig.12.9. Littoral-Zone Military-and-Economic Efficiency of Submarines

This is explained by the fact that in the littoral zone nuclear submarine characteristics become excessive, while the capabilities of non-nuclear submarines, with auxiliary air-independent plants, in terms of stealth features and search potentials measure up closely to those of nuclear ones. Similar results have been obtained by US experts [102].

However, in our opinion, conclusions drawn in [57], [102] may considerably change should one take into account the costs of decommissioned nuclear submarine disposal and the costs of AIP operation, which require highly expensive shore-based infrastructures.

As a particular example of compatibility analysis of staff requirements we suggest performing an exercise on what is more preferable: to improve sonar package performances (through both the installation of a more powerful sonar and the introduction of new design solutions to provide the sonar with more favourable operation conditions) (Fig.12.10), or to install more torpedo tubes with a larger number and a wider choice of reloads [2].

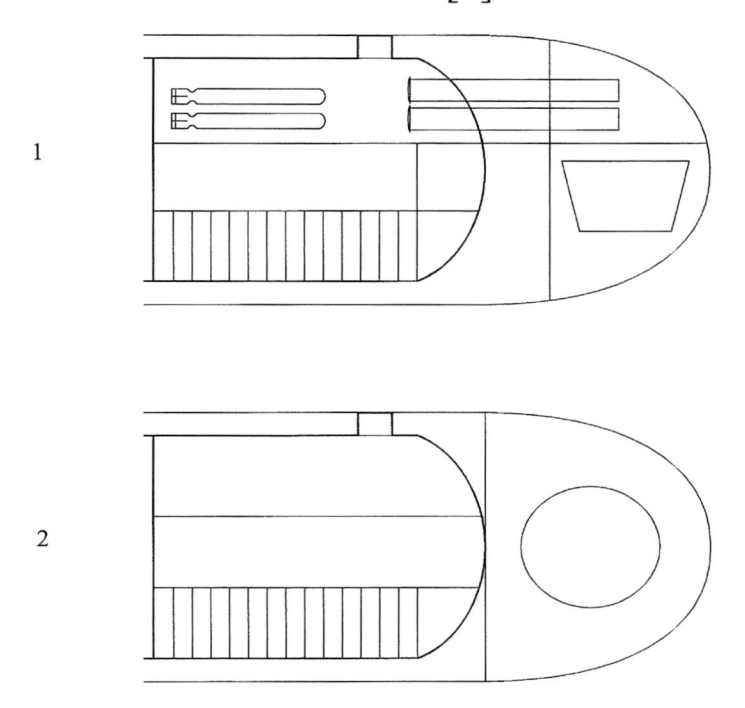

Fig. 12.10. Fore End Arrangement Options
1 – Limitations on sonar performance 2 – optimum sonar performance

The preference of this, or another, solution is estimated through the efficiency of submarine operation with different combinations of sonar packages with torpedo tube and reload torpedo numbers in various tactical situations.

RELIABILITY AND EXPERIENCE — TRADEMARK OF AMUR SHIPBUILDING PLANT

JSC Amur Shipbuilding Plant (former Leninsky Komsomol Plant) is the biggest shipbuilding plant of the Russian Far East on the production facilities of which the complex floating constructions for the Russian Navy and civil customers are manufactured.

Within 65 years of the fruitful cooperation between JSC Amur Shipbuilding Plant and Central Design Bureau for Marine Engineering RUBIN a great number of ships has gone off the slipways. Since 1935 nuclear-powered and diesel submarines of the following Projects were constructed: 659 (5 units), 675 (13 units), 667A (10 units), 667B (7 units), 877 (15 units).

Nowadays the cooperation between enterprises continuous successfully.

Within 1997-1998 Amur Shipbuilding Plant manufactured the Spacer for "Molikpaq" Drilling Installation under the order of International Consortium Sakhalin Energy formed by Shell, Mitsui, Mitsubishi. The total weight of construction is 15,000 tones. The design works were performed by Central Design Bureau for Marine Engineering RUBIN.

OUR ADDRESS:

JSC «Amur Shipbuilding Plant»
Alleya Truda, 1,
Komsomolsk-on-Amur, 681000, Russia
market_department@amurshipyard.ru
phone: (42172) 4 30 19; 4 32 24
fax: (42172) 4 50 26

Joint Stock Company
BALTIC-RUSSIAN SHIP SERVICE

БАРС

- **SHIPBUILDING**
- **SHIPREPAIR**
- **UPGRADE**

1/3, Dekabristov str.
St. Petersburg, 193000, RUSSIA

Tel.: +7-812-314-8080
Tel/Fax: +7-812-312-4084

REFERENCES

1. *Aleksandrov V.L., Glozman M.K., Rostovtsev D.M. and Sivers N.L.* Main Hull Structures Design of Submersible Vehicles. St.Petersburg: published by St.Petersburg Maritime Technical University, 1994.

2. *Aleksandrov Yu.I., Shevchenko P.P. and Khorenko V.I.* Multilevel Justification System of Submarine Appearance Intended for Export. St.Petersburg: Morintech-96.

3. *Alekseev G.N. and Murugov V.S.* Marine Underwater Engines. Moscow: Transport, 1964.

4. *Anishchenko V.V., Zhenishek M.M. and Pen A.V.* Development of Cost Prediction Methods for Prospective Ships and Vessels. St.Petersburg: Morintech-96.

5. *Antonov A.M., Dronov B.F., Kuteinikov A.V., Baratsev V.I. and Vinogradov V.P.* Submarine Architecture: Textbook, Part 2. St.Petersburg, published by Design Bureau «Malakhit», 1997.

6. *Ashik V.V.* Ship Design. Leningrad: Sudostroenie, 1985.

7. *Ashik V.V., Bogdanov A.A., Maraeva I.B. and Shebalov A.N.* Methods of Construction and Ship Surface Mating Using Computers. Leningrad: Sudostroenie, 1978.

8. *Babakov V.V.* Surface Design Using Quadratic Curves. Moscow: Mashinostroenie, 1969.

9. *Bazilevsky S.A.* Submarine Design: Synopsis of Lectures. Leningrad: published by Krylov Naval Academy, 1954.

10. *Batyrov A.N., Kosheverov V.D. and Leikin O.Yu.* Foreign Ships' Nuclear Power Plants. St.Petersburg: Sudostroenie, 1994.

11. *Breslav L.B.* Feasibility Study of Facilities for World Ocean Development. Leningrad: Sudostroenie, 1982.

12. *Breslav L.B. and Khalisev O.A.* Military-and-Economic Study of Submarine Designs. Manual. Leningrad: published by Leningrad Shipbuilding Institute, 1978.

13. *Breslav L.B., Zarubin E.P. and Khalisev O.A.* Evaluation of Submersible Vehicle Cost During Design. Manual. Leningrad: published by Leningrad Shipbuilding Institute, 1982.

14. *Bronnikov A.V.* Ship Design. Leningrad: Sudostroenie, 1991.

15. *Bubnov I.G.* About One Method of Designed Ship Main Dimensions Determination. Annual book of the Naval Architect Society, Vol.1, 1916.

16. *Bukalov V.M. and Narusbaev A.A.* Nuclear Submarine Design. Leningrad: Sudostroenie, 1968.

17. *Burov V.N.* National Naval Shipbuilding in the Third Century of Its History. St.Petersburg, Sudostroenie, 1995.

18. *Bykhovsky I.A.* Nuclear Submarines. Leningrad: Sudpromgiz, 1963.

19. *Bystrov A.I. and Levko A.F.* Ship Diesel Power Plants. Leningrad, 1989.

20. *Vekslyar V.Ya.* Methodology of Hullform and Appendages Design. St.Petersburg: 2nd International Shipbuilding Conference, Sudostroenie, 1998.

21. *Vlasov V.G.* Ship Statics. Moscow, Voenizdat, 1948.

22. *Vlasov V.G.* Collected Works, volumes 1, 2: Leningrad, Sudpromgiz, 1959 through 1966.

23. *Gilmer T.K.* Modern Ship Design. Leningrad: Sudostroenie, 1974.

24. *Dasoyan M.A. and Aguf I.A.* Modern Theory of Lead Cell. Moscow: Energia, 1975.

25. *George C. Maning.* Theory and Technique of Ship Design. Moscow: Voenizdat, 1960.

26. *Dolgov V.N.* Optimisation of Ship Nuclear Power Plant Parameters. Leningrad: Sudostroenie, 1980.

27. *Dronov B.F., Kuteinikov A.V. and Thernousov V.V.* About Normand's Number for Submarines: Textbook. St.Petersburg: published by Design Bureau «Malakhit», 1997.

28. *Dronov B.F., Antonov A.M., Kuteinikov A.V. et al.* Submarine Architecture: Textbook, Part 1. Published by Design Bureau «Malakhit», 1997.

29. *Droblenkov A.I., Ermolaev A.I., Muru N.P., et al.* Ship Theory Reference Book. Moscow: Voenizdat, 1984.

30. *Yefimiev N.N.* Basics of Submarine Theory. Moscow: Voenizdat, 1965.

31. *Yefremov K.P.* Submarine Design. Leningrad: published by Dzerzhinsky Naval Colledge, 1959.

32. *Zabie, A.V.* Torpedo Weapon. Moscow: Voenizdat, 1984.

33. The Classifier of the Unified Design Document System, Class 36 (ships, ship equipment). Arrangement and Rules of Loading. Regulations, 1988.

34. *Karpenko A.V.* Russian Missile Weapon 1943 - 1993. Reference Book. St.Petersburg: Pika, 1993.

35. *Kvasnikov V.N. and Saveliev M.V.* Basics of Submarine Design. Manual. Leningrad: published by Leningrad Shipbuilding Institute, 1954.

36. *Kvasnikov V.N. and Khalisev O.A.* Submarine Design. Issue 4. Manual. Leningrad: published by Leningrad Shipbuilding Institute, 1983.

37. *Kitainov D.S.* Radius-Graphical Method of Design and Analytical Analysis of Complex Curvilinear Surfaces. Published by Rostov NIITM, 1962.

38. *Kormilitsin Yu.N.* Russian Experience of Submarine Creation with Air-Independent Propulsion Plants.// Military Parade, 1997.

39. *Kormilitsin Yu.N. and Baranov I.L.* Automatisation of Design, Construction and Ship Production Preparation.// Shipbuilding Technology, 1970. No.4

40. *Kormilitsin Yu.N.* Project 877 Submarine.// Military Parade, July-August, 1994.

41. *Kormilitsin Yu.N.* Project 636 Submarine.// Military Parade, September-October, 1994.

42. *Kormilitsin Yu.N.* Submarines of «Amur» Family.// Military Parade, January-February, 1995.

43. *Kormilitsin Yu.N.* Status and Prospects of Development of Russian Non-Nuclear Submarines.// Submarine Design Problems. Issue 10. St.Petersburg: published by CDB ME «Rubin», 1996.

44. *Kormilitsin Yu.N.* KILO Class Optimum Performance for All Naval Scenarios.// Naval Forces. No.6. FRG, 1995.

45. *Kormilitsin Yu.N.* «KILO» Class Submarines.// African Defence, No.2, RSA, 1996.

46. *Kormilitsin Yu.N.* Experience of Russia in the Creation of Submarines with AIP.// Asian Defence Journal, May, 1997.

47. *Kormilitsin Yu.N.* Layout Diagrams and Procedure of Variable Load Compensation on Foreign and Soviet Submarines. Leningrad: CDB ME «Rubin», 1977.

48. *Kormilitsin Yu.N.* Project 877 Submarine. //Military Parade, July-August, 1996.

49. *Kormilitsin Yu.N., Pinegin A.N. and Khalisev O.A.* Methods of the Submarine Lines Drawing Design. Manual. Published by St.Petersburg Maritime Technical University, 1999.

50. *Kormilitsin Yu.N. and Khalisev O.A.* Submarine Design. Manual. Part 1. Published by St.Petersburg Maritime Technical University, 1998.

51. *Kormilitsin Yu.N. and Khalisev O.A.* Design. Published by St.Petersburg Maritime Technical University, 1999.

52. *Korshunov Yu.L. and Strokov A.A.* Torpedoes of the USSR Navy. St.Petersburg: Gangut, 1994.

53. *Kuzin V.P. and Nikolsky V.I.* The USSR Navy in 1945 - 1991. St.Petersburg: published by Elite Marine Society, 1996.

54. *Kuenn R.* Torpedo Submarines. Moscow: Voenizdat, 1970.

55. *Malinin B.M.* Modern Trends in Submarine Development. Leningrad, 1947.

56. *Malyutin A.A. and Pinegin A.N.* Experience of Solving Ship Theory Problems during Creation of the 2nd and the 3rd Submarine Generations. Submarine Design Problems. St.Petersburg: CDB ME «Rubin», 1996.

57. *Muradian V.A., Proniakin V.I. and Seliakov S.I.* Non-Nuclear Submarines with Air-Independent Propulsion Plants – a New Stage in the Underwater Shipbuilding. St.Petersburg: Morintech-96.

58. *Narusbaev A.A.* Introduction into the Theory of Substantiated Technical Solutions. Leningrad: Sudostroenie, 1976.

59. *Nogid L.M.* Ship Design Theory. Leningrad: Sudpromgiz, 1955.

60. *Nogid L.M.* Hullform and Lines drawing Design. Leningrad: Sudpromgiz, 1962.

61. *Pashin V.M.* Ship Optimisation. Leningrad: Sudostroenie, 1983.

62. *Perlovsky V.* Submarine Statics. Leningrad: published by Leningrad Shipbuilding Institute, 1983.

63. *Pinegin A.N.* Modern Problems of the Outer Hull Shape Optimisation. St.Petersburg: published by CDB ME «Rubin», 1998.

64. *Pravdin A.A.* Submarine Structure. Moscow: Oborongiz, 1947.

65. *Prasolov S.N. and Amitin M.B.* Submarine Arrangement. Moscow: Voenizdat, 1973.

66. *Popov G.I. and Zakharov I.G.* Theory and Methods of Ship Design. Leningrad: published the Naval Academy named after A.A.Grechko, 1985.

67. *Pozdyunin V.L.* Ship Design Theory. Leningrad: published by Leningrad Shipbuilding Institute, 1938, part 1, 1939, Part 2.

68. Submarines and Their Armament. Leningrad: Sudostroenie, 1983.

69. Submarines of Foreign Navies. Leningrad: published by Research Institute «Rumb», 1984.

70. *Pronyakin V.A.* Nuclear Submarines of the XXI Century. St.Petersburg: Morintech-96.

71. *Rakitsky V.V.* Ship Nuclear Power Plants. Leningrad: Sudostroenie, 1976.

72. *Rozhdestvensky V.V.* Submarine Dynamics. Leningrad: Sudostroenie, 1970.

73. *Rudenko V.M., Gorbunov N.M. and Soloviev I.P.* Power Plants for Submarines. Moscow: Voenizdat, 1962.

74. *Saveliev M.V.* Submarine Design: Manual. Issue 3. Leningrad: published by Leningrad Shipbuilding Institute, 1981.

75. *Saveliev M.V. and Khalisev O.A.* Military-and-Economic Analysis and Its Role in Submarine Design. Manual. Leningrad: published by Leningrad Shipbuilding Institute, 1980.

76. *Sagaidakov F.R., Thekalov Yu.N. and Thernetsov N.A.* Status and Development of the World Market for Submarines with Non-Nuclear Power Plants. St.Petersburg: Morintech-96.

77. *Solomenko N.S. and Rumyantsev Yu.N.* Submarine Structural Mechanics. Leningrad: published by Dzerzhinsky Naval College, 1962.

78. *Spassky I.D.* From «Dolphin» to «Typhoon». // Military Parade, May-June, 1996.

79. *Spassky I.D.* Submarines of the XXI century. // Military Parade, September-October, 1998.

80. Reference book of a Submarine Designer./ Edited by B.M.Malinin. Vol.1. Leningrad: Sudpromgiz, 1949.

81. *Tomashevsky V.T., Astashenko O.G. and Yakovlev V.S.* Submarine Strength. St.Petersburg: published by Grechko Naval Academy, 1994.

82. *Khalisev O.A. and Kvasnikov V.N.* Submarine Design. Issue 7. Manual. Leningrad: published by Leningrad Shipbuilding Institute, 1986.

83. *Khalisev O.A. and Kvasnikov V.N.* Submarine Control Surfaces Design. Manual. Leningrad: published by Leningrad Shipbuilding Institute, 1984.

84. *Khalisev O.A. and Kvasnikov V.N.* Variable Loads Compensation. Manual. Leningrad: published by Leningrad Shipbuilding Institute, 1985.

85. *Khiyaynen L.P.* Development of Foreign Submarines and their Tactics. Moscow: Voenizdat, 1988.

86. *Khmelnov I.N., Kozhevnikov V.A., et al.* Russian Submarines: History and Present. Vladivostok: Ussuri, 1996.

87. *Khudyakov L.Yu.* Ship Conceptual Design. Leningrad: Sudostroenie, 1980.

88. *Khudyakov L.Yu.* Submarines of the XXI Century. St.Petersburg: published by Design Bureau «Malakhit», 1994.

89. *Khudyakov L.Yu.* Special Features of the Surface Survivability of Submarines without Kingston Valves. St.Petersburg: Elmor, 1994.

90. *Shimansky Yu.A.* Structural Mechanics of Submarines. Leningrad: Sudpromgiz, 1948.

91. *Shirokorad, A.B.* Soviet Post-War Submarines. Moscow: «Arsenal Press», 1997.

92. *Shcheglov A.N.* Design of Submarine. Leningrad: Sudpromgiz, 1940.

93. *Yashenkin L.N.* Concept of Modification Process in the Submarine Life Cycle. St.Petersburg: Mirintech-97. Vol. 1, pp.193-195.

94. *Polmar N. and Noot J.* Submarines of the Russian and Soviet Navies 1718-1990. Naval Institute Press. Annapolis, 1991.

95. Jane's Fighting Ships, 1945-1994.

96. *Freidman N.* Submarine Design and Development. London. Conwey Maritime Ltd., 1983.

97. *Cotney Arnold.* The Development of a Computer-aided Conceptual Submarine Design Evaluation Tool. Auburn University, Alabama, 1973.

98. *Poy Burcker and Louis Rydill.* Concepts in Submarine Design. Cambridge University Press, 1994.

99. Warship'88 International Symposium on Conventional Naval Submarines. London, May 1988.

100. Warship'91 International Symposium on Naval Submarines 5. Vol.1, London, May 13-15, 1991.

101. Warship'93 International Symposium on Naval Submarines. London, May 1993.

102. *Dean A.Rains, Kenneth A.Mitchell* «Nuclear vs. Non-Nuclear Attack Submarine Powerplants», Naval Engineers Journal, May 1993, pp. 224-231.

103. *Diner I.Ya.* Operation Study. Leningrad: Naval Academy, 1969.

104. *Glozman M.K.* Constructability of Sea Ship Hulls. Leningrad: Sudostroenie, 1984.

105. *Tokmakov A.A.* Underwater Transport Ships. Leningrad: Sudostroenie, 1965.

106. Propulsion Batteries for Submarines. Advertising brochure of company «The Standard Batteries Limited». India, 1996.

107. *Serditov G.D.* Russian Submarines of the IV-th Generation «Amur 1650». «Defence International», August 1998.

108. *Badanin V.A.* Submarines with a Single Engine. St.Petersburg: «Gangut», 1998.

109. *Braiton Harris* Edited by *Walter J.Boine.* The Navy Times Book of Submarines. A Political, Social and Military History. Berkley Books, New York, 1997.

Yury Nikolaevich KORMILITSIN
Oleg Anatolievich KHALIZEV

THEORY OF SUBMARINE DESIGN